Advances in
ORGANOMETALLIC CHEMISTRY

VOLUME 30

Advances in Organometallic Chemistry

EDITED BY

F. G. A. STONE

DEPARTMENT OF INORGANIC CHEMISTRY

THE UNIVERSITY

BRISTOL, ENGLAND

ROBERT WEST

DEPARTMENT OF CHEMISTRY

UNIVERSITY OF WISCONSIN

MADISON, WISCONSIN

VOLUME 30

ACADEMIC PRESS, INC.

Harcourt Brace Jovanovich, Publishers

San Diego   New York   Boston
London   Sydney   Tokyo   Toronto

ACADEMIC PRESS, INC.
San Diego, California 92101

United Kingdom Edition published by
ACADEMIC PRESS LIMITED
24-28 Oval Road, London NW1 7DX

LIBRARY OF CONGRESS CATALOG CARD NUMBER: 64-16030

ISBN 0-12-031130-5 (alk. paper)

PRINTED IN THE UNITED STATES OF AMERICA
90 91 92 93   9 8 7 6 5 4 3 2 1

Contents

Chemistry of the Cyclopentadienyl Bisphosphine Ruthenium Auxiliary

STEPHEN G. DAVIES, JOHN P. MCNALLY, and ANDREW J. SMALLRIDGE

Thermochromic Distibines and Dibismuthines

ARTHUR J. ASHE III

Syntheses, Structures, Bonding, and Reactivity of Main Group Heterocarboranes

NARAYAN S. HOSMANE and JOHN A. MAGUIRE

η^2 Coordination of Si — H σ Bonds to Transition Metals

ULRICH SCHUBERT

Isoelectronic Organometallic Molecules

ALLEN A. ARADI and THOMAS P. FEHLNER

Dissociative Pathways in Substitution at Silicon in Solution: Silicon Cations R_3Si^+, $R_3Si^+ \leftarrow$ Nu, and Silene-Type Species $R_2Si{=}X$ as Intermediates

JULIAN CHOJNOWSKI and WŁODZIMIERZ STAŃCZYK

Chemistry of the Cyclopentadienyl Bisphosphine Ruthenium Auxiliary

STEPHEN G. DAVIES, JOHN P. MCNALLY, and
ANDREW J. SMALLRIDGE

The Dyson Perrins Laboratory
University of Oxford
Oxford OX1 3QY, England

I

INTRODUCTION

Organotransition metal chemistry is being utilized with increasing success in the development of highly regio- and stereoselective elaboration methodology for bound organic fragments. Particular attention has recently focused on the potential of the cyclopentadienyl bisphosphine ruthenium system to act as an organometallic auxiliary for such elaborations. The attractions of this system are quickly apparent. Early work by Stone and colleagues (1) has shown that both the chloride and the phosphines in $(\eta^5\text{-}C_5H_5)(PPh_3)_2RuCl$ (**1**) are readily replaced with a variety of other anionic or neutral ligands to give complexes of the form $(\eta^5\text{-}C_5H_5)L_2RuX$ and $(\eta^5\text{-}C_5H_5)L_2RuL'^+X^-$. The ease and generality of these syntheses allow for the convenient preparation of a wide range of derivatives of the ruthenium auxiliary and enable the chemist to influence the reactivity of the bound organic moiety by controlling both the steric and electronic environment in the organometallic complex. Furthermore, the well-defined geometry and configurational stability of the ruthenium system allow for the synthesis of complexes with a chiral ruthenium center. This chirality, properly harnessed, can provide the means to conduct stereoselective elaborations on the organic fragment.

Recent efforts to explore the synthesis and reactivity of organotransition metal complexes with the cyclopentadienyl bisphosphine ruthenium auxiliary have emphasized the potential of this system for the development of new organic synthetic methods, but they have also uncovered a number of stumbling blocks which must be overcome to achieve real success in this area. Future developments in the chemistry of these ruthenium complexes will be based on the successful application of the reactivity trends uncovered in the past work. The rapid expansion of studies in this field since the mid-1980s has made it increasingly difficult for the synthetic chemist to keep abreast of recent developments, and it is the purpose of this article to provide a broad review of the major synthetic and elaborative reactions demonstrated for the $(\eta^5\text{-}C_5H_5)(PR_3)_2RuX$ system.

The review is split into two major parts. In the first part (Sections II–V), ligand exchange reactions, which provide the means to control the steric and electronic environment about the metal center, are classified and described. Special attention is paid to the influences of geometry and steric constraints on the reactivity of the ruthenium complexes. The second part of the review (Sections VI–X) discusses the reactivity of the organic fragment bound to the ruthenium auxiliary.

II

PREPARATION OF CYCLOPENTADIENYL BISPHOSPHINE RUTHENIUM COMPLEXES

Gilbert and Wilkinson first prepared $(\eta^5\text{-}C_5H_5)(PPh_3)_2RuCl$ (1) in 1969 by reacting cyclopentadiene and tris(triphenylphosphine)ruthenium dichloride (2) over a period of 2 days [Eq. (1)] (2). This preparation, however, suffers from a competing dimerization reaction to form the unreactive $[(PPh_3)_2RuCl_2]_2$.

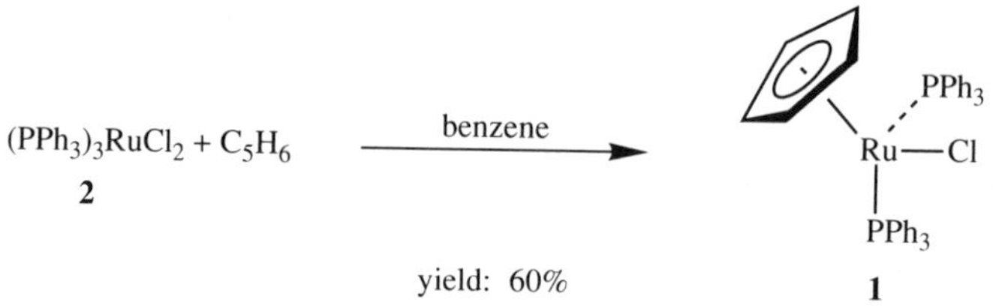

$$(PPh_3)_3RuCl_2 + C_5H_6 \xrightarrow{\text{benzene}} \underset{\textbf{1}}{\text{(complex)}} \quad (1)$$

2

yield: 60%

In 1971, a preparation of **1** from **2** using thallium cyclopentadienide was reported (1) but the toxicity of thallium and the mass of the reagent needed render this procedure unsuitable for large-scale preparations. An improved method was reported by Bruce *et al.* (3,4), using cyclopentadiene, ruthenium trichloride hydrate (3), and triphenylphosphine, which gives the desired complex in high yield [Eq. (2)]. The primary advantage of this latter method is formation of the complex in one pot.

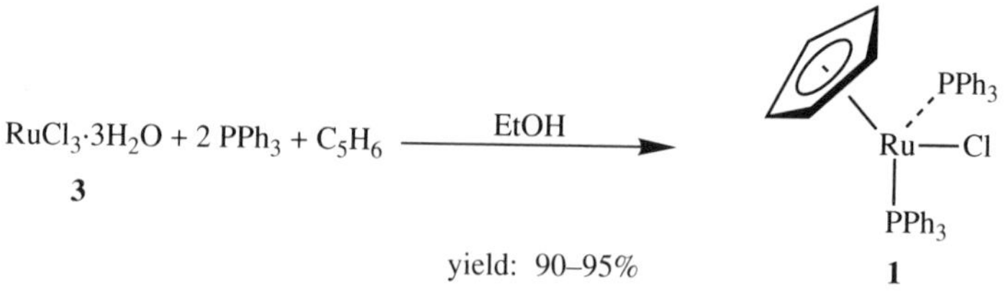

$$RuCl_3{\cdot}3H_2O + 2\,PPh_3 + C_5H_6 \xrightarrow{\text{EtOH}} \underset{\textbf{1}}{\text{(complex)}} \quad (2)$$

3

yield: 90–95%

The pentamethylcyclopentadienyl analog **4** can be prepared in good yield (77%) using a similar procedure, although considerably longer reaction times are required (60 hours) (*5*). Complex **4** can be prepared using less strenuous conditions if **3** and pentamethylcyclopentadiene are first reacted to give polymeric pentamethylcyclopentadienylruthenium dichloride, which is then treated with excess triphenylphosphine to generate the required ruthenium product **4** [Eq. (3)] (*6,7*). The corresponding indenyl

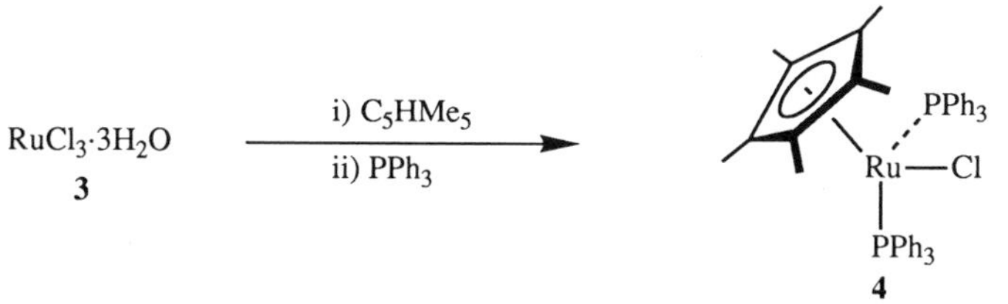
$RuCl_3 \cdot 3H_2O$
3
i) C_5HMe_5
ii) PPh_3
PPh_3
Ru—Cl
PPh_3
4

(3)

ruthenium complex **5** can be prepared from indene, ruthenium trichloride, and triphenylphosphine, using a procedure similar to that used for $(\eta^5\text{-}C_5H_5)(PPh_3)_2RuCl$, if potassium hydroxide is added to the reaction [Eq. (4)] (*8*). The fulvalene analog **6** can also be synthesized (*9,10*) and can

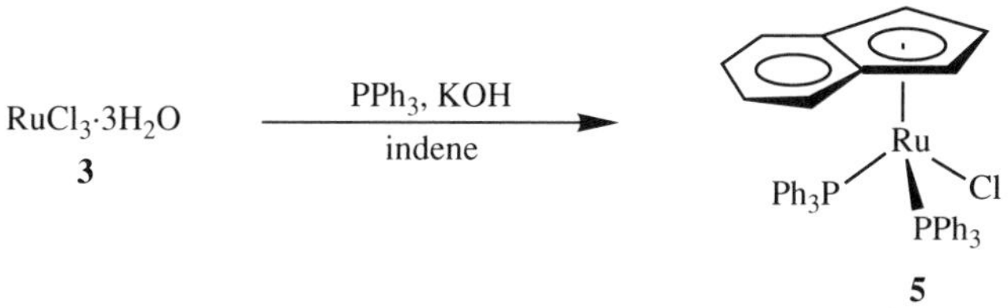
$RuCl_3 \cdot 3H_2O$
3
PPh_3, KOH
indene
Ru
Ph_3P
Cl
PPh_3
5

(4)

subsequently be converted to the heterobimetallic fulvalenyl complex **7** [Eq. (5)] (*11*).

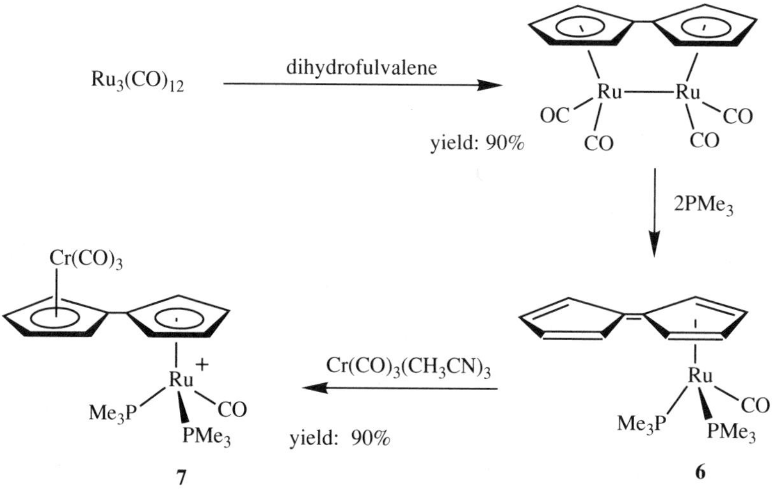
$Ru_3(CO)_{12}$
dihydrofulvalene
yield: 90%
Ru—Ru
OC
CO
CO
CO
$2PMe_3$
$Cr(CO)_3$
Ru +
Me_3P
CO
PMe_3
7
$Cr(CO)_3(CH_3CN)_3$
yield: 90%
Ru
CO
Me_3P
PMe_3
6

(5)

An interesting synthesis of the cyclopentadienyl bisphosphine ruthenium nitrosyl complex 9 involves the thermal displacement of both phenyl groups from $(\eta^5\text{-}C_5H_5)Ru(NO)Ph_2$ (8) with a chelating diphosphine [Eq. (6)] (12). Infrared data indicate that the nitrosyl ligand is linear (3 e^-

$$(6)$$

8   9

donor) in complex 8 and bent (1 e^- donor) in the product 9, as is expected to satisfy the 18-electron rule for each complex. The reaction presumably occurs by initial coordination of one end of the diphosphine to the ruthenium with concomitant conversion of the nitrosyl from a linear to a bent geometry, followed by the elimination of the two phenyl rings, presumably as biphenyl, and coordination of the free end of the diphosphine. Addition of the monophosphines PPh_3 or PMe_3 to 8 gives the monophosphine complexes (10) containing linear nitrosyl ligands [Eq. (7)].

$$(7)$$

8   10

III

GEOMETRY OF $(\eta^5\text{-}C_5H_5)(PR_3)_2RuCl$

The geometry of $(\eta^5\text{-}C_5H_5)(PMe_3)_2RuCl$ (11) is octahedral about the metal center, with the cyclopentadienyl ligand occupying three coordination sites. This is evidenced by the near 90° bond angles between the noncyclopentadienyl ligands and the metal center seen in the crystal structure of this compound (Fig. 1, Table I) (13). In 1 the increased bulk of the triphenylphosphine ligands relative to trimethylphosphine leads to a dis-

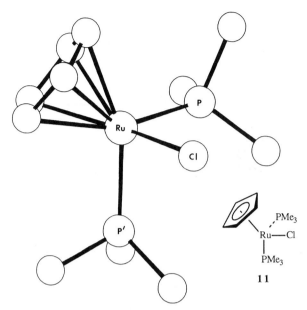

FIG. 1.   X-Ray crystal structure of $[(\eta^5\text{-}C_5H_5)(PMe_3)_2RuCl]$ (**11**) (hydrogens removed for clarity).

tortion from idealized octahedral geometry owing to severe steric inter-actions between the two phosphine ligands. Although the P—Ru—P bond angle has increased to 104°, the P—Ru—Cl bond angles are still approxi-mately 90° (Fig. 2, Table II) (*13*). The instability resulting from the steric strain about the metal center in bistriphenylphosphine complexes leads to facile dissociation of a phosphine ligand, which is a key step in many of the reactions of this system (*vide infra*).

TABLE I

BOND ANGLES IN
$[(\eta^5\text{-}C_5H_5)(PMe_3)_2RuCl]$ (**11**)[a]

Angle	θ (degrees)
P—Ru—P′	94.68
P—Ru—Cl	89.75
P—Ru—Cen	122.67
P′—Ru—Cl	90.11
P′—Ru—Cen	126.35
Cl—Ru—Cen	123.32

[a] Cen, C_5H_5 centroid.

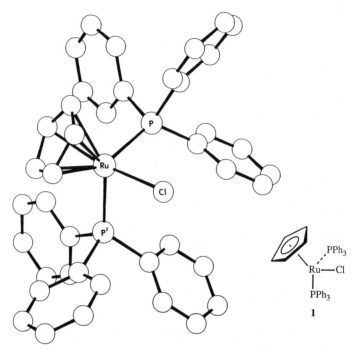

FIG. 2. X-Ray crystal structure of $[(\eta^5\text{-}C_5H_5)(PPh_3)_2RuCl]$ (**1**) (hydrogens removed for clarity).

Sterically bulky chelating diphosphines such as (R)-Ph$_2$PCHMeCH$_2$PPh$_2$ [(R)-PROPHOS] form ruthenium complexes with P—Ru—P bond angles considerably smaller than those exhibited for the analogous triphenylphosphine complexes, often less than the idealized 90°. This is a consequence of constraints on the bite angle of the two phosphorus atoms

TABLE II

BOND ANGLES IN
$[(\eta^5\text{-}C_5H_5)(PPh_3)_2RuCl]$ (**1**)^a

Angle	θ (degrees)
P—Ru—P′	103.99
P—Ru—Cl	89.05
P—Ru—Cen	121.56
P′—Ru—Cl	90.41
P′—Ru—Cen	121.36
Cl—Ru—Cen	122.49

^{a} Cen, C$_5$H$_5$ centroid.

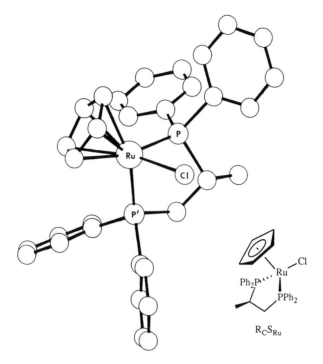

$R_C S_{Ru}$

FIG. 3. X-Ray crystal structure of (R_C,S_{Ru})-$[(\eta^5\text{-}C_5H_5)(PROPHOS)RuCl]$ (hydrogens removed for clarity).

imposed by the carbon backbone (Fig. 3, Table III) (*14*). The complete dissociation of chelating phosphines is not a facile process for both kinetic and thermodynamic reasons (*15*), and, consequently, reactivity of these complexes is often dissimilar to that of the bisphosphine analogs.

TABLE III

BOND ANGLES IN (R_C, S_{Ru})-
$[(\eta^5\text{-}C_5H_5)(PROPHOS)RuCl]^a$

Angle	θ (degrees)
P—Ru—P'	82.90
P—Ru—Cl	84.05
P—Ru—Cen	131.35
P'—Ru—Cl	92.93
P'—Ru—Cen	129.47
Cl—Ru—Cen	122.05

^{a} Cen, C_5H_5 centroid.

IV

EXCHANGE REACTIONS OF PPh$_3$ IN (η^5-C$_5$H$_5$)(PPh$_3$)$_2$RuX

The steric strain between the two bulky triphenylphosphine ligands together with the high electron density localized at the ruthenium center result in the ready dissociation of one triphenylphosphine ligand from (η^5-C$_5$H$_5$)(PPh$_3$)$_2$RuCl (1). Supporting evidence for this lability can be found in the analysis of mass spectrometric data for the complexes (η^5-C$_5$H$_5$)(PPh$_3$)$_2$RuX, which show a high abundance of the [(η^5-C$_5$H$_5$)-Ru(PPh$_3$)]$^+$ ion (13). Trapping the coordinatively unsaturated intermediate "(η^5-C$_5$H$_5$)Ru(PPh$_3$)Cl" with a variety of two-electron donor ligands provides a convenient route to phosphine-substituted ruthenium complexes (η^5-C$_5$H$_5$)(PPh$_3$)LRuCl. These complexes generally undergo a second substitution for the remaining bulky triphenylphosphine ligand [Eq. (8)], unless L is both sterically undemanding and electron withdrawing.

A. Exchange of PPh$_3$ for PR$_3$

Heating 1 in a nonpolar solvent (e.g., toluene or decalin) in the presence of excess phosphine or phosphite results in the smooth displacement of both triphenylphosphine ligands to give the bis-substituted ruthenium complexes in moderate to high yield [Eq. (9)] (5). This reaction repeated in

L	Yield(%)
PPh$_2$Me	77
PPhMe$_2$	30
PMe$_3$	65
PPh$_2$OMe	82
P(O^{i}Pr)$_3$	90

polar solvents promotes halide ionization and the eventual generation of trisphosphine ruthenium salts (*vide infra*). The mixed phosphine derivative (**12**) is formed in high yield as a racemic mixture of the two possible enantiomers by refluxing **1** with an equimolar amount of PMe_3 [Eq. (10)] (*13*).

$$\text{(10)}$$

Thermolysis of **1** in the presence of diphosphines such as $Ph_2PCH_2PPh_2$ (dppm) results in displacement of both triphenylphosphine groups and formation of the chelated phosphine derivative (**13**) [Eq. (11)] (*16*). The

$$\text{(11)}$$

reaction presumably proceeds via an intermediate monodentate dppm complex, which then ultimately displaces the second PPh_3 ligand to form the bidentate product. The monodentate intermediate (**14**) can be isolated if the reaction is conducted in a limited amount of benzene since precipitation of the complex occurs before displacement of the second triphenylphosphine ligand [Eq. (12)] (*17*). Conversion to the bidentate complex (**13**) is achieved simply by dissolution in additional benzene and heating.

$$\text{(12)}$$

Displacement of the phosphine groups in **1** by unsymmetrical bidentate phosphine ligands containing a chiral backbone leads to the formation of two diastereomers epimeric at the ruthenium center. Studies using the phosphines **15** (R = Me, Ph, c-C_6H_{11}) have shown that these ligands readily displace the triphenylphosphine groups at 80°C to give a 1:1 ratio of diastereomers **16** and **17** [Eq. (13)] (*18*). At lower temperatures (20°C) the diastereomeric selectivity increases to a maximum of approximately 4:1

$$
\begin{array}{c}
\mathbf{1} \xrightarrow[\substack{R = Me, Ph \\ c\text{-}C_6H_{11}}]{\mathbf{15}\ R} \mathbf{16} + \mathbf{17}
\end{array}
\tag{13}
$$

but with a considerable drop in yield (25%), even after stirring for 5 days. The two diastereomeric ruthenium complexes can be separated by fractional recrystallization to obtain the pure diastereomers in 30–40% yield (*14*). Epimerization of the pure diastereomers occurs at 80°C in C_6D_5Cl to give an equilibrium diastereomeric ratio of approximately 2:1, depending on the phosphine. This low asymmetric induction of the chiral ligand on the ruthenium center is presumably a consequence of the significant separation of the two chiral centers in the diastereomers (see Fig. 3).

Phosphines containing other metal-binding functional groups can also react with (η^5-C_5H_5)(PPh_3)$_2$RuCl (**1**) to form bidentate complexes. For example, 2-styryldiphenylphosphine displaces both triphenylphosphine groups to give the chelated complex **18** [Eq. (14)] (*19*). Reduction of **18** to

$$
\mathbf{1} \xrightarrow{\text{yield: 93\%}} \mathbf{18}
\tag{14}
$$

the hydrido complex **19** with sodium methoxide in methanol occurs in 87% isolated yield. Addition of carbon disulfide to **19** yields the insertion product **20**. An X-ray crystal structure suggests an unusual η^3-bonding mode for the dithiocarboxylate rather than the anticipated η^2-S,S' bonding [Eq. (15)].

$$(15)$$

B. *Exchange of PPh₃ for CO*

Stone and colleagues (*1*) have reported that one triphenylphosphine ligand in $(\eta^5\text{-}C_5H_5)(PPh_3)_2RuCl$ (**1**) can be replaced by CO either under forcing carbonylation conditions (150 atm CO) or via the addition of $Fe_2(CO)_9$ in tetrahydrofuran [Eq. (16)]. The latter reaction suffers from

$$(16)$$

occasional poor yields and contaminated product (*20*). Direct replacement of a phosphine group with carbon monoxide can be conducted under much milder reaction conditions (2 atm CO) if sulfur is used to drive the equilibrium to the right by removing the free triphenylphosphine as the sulfide [Eq. (17)] (*20*).

$$(17)$$

Decreases in electron density and steric strain at the ruthenium center on replacement of one PPh_3 for CO in $(\eta^5\text{-}C_5H_5)(PPh_3)_2RuX$ stabilizes the monophosphine complex, and, indeed, all attempts to prepare the dicarbonyl species $(\eta^5\text{-}C_5H_5)(CO)_2RuX$ via phosphine substitution have been unsuccessful. However, a mixture of both the mono- and disubstituted carbonyl complexes can be prepared under moderately forcing conditions (5 atm CO) from the corresponding pentamethylcyclopentadienylruthenium complex **4** [Eq. (18)] (21). This difference in reac-

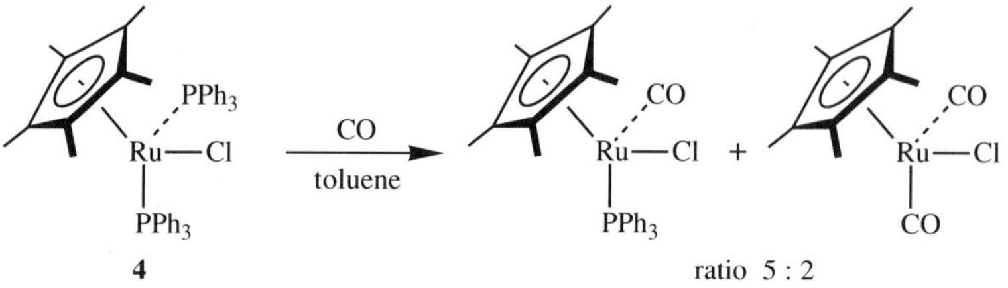

$$\text{(18)}$$

4   ratio 5 : 2

tivity is most likely a consequence of increases in both electron density and steric congestion at the metal center. Pure monosubstituted complex can be prepared in good yield by heating **4** in the presence of formic acid [Eq. (19)]. Presumably, the carbonyl hydride is the initial product which is converted to the chloride complex on work-up in CH_2Cl_2.

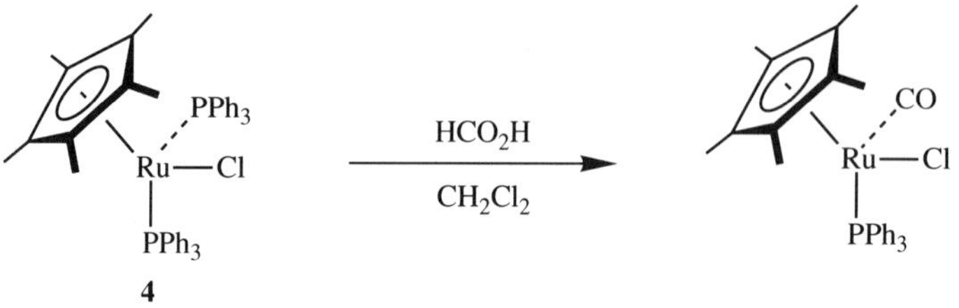

$$\text{(19)}$$

4

C. Exchange of PPh₃ for C, N, and S Donor Ligands

Bruce and Wallis (22) have reported that heating **1** in the presence of isocyanides results in the replacement of one of the triphenylphosphine ligands in high yield [Eq. (20)]. Isocyanides have steric and electron-with-

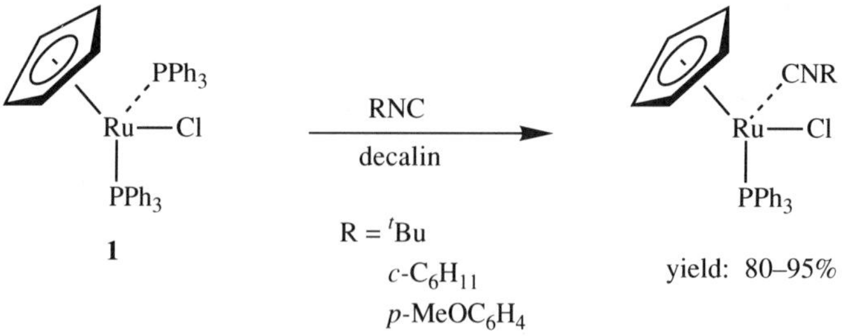

$$\text{(20)}$$

1

R = ^{t}Bu
    $c\text{-}C_6H_{11}$
    $p\text{-}MeOC_6H_4$

yield: 80–95%

drawing properties similar to carbon monoxide, and, as in the CO substitution chemistry described above, only monosubstitution is generally observed. Forcing conditions (180°C, 12 hours) did lead to the disubstituted product (21) for *tert*-butyl isocyanide, but only in low yield [Eq. (21)] (22). Again paralleling CO substitution chemistry, the doubly

$$(21)$$

substituted isocyanide complex 23 can be formed in the pentamethylcyclopentadienylruthenium system (21). Experimentally, this disubstitution proves most convenient via the intermediate formation of the chelating norbornadiene complex 22 [Eq. (22)].

$$(22)$$

Reaction of $(\eta^5\text{-}C_5H_5)(PPh_3)_2RuX$ with a wide range of N-heterocyclic bases results in the displacement of only one of the triphenylphosphine ligands in good yield. Reactions with bicyclic compounds (e.g., 2,2′-dipyridine) yield bridged diruthenium complexes (24) when X is I, CN, NCS, or $SnCl_3$ [Eq. (23)]. A cationic mononuclear species (25) is formed, however, from the displacement of both the anion X^- and one triphenylphosphine ligand when X is Br or Cl [Eq. (24)] (23).

Thiocarboxamides also displace a triphenylphosphine ligand from $(\eta^5\text{-}C_5H_5)(PPh_3)_2RuCl$ (1). Detailed infrared analyses indicate that the

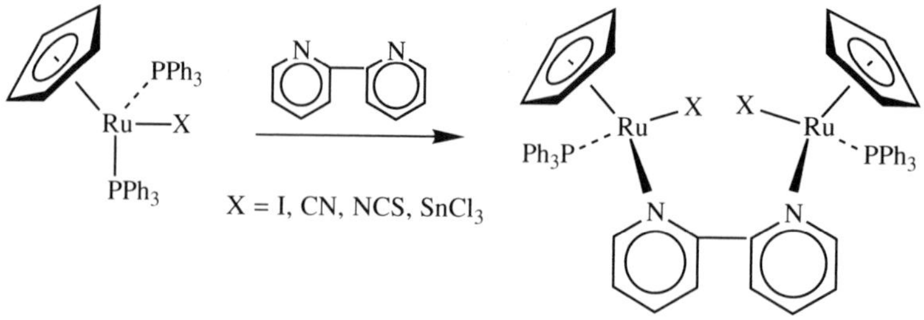

(23)

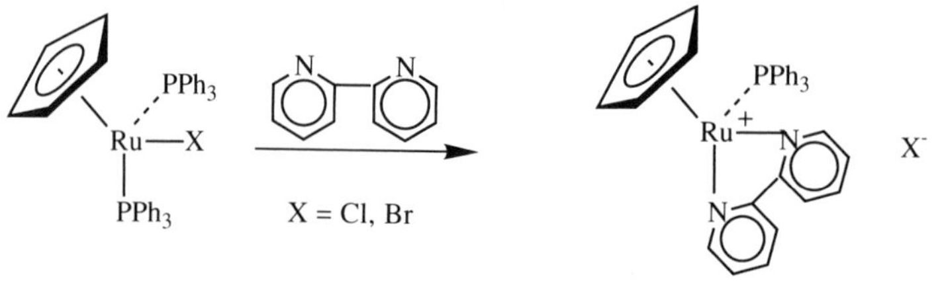

(24)

products can be broken into two classes: the first class binds to ruthenium via sulfur, whereas the second class binds via nitrogen [Eq. (25)] (24).

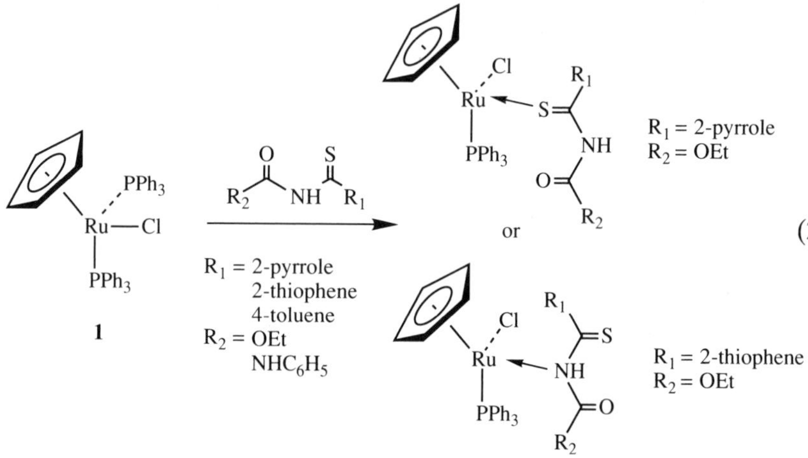

(25)

yield: 50–60%

Dithiocarboxylate anions with a range of electronic and steric properties thermally displace both the chloride ion and one of the triphenylphosphine groups from **1** to afford the η^2-S,S′ bound complexes (**26**) in varying yields [Eq. (26)] (*25*).

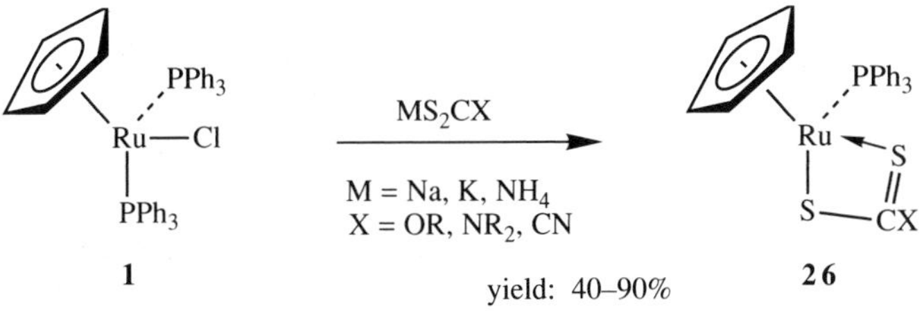

$$(26)$$

M = Na, K, NH$_4$
X = OR, NR$_2$, CN

yield: 40–90%

The tetrathiometallates $[MS_4]^{2-}$ (M = Mo, W) combine with **1** to form the air-stable, sulfur-bridged trimetallic species **27** [Eq. (27)] (*26*). The

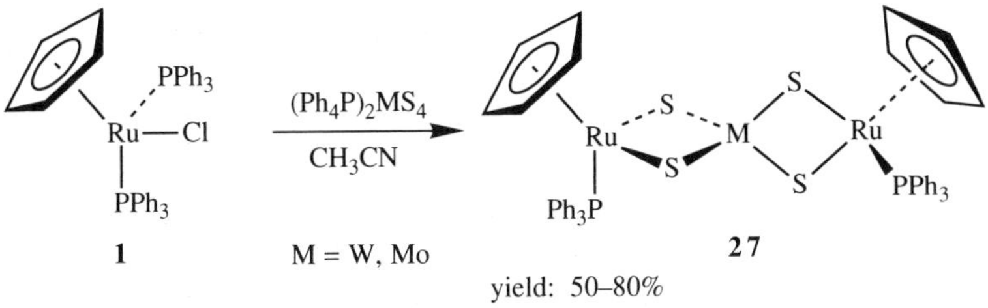

$$(27)$$

M = W, Mo

yield: 50–80%

corresponding tetraselenometallates can be formed in a similar fashion. The triphenylphosphine groups in **27** are readily exchanged for other phosphines in a manner similar to that of $(\eta^5$-C$_5$H$_5)$Ru(PPh$_3$)$_2$Cl. Surprisingly, only one of the two phosphine ligands can be exchanged for CO [(Eq. (28)]. This would imply that the electron-withdrawing effect on the substituted ruthenium center may be transmitted through the metal thiolate bridge to the second center.

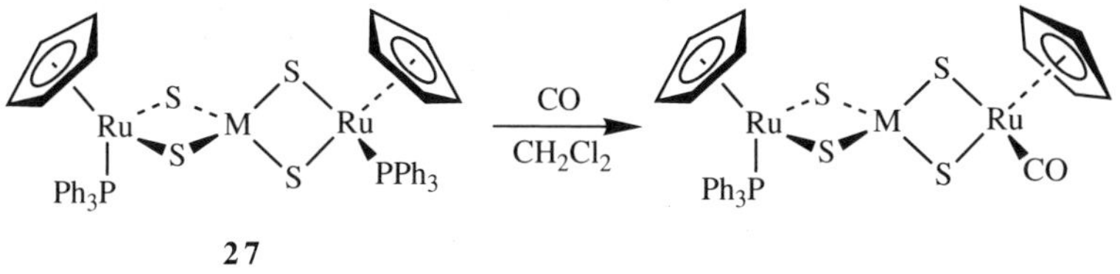

$$(28)$$

Displacement of both phosphine ligands and the halide ion in $(\eta^5$-C$_5$H$_5)$-(PPh$_3$)$_2$RuX can occur under certain conditions. For example, the addition of sodium tetraphenylborate to **1** in refluxing methanol affords the interesting sandwich compound (**28**) with one of the phenyl groups of BPh$_4^-$ η^6-bonded to the ruthenium atom [Eq. (29)] (*27*). Similarly, reaction of **1**

$$(29)$$

with thiophene in the presence of $AgBF_4$ gives complex **29** in moderate yield [Eq. (30)] *(28)*.

$$(30)$$

V

EXCHANGE REACTIONS OF X IN $(\eta^5\text{-}C_5H_5)(PR_3)_2RuX$

Whereas exchange reactions of $(\eta^5\text{-}C_5H_5)(PPh_3)_2RuX$ in nonpolar solvents generally result in the substitution of phosphine ligands, reactions in polar solvents lead to halide exchange. This section reviews general methods for ligand substitution of the halide in $(\eta^5\text{-}C_5H_5)L_2RuX$. Although the majority of examples given involve ligand substitution within the "parent compound" $(\eta^5\text{-}C_5H_5)(PPh_3)_2RuCl$ (**1**), it should be emphasized that similar reactivity is generally observed for other $(\eta^5\text{-}C_5H_5)L_2RuX$ compounds.

A. Exchange of X for X'

Treatment of **1** with a variety of inorganic salts MX in polar solvents leads to the displacement of Cl^- for X^-. The reaction most likely involves predissociation of the chloride ion followed by attack of the exchanging anion X^- at the coordinatively unsaturated ruthenium center [Eq. (31)]. Conductivity tests indicate that although $(\eta^5\text{-}C_5H_5)(PPh_3)_2RuCl$ is a non-electrolyte in acetone, appreciable ionic behavior develops in donor

$$\underset{\mathbf{1}}{\text{Cp(PPh}_3)_2\text{Ru}-\text{Cl}} \rightleftharpoons \underset{}{\text{Cp(PPh}_3)_2\overset{+}{\text{Ru}}-\text{S} \ \text{Cl}^-} \underset{-MX}{\overset{+ MX}{\rightleftharpoons}} \underset{+ MCl}{\text{Cp(PPh}_3)_2\overset{+}{\text{Ru}}-\text{X}} \quad (31)$$

S = solvent

solvents such as methanol, indicating that dissociation to the solvent-stabilized salt is a significant reaction (27).

The electron-rich ruthenium center can render the bound X ligand nucleophilic as has been demonstrated by the reactions of $(\eta^5\text{-}C_5H_5)$-$(PPh_3)_2Ru-C\equiv N$ (**30**) with a variety of electrophiles. Baird and Davies have shown, for example, that addition of alkyl halides to **30** gives the corresponding isocyanides in moderate yield [Eq. (32)] (29). Other

$$\underset{\mathbf{30}}{\text{Cp(PPh}_3)_2\text{Ru}-\text{CN}} \xrightarrow{\text{RX}} \text{Cp(PPh}_3)_2\overset{+}{\text{Ru}}-\text{CNR} \ \text{X}^- \quad (32)$$

RX	Yield(%)
EtI	45
$CH_2=CHCH_2Br$	44
$PhCH_2Br$	45
ICH_2CH_2OH	38

electrophiles, including $[Me_3O]BF_4$ (22), MeI (30), BF_3, (31), and HBF_4 (31), react similarly to give isonitrile-type derivatives. Interesting cyano-bridged bimetallic complexes can also be formed by reaction of $(\eta^5\text{-}C_5H_5)$-$(PR_3)_2Ru-C\equiv N$ with $(\eta^5\text{-}C_5H_5)(PR'_3)_2MCl$ [Eq. (33)] (32).

$$\underset{\mathbf{30}}{\text{CpL}_2\text{Ru}-\text{CN}} + \text{Cl}-\text{ML'}_2\text{Cp} \longrightarrow \left[\text{CpL}_2\text{Ru}-\text{CN}-\text{ML'}_2\text{Cp} \right]^+ \text{Cl}^- \quad (33)$$

M = Ru, Fe
L, L'=PPh$_3$, dppe

B. Exchange of X for R

The reaction of **1** with primary Grignard reagents is the most versatile route to alkyl complexes of type **31** [Eq. (34)] (33). More hindered

$$(34)$$

R	Yield (%)
Me	52
Et	31
n-Pr	35
n-Bu	70
CH$_2$CHMe$_2$	52

secondary and tertiary alkyl Grignard reagents, however, generally fail to give the anticipated ruthenium alkyl complexes, forming instead the hydride complex **32** and free alkene [Eq. (35)]. One exception to this observa-

$$(35)$$

tion is the reaction of **1** with cyclopropylmagnesium bromide which affords the cyclopropylruthenium complex **33** in high yield [Eq. (36)] (*33*).

$$(36)$$

yield: 87%   7%

The mechanism for alkylation with Grignards has been proposed by Lehmkuhl *et al.* (*33*) to involve a concerted halide and alkyl transfer between Mg and Ru (Scheme 1, Path A). The propensity of bulky, non-strained secondary and tertiary alkyl Grignard reagents with β hydrogens to act as hydride donors (*34*) would explain the formation of **32** in these reactions, invoking a similar concerted exchange (Scheme 1, Path B). The strained cyclopropyl Grignard reagent is both a poor hydride source and relatively small, which explains the formation of the secondary alkylruthenium product in this case.

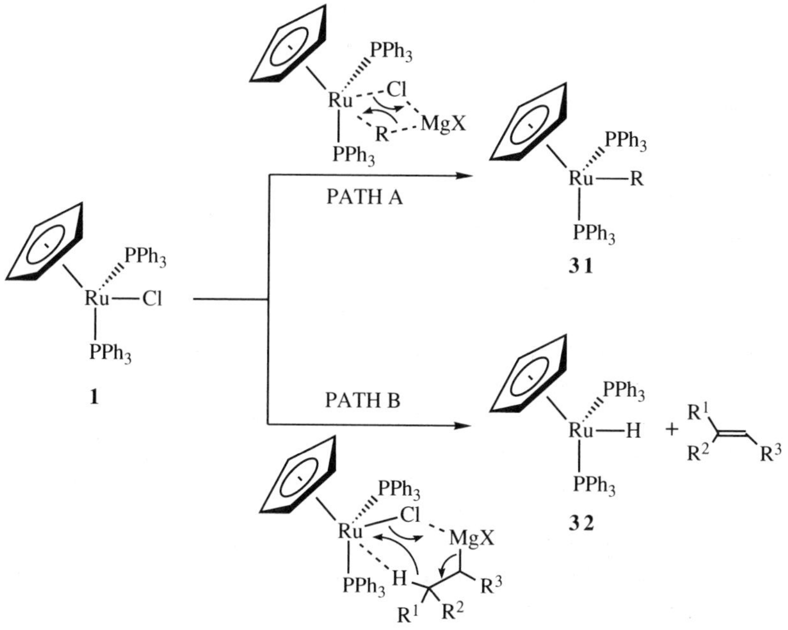

SCHEME 1

Stereochemical studies by Morandini and Consiglio and colleagues, on the reaction of Grignard reagents with the separate epimers of (η^5-C_5H_5)-[(*R*)-PROPHOS]RuCl (**34** and **35**) support the mechanisms shown in Scheme 1 since they indicate that the formation of both alkyl and hydride products proceeds with retention of configuration at the ruthenium center [Eqs. (37) and (38)] (*35*). It should be noted, however, that dissociative pathways to the alkyl complexes via coordinatively unsaturated ruthenium

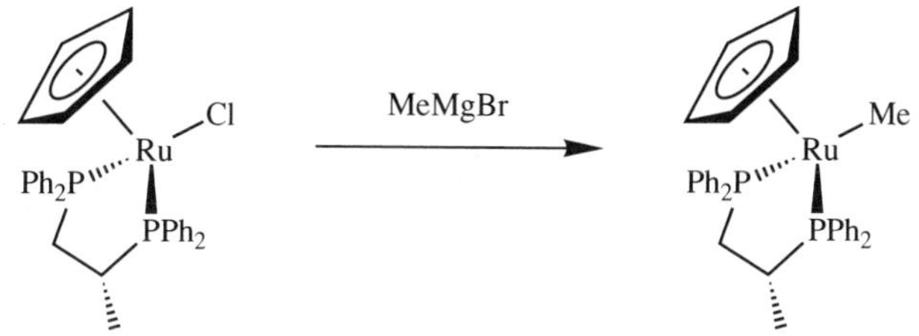

R_CR_{Ru} $\quad$ R_CR_{Ru} $\quad$ (37)

34

Similarly, R_CS_{Ru} (**35**) ⟶ R_CS_{Ru}

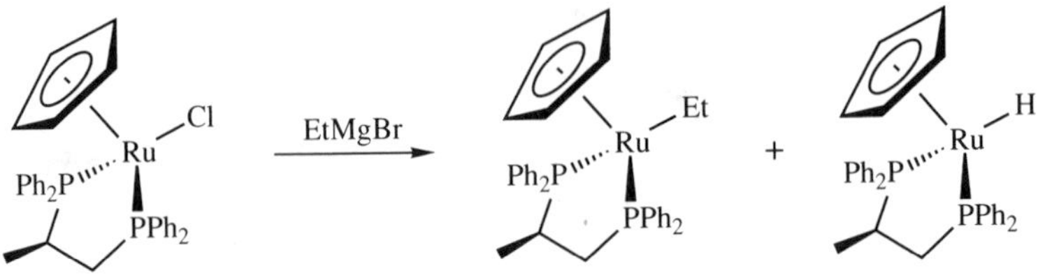

$$(38)$$

R_CS_{Ru}
35

R_CS_{Ru}

R_CS_{Ru}

Similarly, R_CR_{Ru} (**34**) $\longrightarrow$ R_CR_{Ru} + R_CR_{Ru}

intermediates cannot be ruled out as a consequence of the observed stereo-selectivity of this reactions since there is growing evidence that these unsaturated intermediates can retain their stereochemical integrity at the metal center (*vide infra*). Another potential route to the hydride complex **32**, involving initial alkylation at ruthenium followed by loss of phosphine, β-hydrogen elimination, and finally phosphine substitution for alkene, is apparently incompatible with these stereochemical studies as this mechanism would result either in overall inversion of configuration at the ruthenium center or in epimerization (*33*).

Grignard reagents derived from the alkyl chloride are generally preferred over those from the bromide or iodide. For example, MeMgI reacts with $(\eta^5\text{-}C_5H_5)(PPh_3)_2RuCl$ to give $(\eta^5\text{-}C_5H_5)(PPh_3)_2RuI$ rather than the anticipated methyl complex $(\eta^5\text{-}C_5H_5)(PPh_3)_2RuMe$ (*1*). Also, Tilley *et al.* have found that $(\eta^5\text{-}C_5Me_5)(PMe_3)_2RuCl$ reacts more rapidly with RMgCl than RMgBr to give ruthenium alkyl complexes in high yield [Eq. (39)] (*36*).

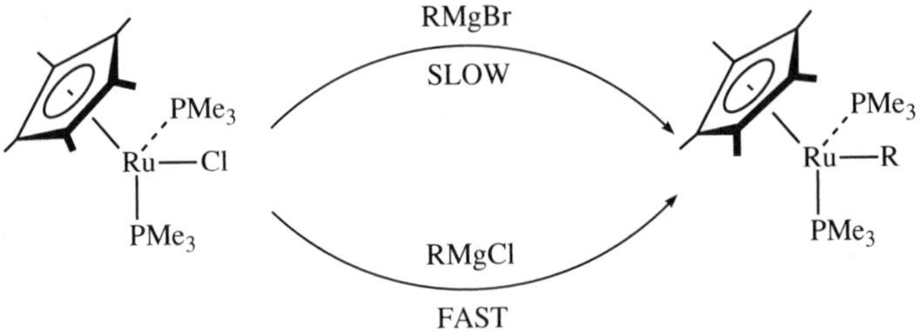

$$(39)$$

This trend reflects the relative Lewis acidity of the Mg center in the Grignard reagents and presumably their relative ability to abstract chloride from the ruthenium complex.

Although not generally as useful as the Grignard reagents, alkyllithium reagents have been used with some success in alkylation reactions. For example addition of MeLi [Eq. (40)] or C_6F_5Li [Eq. (41)] to **1** gives the corresponding methyl- and perfluorophenylruthenium complexes **36** and **37** in good and marginal yield, respectively (*1*).

$$ (40) $$

$$ (41) $$

C. *Exchange of X for H*

As mentioned above, reducing Grignards provide one method for the conversion of $(\eta^5\text{-}C_5H_5)(PPh_3)_2RuX$ to the air-stable yellow crystalline hydride **32**; this route, however, is neither particularly clean nor cost effective. Stone and colleagues found that the reaction of $LiAlH_4$ with $(\eta^5\text{-}C_5H_5)(PPh_3)_2RuX$ gives **32** (*1*), but this synthesis is hampered by variable yields and the generation of a colorless trihydride side product (**38**) on displacement of a phosphine ligand [Eq. (42)] (*37*). Since the

$$ (42) $$

monohydride converts to the trihydride only very slowly in hot THF in the presence of $LiAlH_4$, it is unlikely that **32** is a precursor to **38**. The further observation that treatment of $(\eta^5\text{-}C_5H_5)(PPh_3)_2RuCl$ (**1**) with $LiAlD_4$ followed by water gave **38** containing approximately 30% H at the metal

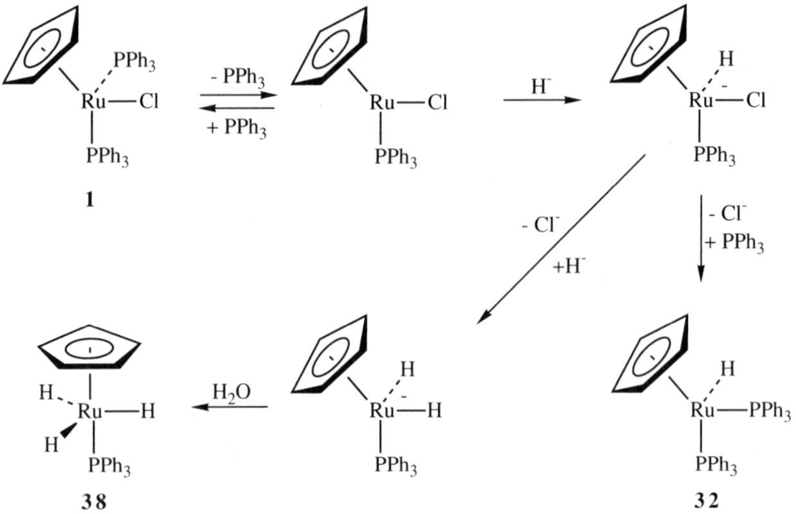

SCHEME 2

center suggests that the mechanism shown in Scheme 2 is the most plausi-ble for the formation of the two products (37).

The reaction of NaOMe on $(\eta^5\text{-}C_5H_5)(PPh_3)_2RuX$ in refluxing MeOH provides the best synthesis of **32** (38,39). The most likely mechanism involves S_N1 displacement of chloride for methoxide at the metal center followed by rapid β-hydrogen elimination and concurrent loss of for-maldehyde to give $(\eta^5\text{-}C_5H_5)(PPh_3)_2RuH$ [Eq. (43)]. Conversion of the

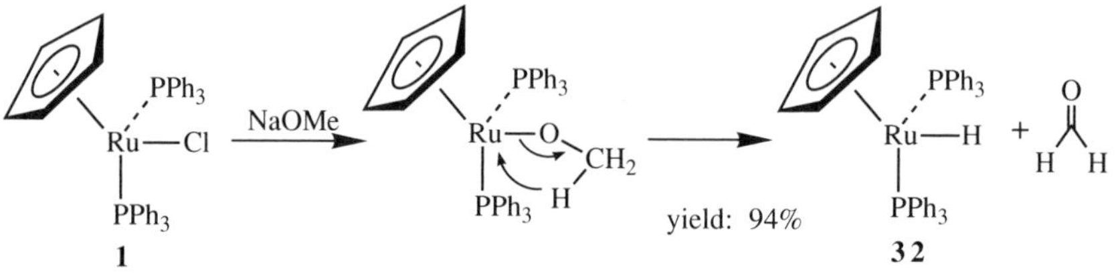

hydride complexes $(\eta^5\text{-}C_5H_5)(PR_3)_2RuH$ back to the corresponding halides $(\eta^5\text{-}C_5H_5)(PR_3)_2RuX$ is easily accomplished by treatment with a suitable polyhalogenated alkane or HX. For example, treatment of **32** with Cl_3CCH_2OH smoothly affords the original chloride complex **1** in near quantitative yield [Eq. (44)] (38).

$$\text{(44)}$$

D. Exchange of X for L

Exchange of X for a neutral two-electron donor ligand L, including, for example, CO (13,27,40), MeOH (27), RCN (1,13,16,40–43), Py (40,44), PR_3 (13,16), $P(OR)_3$ (16), alkenes (40,45,46), alkynes (45,46), and CS_2 (46), generates ruthenium salts of the form $(\eta^5\text{-}C_5H_5)(PR_3)_2RuL^+X^-$. The general mechanism involves initial ionization of the complex to form the coordinatively unsaturated intermediate 39. Competitive attack on 39 by L or X^- affords the substituted product or regenerates the starting material, respectively [Eq. (45)]. The relative nucleophilicity of X^- and L, the polarity of the solvent, and the steric constraints imposed by the two phosphine ligands are all important factors influencing the equilibrium between the neutral and charged species.

$$\text{(45)}$$

Treichel and colleagues have measured the relative rates of substitution of X^- for CD_3CN (41) and DMSO-d_6 (47) in a number of complexes $(\eta^5\text{-}C_5H_5)(PPh_3)_2RuX$, and they have been able to correlate the dissociation rate of X^- to the relative electron-donating properties of the phosphines. It was found, for example, that the rate of chloride substitution for DMSO-d_6 is around 100 times faster in $(\eta^5\text{-}C_5H_5)(PMe_3)_2$-RuCl than in $(\eta^5\text{-}C_5H_5)(PPh_2OMe)_2RuCl$, presumably owing to the greater stability of the unsaturated cationic intermediate with the more electron-donating PMe_3 ligands. Replacement of the counterion X^- with a less nucleophilic anion such as PF_6^- or BF_4^- may be necessary to drive the reaction completely over to the salt. The removal of the free halide anion is most easily accomplished by precipitation as the Ag(I) or Tl(I) salt. Precipitation of the ruthenium complex as the halide salt will also push the equilibrium to the right.

The stereochemical consequences of L for X^- exchange at the ruthenium center have been studied by Morandini *et al.* (*48*). Using the pure epimers **34** and **35**, it was determined that the exchange of chloride for acetonitrile proceeds stereospecifically with retention of configuration at ruthenium [Eq. (46)] (*48*). This study provides good evidence that the stereochemical integrity of the ruthenium center is maintained in the coordinatively unsaturated intermediate formed on loss of Cl^-.

$$\text{(46)}$$

Similarly, $R_C S_{Ru}$ (**35**) $\longrightarrow$ $R_C S_{Ru}$

Intramolecular displacement of halide ions has also been observed in several appropriately substituted complexes. For example, the thiophene unit tethered to the cyclopentadienyl ring in complex **40** can be induced to displace the chloride and form the sulfur-bound ruthenium cation **41** shown in Eq. (47) (*49*).

$$\text{(47)}$$

Treatment of $(\eta^5\text{-}C_5H_5)(PPh_3)_2RuCl$ with the tridentate ligands TRIPOD [$(Ph_2PCH_2)_3CMe$] or TRIPHOS [$(Ph_2PCH_2CH_2)_2PPh$] affords the tridentate ruthenium salts [$(\eta^5\text{-}C_5H_5)(TRIPOD)Ru$]Cl and [$(\eta^5\text{-}C_5H_5)$-(TRIPHOS)Ru]Cl, respectively (*50,51*). The mechanism presumably involves the chelation-aided systematic replacement of the two triphenylphosphine ligands and the chloride by the tridentate phosphine as shown in Eq. (48) for the TRIPHOS example.

$$(48)$$

An unusual route to the formation of $(\eta^5\text{-}C_5H_5)(PR_3)_2RuL^+$ from $(\eta^5\text{-}C_5H_5)(PR_3)_2RuX$ involves the transformation of the X ligand to the L ligand. Methylation of **42** with $MeSO_3CF_3$ yields the highly reactive salt **43** which can be used as a potent methylating reagent for a wide range of nucleophiles [Eq. (49)] (52).

$$(49)$$

E. Reactions of $(\eta^5\text{-}C_5H_5)(PR_3)_2RuX$ with Electrophiles

The electron-rich ruthenium center of $(\eta^5\text{-}C_5H_5)(PR_3)_2RuX$ combines with certain electrophiles E^+ to generate highly reactive cationic addition complexes susceptible to elimination–substitution reactions, as shown in Eq. (50).

$$\text{(50)}$$

1. Reactions of $(\eta^5\text{-}C_5H_5)(PPh_3)_2RuX$ with Acids

Treatment of $(\eta^5\text{-}C_5H_5)(PMe_3)_2RuCl$ (**11**) (*53*) or $(\eta^5\text{-}C_5Me_5)(PMe_3)_2\text{-}$RuMe (*54*) with noncomplexing acids such as HPF_6 or HBF_4 leads to the protonated complexes **44** and **45**, respectively [Eqs. (51) and (52)]. Unfortunately, the geometry and stereochemistry of these species are unknown.

$$\text{(51)}$$

$$\text{(52)}$$

A similar reaction of the basic $(\eta^5\text{-}C_5H_5)(PMe_3)_2RuX$ with complexing acids HA, however, results in ultimate formation of the substitution product $(\eta^5\text{-}C_5H_5)(PMe_3)_2RuA$. Bryndza *et al.* (*54*) have taken advantage of the generality of this reaction to measure the equilibrium constants for the reaction of $(\eta^5\text{-}C_5Me_5)(PMe_3)_2RuA$ with HA′ for a wide range of A and A′ [Eq. (53)]. The electron-rich $(\eta^5\text{-}C_5Me_5)(PMe_3)_2RuA$ system was chosen for the study to facilitate formation of cationic species during the course of the reaction. The equilibria data, together with the K_a values for HA and HA′, were used to estimate the relative homolytic and heterolytic dissociation energies for the $[(\eta^5\text{-}C_5Me_5)(PMe_3)_2Ru]$—A bond. The observed homolytic bond strengths were of the order Ru—(sp)C >

$$\text{Ru—A} + \text{HA'} \underset{}{\overset{K_{eq}}{\rightleftharpoons}} \text{Ru—A'} + \text{HA} \quad (53)$$

$Ru—O > Ru—H > Ru—(sp^3)C > Ru—NR_2$, and can be correlated to the bond dissociation energies of H—A.

Addition of HPF_6 or HBF_4 to $(\eta^5\text{-}C_5H_5)(L \triangle L)RuH$ [$L \triangle L$ = chelating diphosphines $Ph_2P(CH_2)_nPPh_2$, $n = 1, 2, 3$] generates the anticipated dihydride species **46** in equilibrium with the η^2-dihydrogen complexes **47** [Eq. (54)] (55,56). Molecular hydrogen bound to a metal center represents

46   **47**

an important potential intermediate in the activation of dihydrogen by transition metal complexes, and work done with the ruthenium examples described above, including deprotonation and H/D exchange reactions, provides information on the synthesis and reactivity of these complexes complementary to that of the Mo (57) and W (57,58) examples studied by Kubas and colleagues.

2. Insertion of Electrophiles into the Ru—X Bond

An interesting class of reactions of $(\eta^5\text{-}C_5H_5)(PR_3)_2RuX$ involves the insertion of certain electrophiles (E) into the Ru—X bond [Eq. (55)].

$$\text{Ru—X} \xrightarrow{E} \text{Ru—E—X} \quad (55)$$

Bruce has shown that **32** reacts with CS_2 over 12 hours to give the $\eta^1\text{-}S$-dithioformate insertion product **48** in high yield [Eq. (56)] (39). Although the mechanism of this reaction has not been determined, a reasonable

$$(56)$$

32   **48**

route involves initial loss of a phosphine ligand, followed by coordination of the CS_2 unit, which then inserts into the Ru—H bond with recoordination of phosphine.

A more recent example by Bruce et al. (59) of CS_2 insertion into the Ru—C bond of an acetylide complex yields the η^2-S,S-dithiocarboxylate complex **49**. Again, the most likely mechanism involves initial replacement of a phosphine ligand for CS_2 followed by insertion into the Ru—C bond. In contrast to the dithioformate example, however, the η^1-dithiocarboxylate complex rearranges to the η^2- complex rather than recoordinating phosphine [Eq. (57)].

$$(57)$$

Similarly, stannous chloride can insert into the Ru—Cl bond of **1** to yield the bimetallic complex **50** [Eq. (58)] (1). Consiglio and Morandini and

$$(58)$$

1   **50**

colleagues have conducted a stereochemical study of the $SnCl_2$ insertion with the separate epimers **34** and **35** and found that the reaction proceeds stereospecifically with net retention of configuration at the metal center [Eq. (59)] (*60*). The high stereospecificity of the insertion rules out phos-

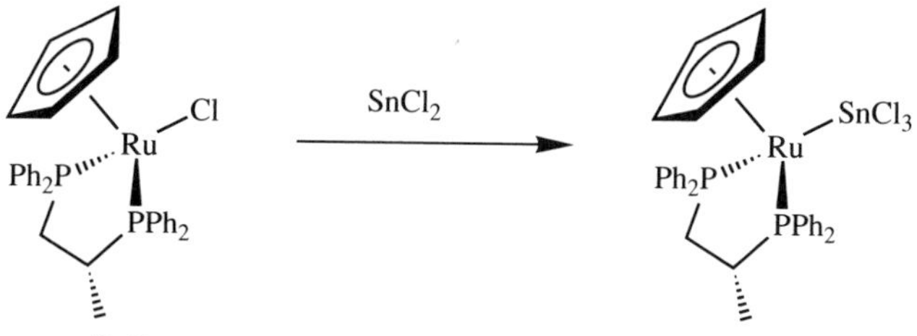

R_CR_{Ru} **34** $\qquad$ R_CS_{Ru} (59)

Similarly, R_CS_{Ru} (**35**) ⟶ R_CR_{Ru}

phine dissociation during the reaction, and a plausible mechanism involves initial formation of the tight ion pair **51**, retaining stereochemical integrity at the metal center, which eventually collapses to **52** (Scheme 3).

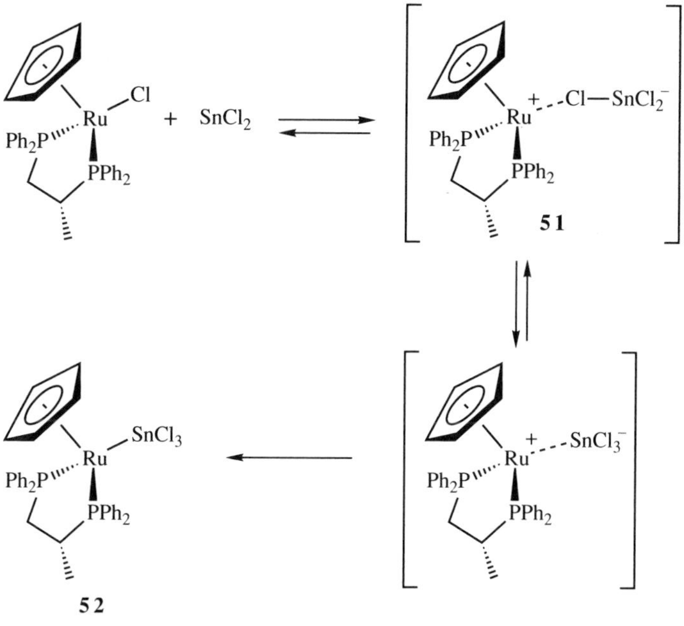

SCHEME 3

VI

RUTHENIUM ACETYLIDE, VINYLIDENE, AND CARBENE COMPLEXES

A wide range of complexes can be readily derived from $(\eta^5\text{-}C_5H_5)$-$(PPh_3)_2RuCl$ (1) and terminal acetylenes. Thus, treatment of methanol solutions of 1 with a terminal acetylene generates, presumably via the η^2 intermediate 53, the vinylidene cation (54). Deprotonation of 54 with weak bases generates the η^1-acetylide complexes (55), which may also be prepared directly from 1 with lithium acetylides. The acetylide complexes (55) are nucleophilic at the β carbon, protonating to regenerate 54 or reacting with a variety of electrophiles to give the disubstituted vinylidene cation (56). The cationic vinylidene complexes (56 and 54) are electrophilic at the α carbon and react with nucleophiles to generate the η^1-vinyl complex (57). Complex 57 is, in turn, nucleophilic at the β carbon, reacting with elec-

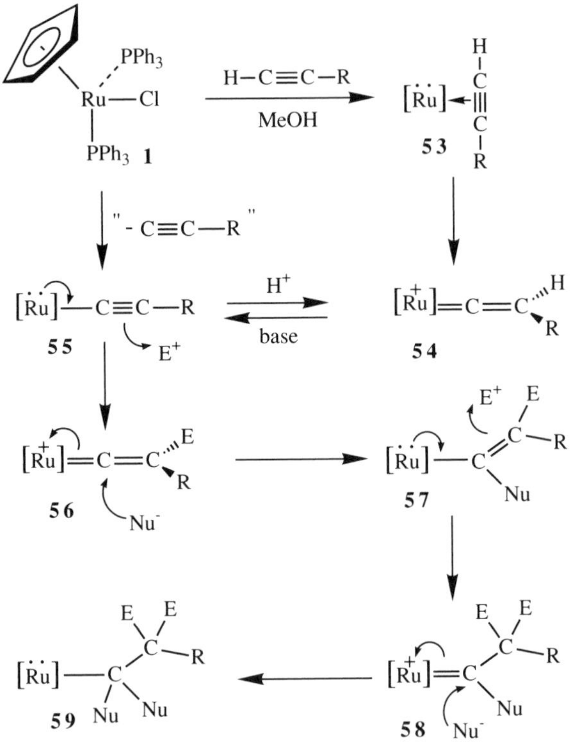

SCHEME 4

trophiles to generate the cationic carbene complex (**58**). Finally, nucleophilic addition to the α carbon of **58** generates the alkyl complexes (**59**) (Scheme 4). The ease and versatility of these interconversions make them potentially extremely useful for synthesis (*vide infra*).

A. Ruthenium Acetylide Complexes

1. Preparation

As described in Section V,B, ruthenium–carbon bonds can be formed by the reaction of $(\eta^5\text{-}C_5H_5)(PPh_3)_2RuCl$ with a variety of metal alkyl reagents. This route can also be applied to the synthesis of acetylide complexes with varying success. Consiglio and Morandini and colleagues have shown, for example, that the addition of lithium phenylacetylide to $(\eta^5\text{-}C_5H_5)[(R)\text{-PROPHOS}]RuCl$ (**34**) affords the acetylide complex (**60**). The substitution was found to occur with retention of configuration at the ruthenium center for both epimers of the ruthenium complex [Eq. (60)] (*61*). This is consistent with the short-lived coordinatively unsaturated intermediate formed on dissociation of Cl$^-$ being configurationally stable on the time scale necessary for the attack of the acetylide to occur (see Section V,D).

$$(60)$$

Similarly, R_CS_{Ru} (**35**) $\longrightarrow$ R_CS_{Ru}

In contrast to the lithium acetylide reaction, addition of copper(I) phenylacetylide to $(\eta^5\text{-}C_5H_5)(PPh_3)_2RuCl$ (**1**) affords the monomeric ruthenium acetylide–copper chloride adduct (**62**) as the major product. An X-ray crystal structure of this complex reveals an η^1,η^2-bridging acetylide between the ruthenium and copper centers, respectively (*62*). A small amount of the dimeric chloride bridged complex **61** was also isolated. The copper chloride can be removed from the monomeric complex by the

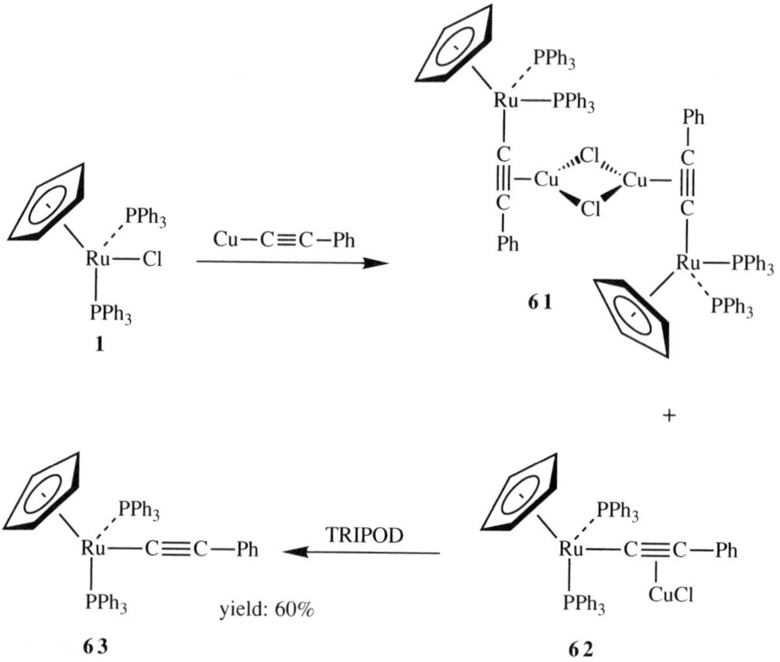

SCHEME 5

addition of TRIPOD to give the ruthenium acetylide complex (63) in moderate yield (Scheme 5). No reaction occurs on addition of a non-chelating phosphine (e.g., PPh$_3$) to the copper complex (63).

Although the reaction of copper acetylides with transition metal halides has been successfully applied to the preparation of a variety of transition metal acetylides (64), the generation of copper-complexed derivatives is not unprecedented (65). A simpler and more general route to ruthenium acetylide complexes involves the deprotonation of ruthenium vinylidene complexes as described in Section VI,C.

2. Geometry

An X-ray crystal structure of (η^5-C$_5$H$_5$)(PPh$_3$)$_2$Ru—C≡CPh (63) (66) (Fig. 4) again shows the octahedral geometry about the ruthenium center with a slight distortion of the P—Ru—P angle owing to the steric bulk of the two adjacent triphenylphosphine ligands (Table IV). A recurring principle in the chemistry of unsaturated organic groups bound to (η^5-C$_5$H$_5$)-(PR$_3$)$_2$Ru is that steric constraints imposed by bulky phosphine ligands

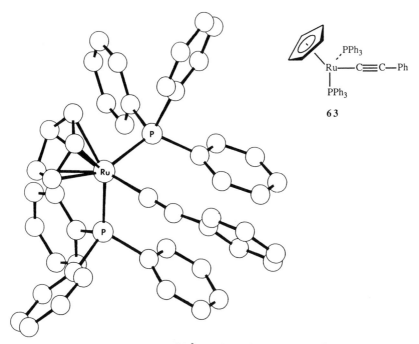

FIG. 4. X-Ray crystal structure of $[(\eta^5\text{-}C_5H_5)(PPh_3)_2Ru\text{—}C\equiv CPh]$ (**63**) (hydrogens removed for clarity).

influence the conformation, and potentially the reactivity, of the unsaturated organic moiety. In the above complex, the acetylenic phenyl group is forced to lie in the plane bisecting the phosphine groups, slotted between one phenyl ring of each phosphine, as best shown in the space filling representation (Fig. 5).

TABLE IV

BOND ANGLES IN
$[(\eta^5\text{-}C_5H_5)(PPh_3)_2Ru\text{—}C\equiv CPh]$ (**63**)[a]

Angle	θ (degrees)
P—Ru—P′	100.84
P—Ru—C	89.17
P—Ru—Cen	123.48
P′—Ru—C	88.54
P′—Ru—Cen	122.53
C—Ru—Cen	123.04

[a] Cen, C_5H_5 centroid.

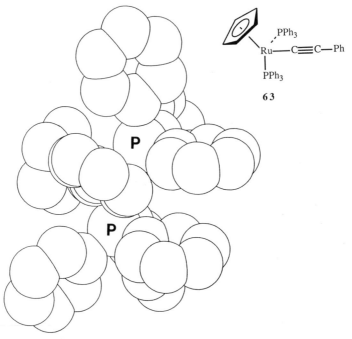

FIG. 5. Space filling representation of $[(\eta^5\text{-}C_5H_5)(PPh_3)_2Ru\text{---}C\equiv CPh]$ **(63)** (hydrogens removed for clarity).

B. Monosubstituted Ruthenium Vinylidene Complexes

1. Preparation

The interaction of an alkyne with $(\eta^5\text{-}C_5H_5)(PR_3)_2RuX$ can result in the formation of a wide variety of ruthenium complexes. The nature of the products formed depends on the conditions used and the type of alkyne reacted. Reactions between **1** and terminal alkynes in the presence of ammonium hexafluorophosphate lead to the formation of cationic mono-substituted ruthenium vinylidene complexes in high yield, as shown for phenylacetylene in Eq. (61) *(4,67,68)*.

Consiglio and Morandini and co-workers (*61*) have investigated the stereochemistry involved in the addition of acetylenes to chiral ruthenium complexes. Reaction of propyne with the separated epimer of the chiral ruthenium phosphine complex **34** at room temperature results in the chemo- and stereospecific formation of the respective propylidene complex **64**. An X-ray structure of the product (**64**) proves that the reaction proceeds with retention of configuration at the ruthenium center. The identical reaction utilizing the epimer with the opposite configuration at ruthenium (**35**) also proceeded with retention of configuration at the metal center, proving that the stereospecificity of the reaction in not under thermodynamic control [Eq. (62)].

$$\text{Similarly, } R_CS_{Ru} \; (\textbf{35}) \longrightarrow R_CS_{Ru}$$

Similar reactions of **34** and **35** with phenylacetylene at room temperature are also stereospecific, and they are presumed to occur with retention of configuration at the metal center by analogy to the propyne reactions. When these reactions are performed in refluxing methanol, both chemoselectivity and stereospecificity are lost, with almost equal amounts of the two benzylidene diastereomers (**65** and **66**) and a small amount (10–15%) of the methanol adducts (**67**) (*vide infra*) being formed from **34** [Eq. (63)]. The individual benzylidene epimers **65** and **66** do not epimerize at the ruthenium center in refluxing methanol, which indicates that the loss of stereochemical integrity occurs prior to addition of the acetylene.

2. Mechanism

Silvestre and Hoffmann have considered the conversion of ruthenium acetylide complexes to the corresponding vinylidene species using extended Hückel molecular orbital calculations (*69*). Although the rearrangement of free acetylene to its vinylidene isomer is thermodynamically disfavored, their results indicate that the transformation becomes thermodynamically

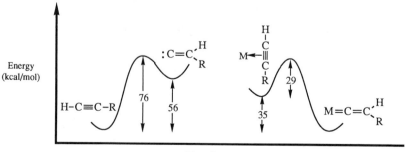

favorable on complexation of the starting acetylene and product vinylidene to an appropriate metal fragment ML_n (Fig. 6).

Analysis of the intimate mechanism for conversion of L_nM-$(\eta^2$-HC≡CH) to $L_nM=C=CH_2$ suggests that the reaction most likely involves initial slippage of the acetylene from an η^2- to an η^1-binding mode. In this geometry, positive charge develops on the β carbon, and hydrogen transfer in a "1,2-hydride shift" type mechanism with concurrent

Fig. 6.

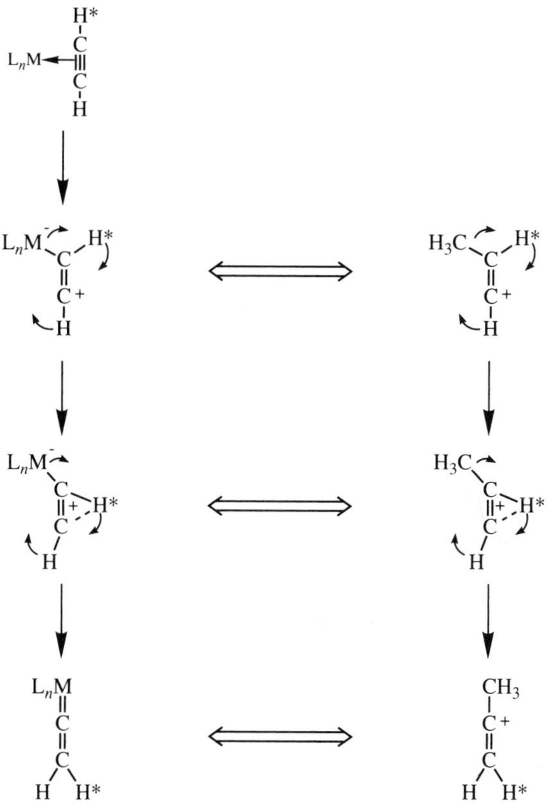

SCHEME 6

further migration of the ML_n fragment onto the C_α–C_β axis gives the product vinylidene species (Scheme 6). Silvestre and Hoffmann (69) suggest that the reaction is analogous to the isomerization of methyl vinyl cation since ML_n is isolobal with CH_3.

An intermolecular version of this rearrangement involving dissociation of the acidic proton on C_α of the slipped acetylene, followed by reprotonation of an intermediate acetylide (discussed in Section VI,C), must also be considered as a potential route to the cationic ruthenium vinylidene species (Scheme 7). Unfortunately, to date this mechanism has not been addressed experimentally or theoretically.

Another mechanism for the rearrangement involving insertion of the ruthenium into the C—H bond of the acetylene, followed by hydride migration to the β carbon, was also explored by Silvestre and Hoffmann

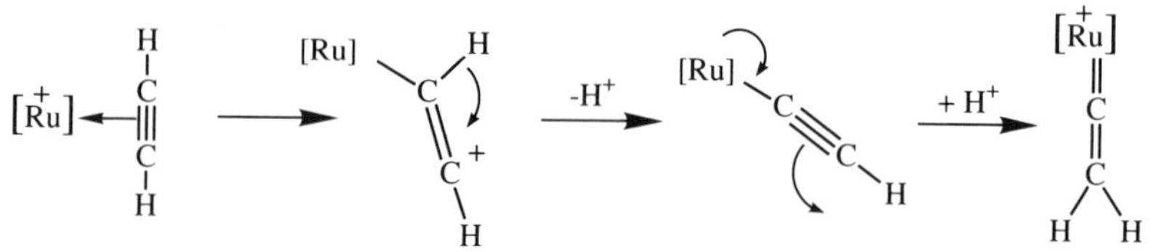

SCHEME 7

(Scheme 8). Although the insertion step forming the intermediate acetylene hydride complex appears feasible, the migration of hydride from the metal to the β carbon is energetically too costly for this to be a significant pathway for the reaction (69).

3. Geometry

Kostic and Fenske, in their general study of the bonding of unsaturated ligand moieties to metal fragments of the form $(\eta^5\text{-}C_5H_5)L_2M$, determined that the preferred geometry of the vinylidene unit has the plane containing the β-carbon constituents perpendicular to the symmetry plane of the molecule (Fig. 7) (70). Fenske–Hall self-consistent field calculations conducted on $(\eta^5\text{-}C_5H_5)(PH_3)_2Ru=C=CH_2$ indicate that the maximum π stabilization between metal and α carbon arises from interaction of the p-type orbital on the unsaturated organic fragment lying in plane with the molecular orbital of $2a''$ symmetry on the metal fragment. In this case, however, the difference in molecular energy for the vertical and horizontal geometries of the vinylidene unit is small (<2 kcal), and steric interactions between the substituents of the vinylidene and the bulky phosphine ligands would be anticipated to stabilize the horizontal geometry shown in Fig. 7 relative to the vertical geometry.

Several X-ray crystallographic studies of $[(\eta^5\text{-}C_5H_5)(PR_3)_2\text{-}Ru=C=CRR']^+$ cations have been reported (61,71), and, in all cases, the vinylidene fragment lies orthogonal to the plane bisecting the

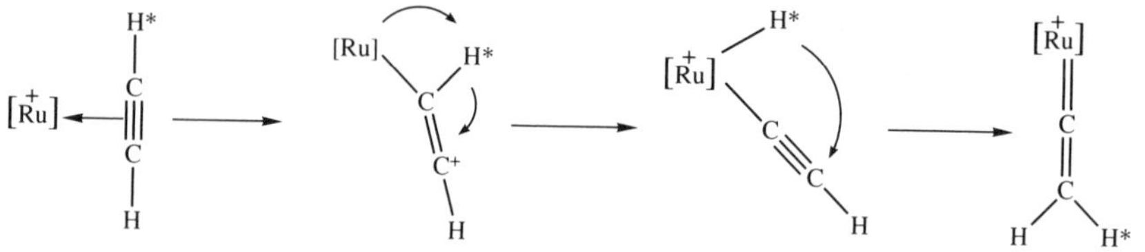

SCHEME 8

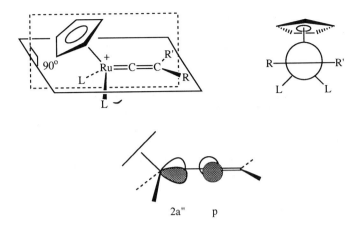

P—Ru—P angle as predicted. The structure of the methylvinylidene cation $[(\eta^5\text{-}C_5H_5)(PMe_3)_2Ru{=}C{=}CHMe]^+$ (**68**) (72) clearly shows the linearity of the vinylidene unit (Fig. 8a) and its orthogonal relationship with the plane bisecting the P—Ru—P angle (Fig. 8b).

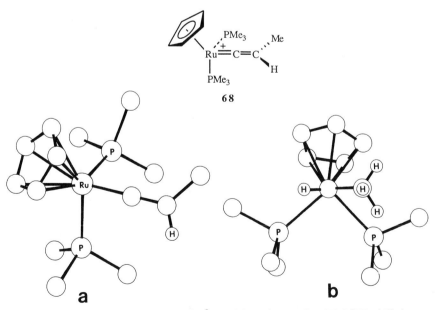

FIG. 8. X-Ray crystal structure of $[(\eta^5\text{-}C_5H_5)(PMe_3)_2Ru{=}C{=}CHMe]PF_6$ (**68**) (some hydrogens and PF_6^- counterion removed for clarity).

Replacement of the two trimethylphosphine ligands for the chelating (R)-PROPHOS results in a considerable decrease of the coordination space available to the vinylidene fragment. As a consequence, the plane of the vinylidene unit is tipped slightly with respect to the idealized orthogonal geometry seen in complex **68** to minimize steric interaction between the terminal methyl group and the phosphine ligand (Fig. 9) (61). In the space filling representation (Fig. 10), two phenyl rings of the diphosphine in this complex are seen to be oriented in a manner which partially protects C_α and C_β from the approach of reactants, and it is not surprising that, as seen by Bruce and Swincer (73), the reactivity of the vinylidene fragment to nucleophiles is inversely dependent on the size of the attached phosphines (*vide infra*). Disubstituted vinylidene complexes exhibit a similar basic geometry to the monosubstituted cases as determined by several crystal structures (66,71,74).

A second π interaction between that π^* orbital of the organic group and a molecular orbital of a' symmetry centered on the metal fragment (Fig. 11), although of less energy than the former π interaction, results in a calculated bond order between M and C_α of between 2 and 3 and, more importantly, indicates that rotation of the vinylidene unit about the M–C_α bond should be facile. This prediction is easily deduced from the lone pair

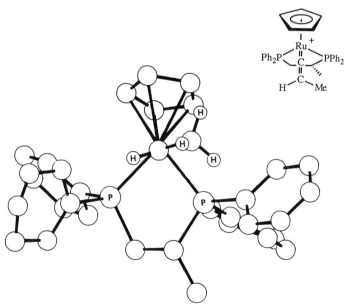

Fig. 9.   X-Ray crystal structure of (R_C, S_{Ru})-[$(\eta^5$-$C_5H_5)$(PROPHOS)Ru=C=CHMe]PF$_6$ (some hydrogens and PF$_6^-$ counterion removed for clarity).

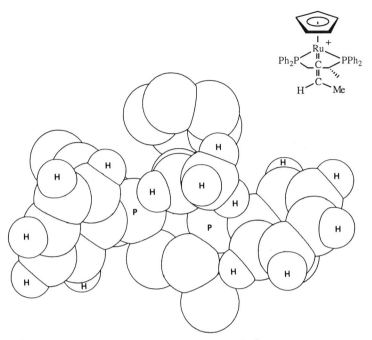

Fig. 10. Space filling representation of (R_C,S_{Ru})-$[(\eta^5\text{-}C_5H_5)(PROPHOS)Ru{=}C{=}CHMe]PF_6$ (some hydrogens and PF_6^- counterion removed for clarity).

counting rules of Davies and colleagues (75) which simply state that for pseudooctahedral systems with at least two available lone pairs of electrons such as $(\eta^5\text{-}C_5H_5)L_2RuX$ rotation about metal–carbon bonds with π–π interactions is stereoelectronically facile.

Dynamic NMR studies by Consiglio and Morandini on $[(\eta^5\text{-}C_5H_5)(dppe)Ru{=}C{=}CHPh]^+$ verify this prediction by determining that the two ^{31}P-NMR signals observed at low temperature coalesce at 194 K, indicating rapid rotation on the NMR time scale about the Ru=C bond [Eq. (64)] (76). A barrier to rotation of 9.1 kcal/mol was calculated for this complex by standard coalescence analysis of the variable-temperature

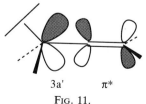

3a' π^*

Fig. 11.

$$(64)$$

NMR data. Similar energy barriers were determined for a number of other vinylidene complexes in this study.

Interestingly, if the separate epimers of the (R)-PROPHOS complexes $\{(\eta^5\text{-}C_5H_5)[(R)\text{-PROPHOS}]Ru=C=CHR\}^+$ are employed, then the two rotamers of the vinylidene are diastereomeric [Eq. (65)]. Consiglio and Morandini (76) found that the difference in population of these interchangeable conformers appears to be controlled by steric factors and presumably reflects the relative ability of the two rotamers to fit into the chiral pockets formed by the phosphines.

$$(65)$$

C. Interconversion between Vinylidenes and Acetylides

The monosubstituted vinylidene complexes are readily deprotonated with a variety of mild bases (e.g., MeO^-, CO_3^{2-}), and this reaction constitutes the most convenient route to ruthenium acetylide complexes. Experimentally the deprotonation is most easily achieved by passing the vinylidene complex through basic alumina. Addition of a noncomplexing acid (e.g., HPF_6) to the acetylide results in the reformation of the vinylidene complex [Eq. (66)]. Reaction of 1 and terminal alkynes such as phenylacetylene in methanol followed by the addition of an excess of

$$(66)$$

sodium to generate sodium methoxide gives a high-yielding one-pot synthesis of ruthenium acetylides [Eq. (67)] (4).

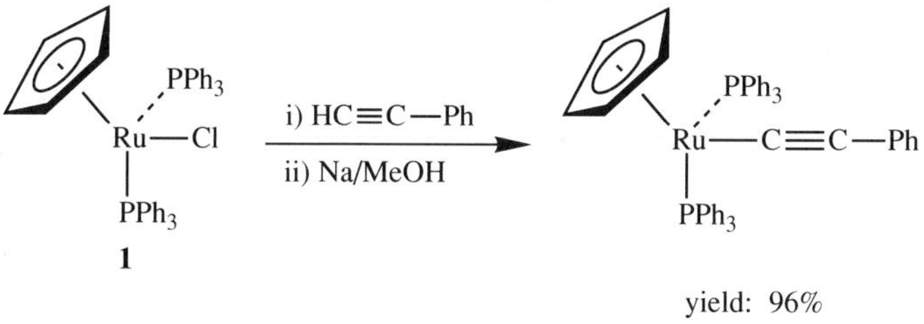

$$\text{(67)}$$

yield: 96%

D. *Disubstituted Ruthenium Vinylidene Complexes*

Addition of internal alkynes to $(\eta^5\text{-}C_5H_5)(PR_3)_2RuCl$ does not lead to the formation of the corresponding disubstituted vinylidene (68). The failure of this reaction could reflect the relative difficulty of a 1,2-alkyl shift for internal alkynes as compared to the 1,2-proton shift for the successful rearrangement of terminal alkynes (Scheme 9). Alternatively, if the deprotonation–reprotonation route is important in the rearrangement of terminal alkynes (*vide supra*), then clearly internal alkynes would not undergo a similar isomerization.

Theoretical studies indicate that electron density in the HOMO of the metal acetylide complex is concentrated on the β carbon, and consequently

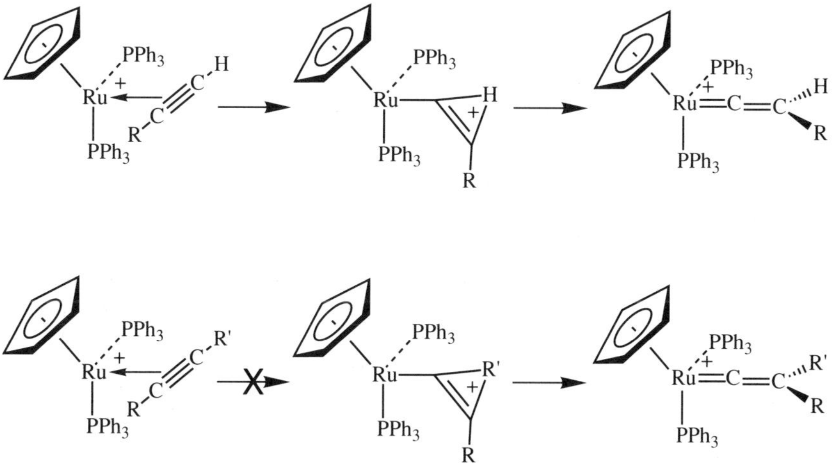

SCHEME 9

this carbon is rendered nucleophilic (70). The protonation of acetylide complexes to generate the monosubstituted vinylidenes demonstrates the nucleophilicity of the β carbon (Section VI,C). Addition of other electrophiles to ruthenium acetylide complexes provides the most convenient route to disubstituted vinylidene complexes. For example, the reaction of [Me$_3$O]PF$_6$ with the phenyl acetylide complex 63 generates the cationic vinylidene species (69) [Eq. (68)] (68). Reaction also occurs under con-

$$\text{(68)}$$

siderably milder conditions than perhaps the above example implies, as evidenced by the formation of the methyl-, ethyl-, and benzyl-substituted vinylidene complexes on treatment of $(\eta^5\text{-C}_5\text{H}_5)(\text{PPh}_3)_2\text{Ru—C}\equiv\text{CMe}$ (70) with MeI, EtI, and PhCH$_2$Br, respectively [Eq. (69)] (77).

RX	Yield (%)
MeI	85
EtI	62
BzBr	87

$$\text{(69)}$$

    Intramolecular attack of the acetylide β carbon on a pendant terminal alkyl halide chain gives cyclic vinylidene complexes (78). Reaction of 6-chlorohex-1-yne with 1 leads to the intermediate vinylidene complex 72, which on passing down an alumina column converts to the cyclic vinylidene complex 73 [Eq. (70)]. The similar reaction with 5-chloropent-1-yne, however, yields the stable chloroacetylide complex 74. Presumably, the steric constraints that would arise from forming a four-membered ring prevent cyclization [Eq. (71)].
    The addition of carbon electrophiles is not limited to simple alkyl halides. Reaction of tropylium hexafluorophosphate with 63, for example,

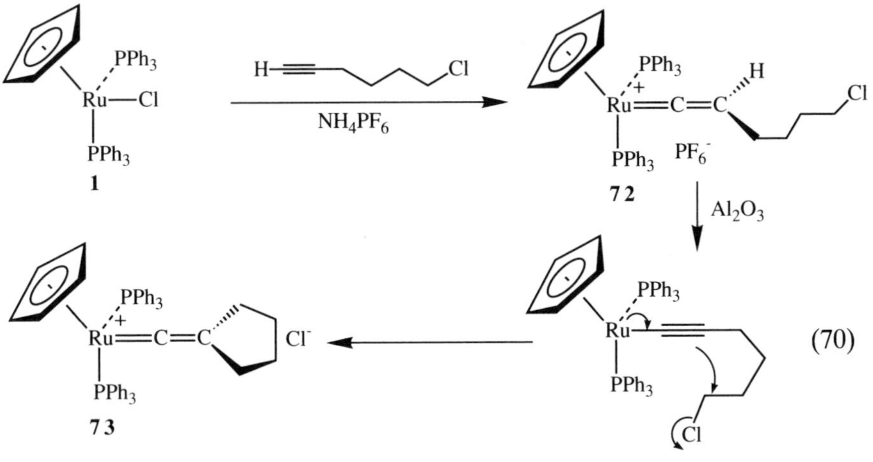

(70)

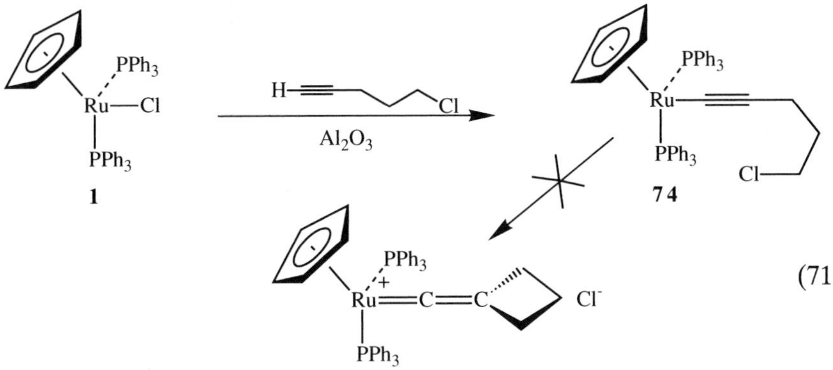

(71)

yields the cycloheptatriene-substituted vinylidene complex **75** in good yield [Eq. (72)] (79).

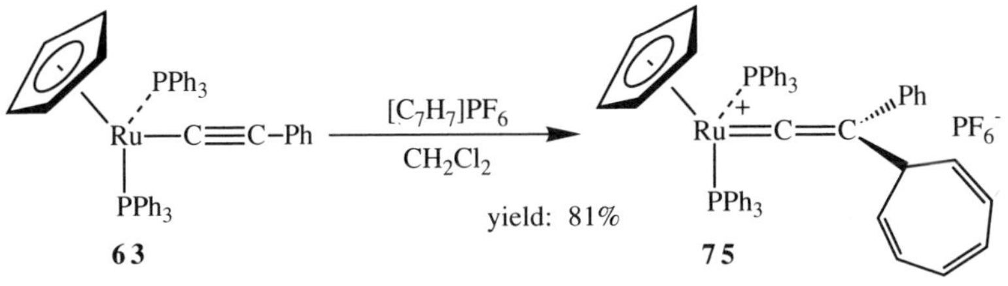

(72)

yield: 81%

Consiglio *et al.* have recently reported a surprising mode of reactivity for the vinylidene complexes of type **76**. Reaction of these complexes with diazomethane results in the insertion of CH_2 into the C_β–H bond in good yield, instead of addition of a methylene unit across the $C_\alpha-C_\beta$ bond or even the $Ru-C_\alpha$ bond as anticipated [Eq. (73)] (80). The acidic nature of

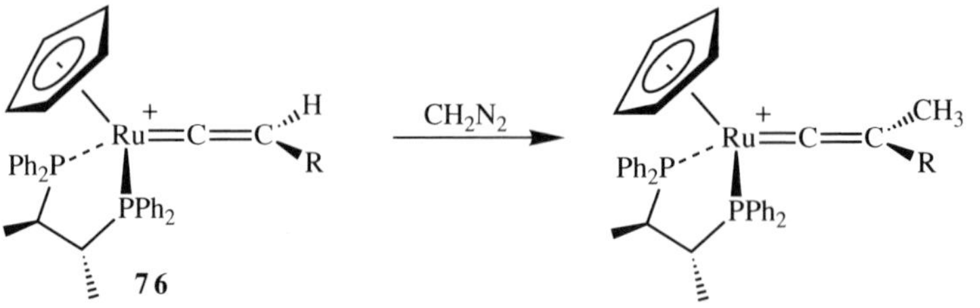

(73)

R	Yield (%)
t-Bu	70
Ph	50

the protons on the β carbon, discussed above, suggests that the mechanism of the reaction most reasonably involves deprotonation of the vinylidene 76 by diazomethane, followed by nucleophilic attack of the resulting acetylide on the electrophilic methylating agent MeN_2^+ [Eq. (74)].

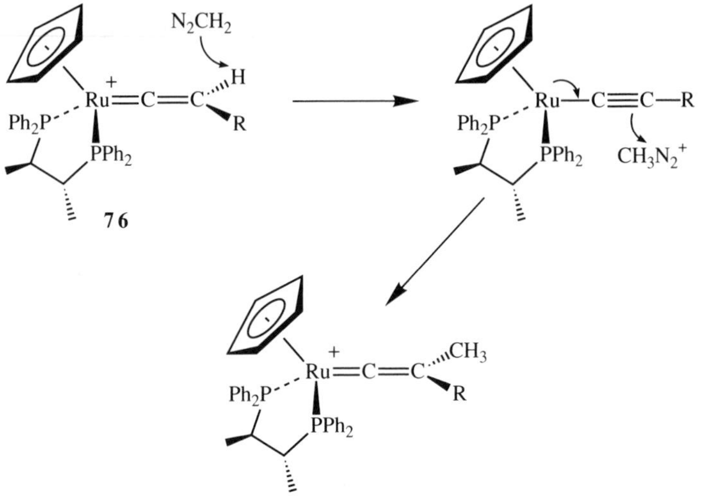

(74)

Reaction between the acetylide complex 70 and carbon disulfide followed by dissolution in methyl iodide gives the thioester vinylidene complex (78) (78). A possible mechanism for the reaction is outlined in Scheme 10. Precedence for the cyclic intermediate 77 comes from the isolation of the iron analog 79 on reaction of $(\eta^5\text{-}C_5H_5)(dppe)Fe-C\equiv CMe$ with carbon disulfide [Eq. (75)] (81).

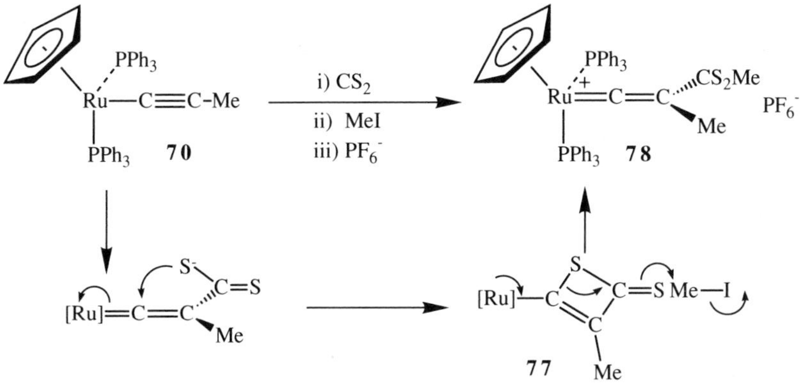

SCHEME 10

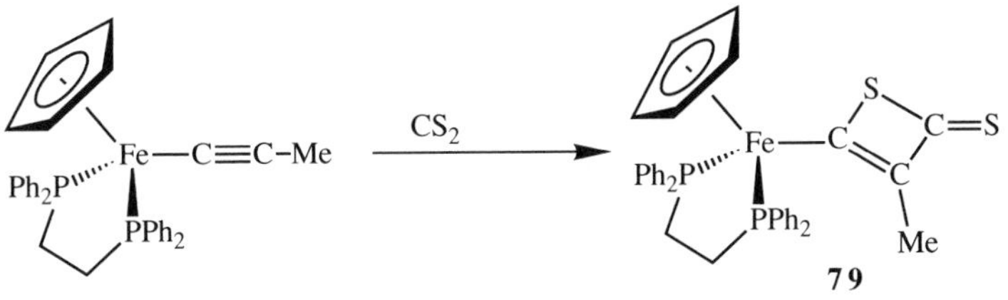

(75)

Noncarbon-based electrophiles also react with ruthenium acetylides to give disubstituted vinylidene complexes. Addition of halogens to the phenyl acetylide complex **63** occurs rapidly and quantitatively to form β-halovinylidene complexes [Eq. (76)] (*71,74*). Facile bromination of the

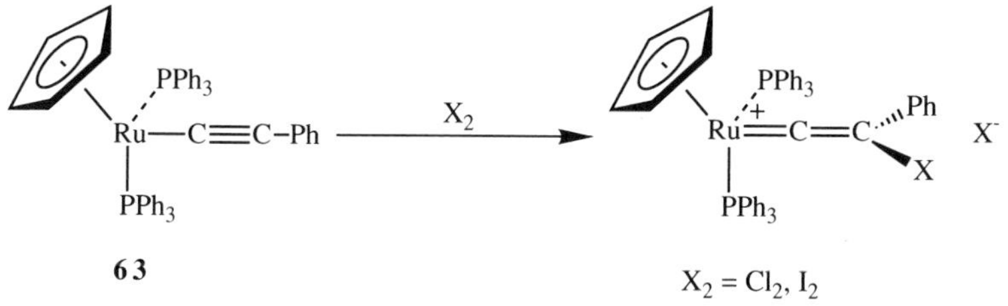

(76)

$X_2 = Cl_2, I_2$

phenyl ring is also observed when X_2 is Br_2, consistent with the electron-releasing nature of the ruthenium acetylide group in the para position [Eq. (77)]. The addition of various arenediazonium salts to ruthenium acetylide complexes can also lead to the formation of substituted vinylidene complexes [Eq. (78)] (*82*). Other ruthenium vinylidene and

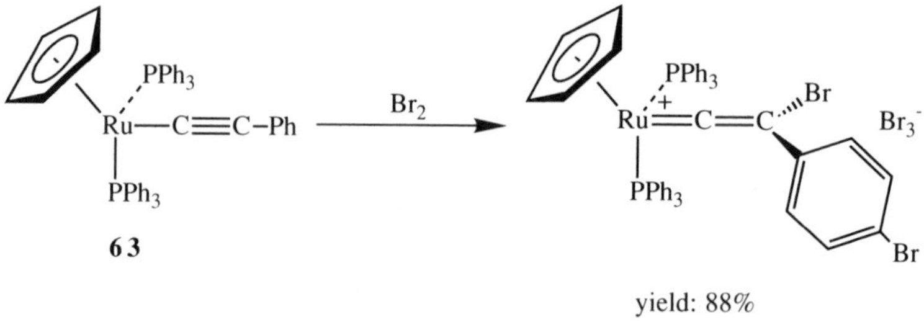

$$\text{(77)}$$

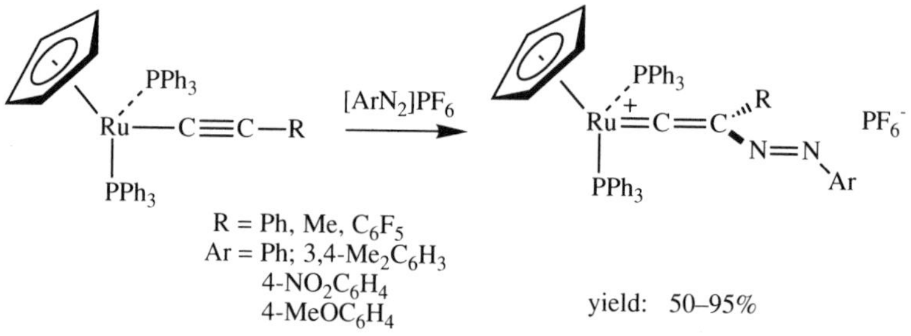

yield: 88%

$$R = Ph, Me, C_6F_5$$
$$Ar = Ph; 3,4\text{-}Me_2C_6H_3$$
$$4\text{-}NO_2C_6H_4$$
$$4\text{-}MeOC_6H_4$$

$$\text{(78)}$$

yield: 50–95%

acetylide complexes containing a variety of phosphine ligands may be prepared either by ligand exchange (*vide supra*) or by reacting the appropriate ruthenium chloride complex with the alkyne (*72,78*).

E. Ruthenium Carbene Complexes

1. Preparation

The vast majority of work exploring the reactivity of ruthenium vinylidene complexes has focused on the attack of alcohols at the electrophilic α carbon of monosubstituted vinylidenes, resulting in the formation of ruthenium alkoxycarbene complexes. Bruce and co-workers have determined, for example, that the phenylvinylidene complex **80** is slowly transformed in refluxing MeOH to the methoxycarbene complex **82** in good yield (*73,83*). The mechanism for this reaction must involve initial attack of the alcohol at the electrophilic C_α to form a transient vinyl intermediate **81** which is rapidly protonated at the nucleophilic C_β, generating the product carbene **82** [Eq. (79)]. In contrast to monosubstituted vinylidene complexes, disubstituted vinylidene complexes are generally unreactive to nucleophiles; even the relatively small dimethylvinylidene complex **83** shows no reaction with MeOH after 70 hours at reflux [Eq. (80)].

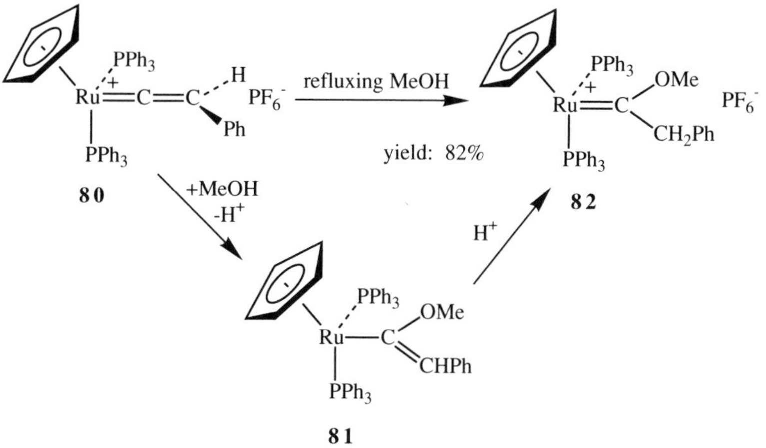

$$\text{Ru}=\text{C}=\text{C}\overset{H}{\underset{Ph}{\diagup}} \;\text{PF}_6^- \quad \xrightarrow[\text{yield: 82\%}]{\text{refluxing MeOH}} \quad \text{Ru}=\text{C}\overset{OMe}{\underset{CH_2Ph}{\diagup}} \;\text{PF}_6^-$$

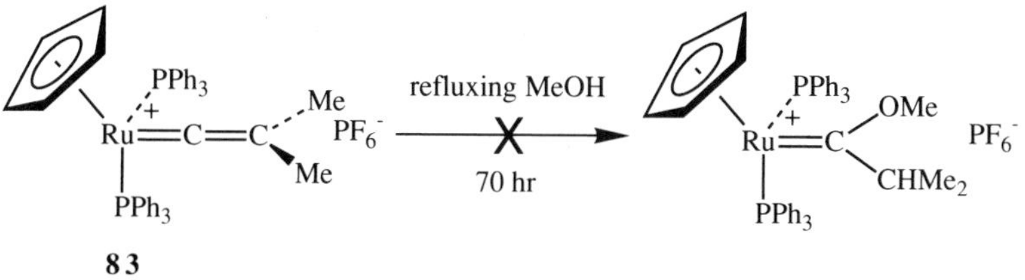

80   **82**

$$\downarrow \begin{array}{c}+\text{MeOH}\\-\text{H}^+\end{array} \qquad \qquad \nearrow \; \text{H}^+$$

(79)

$$\text{Ru}-\text{C}\overset{OMe}{\underset{CHPh}{\diagup}}$$

81

$$\text{Ru}=\text{C}=\text{C}\overset{Me}{\underset{Me}{\diagup}} \;\text{PF}_6^- \quad \xrightarrow[\text{70 hr}]{\overset{\text{refluxing MeOH}}{\quad\times\quad}} \quad \text{Ru}=\text{C}\overset{OMe}{\underset{CHMe_2}{\diagup}} \;\text{PF}_6^- \quad (80)$$

83

As would be anticipated, the unsubstituted vinylidene species $[(\eta^5\text{-}C_5H_5)(PPh_3)_2Ru=C=CH_2]^+$ (**84**) is correspondingly more reactive than the monosubstituted complexes. In fact, the reaction of ethyne with **1** in methanol at room temperature affords only the methoxycarbene **85**, the methanol addition reaction being so rapid that isolation of the unsubstituted vinylidene **84** is impossible [Eq. (81)] (*78*). As the alcohol must

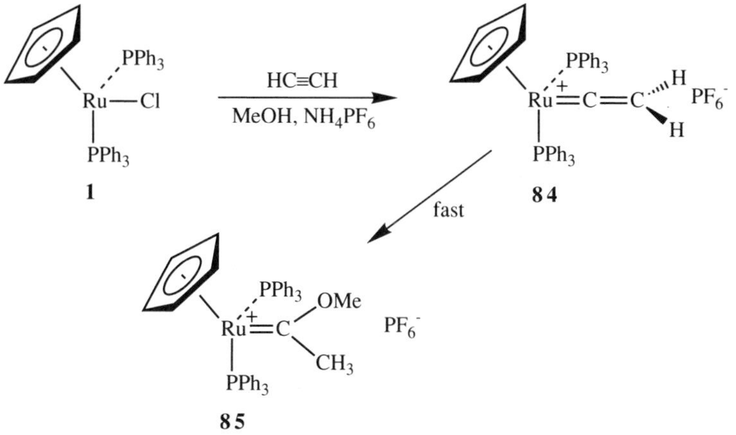

$$\text{Ru}-\text{Cl} \quad \xrightarrow[\text{MeOH, NH}_4\text{PF}_6]{\text{HC}\equiv\text{CH}} \quad \text{Ru}=\text{C}=\text{C}\overset{H}{\underset{H}{\diagup}} \;\text{PF}_6^-$$

1   **84**

$$\swarrow \; \text{fast}$$

$$\text{Ru}=\text{C}\overset{OMe}{\underset{CH_3}{\diagup}} \;\text{PF}_6^- \qquad (81)$$

85

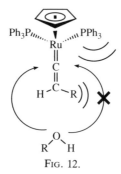

FIG. 12.

attack C_α in the plane of the vinylidene substituents, the observed reactivity may reflect a steric inability of the alcohol to approach past substituents larger than hydrogen (Fig. 12). Reactions conducted under basic conditions with nonacidic work-ups should produce the vinyl species as stable products, and this hypothesis has interesting stereochemical implications for the geometry about the double bond: the kinetic product of the reaction should be the Z isomer, rather than the thermodynamically more stable E isomer [Eq. (82)].

$$\text{(82)}$$

Z

A more detailed study of the general reaction of alcohols with ruthenium vinylidene complexes were conducted in the Bruce and Swincer laboratories (73). It was determined that attack of the alcohol at the α carbon is inhibited by bulky phosphine ligands, as indicated by the inverse relationship between the relative reaction rates of $[(\eta^5\text{-}C_5H_5)\text{-}(PPh_3)LRu{=}C{=}CHPh]^+$ with MeOH and the cone angle of the phosphine L (84). As discussed in Section VI,B,3, attack of a nucleophile at the α carbon can be hindered by the bulk of the flanking phosphine ligands. Electronic factors also influence reactivity, with electron-withdrawing groups on the acetylide unit or on the metal facilitating nucleophilic attack at C_α. This effect can be observed in the relative rates of MeOH reaction with $[(\eta^5\text{-}C_5H_5)(PPh_3)_2Ru{=}C{=}CHR]^+$ ($R = CO_2Me > Ph > Me$). The size of the attacking alcohol is also crucial, with no reaction being observed between $[(\eta^5\text{-}C_5H_5)(PPh_3)_2\text{-}Ru{=}C{=}CHPh]^+$ and refluxing EtOH or i-PrOH after 24 hours. This, again, is almost certainly due to steric constraints imposed by the flanking phosphines and the vinylidene substituents. The more reactive complex $[(\eta^5\text{-}C_5H_5)(PPh_3)(CO)Ru=C=CHPh]^+$ (86), however, converts slowly to the anticipated ethoxy- and isopropoxycarbene complexes 87 in refluxing EtOH and i-PrOH, respectively [Eq. (83)]. The increased reactivity of complex 86 can be explained as a combination of two factors: first, the much smaller steric requirements of the carbonyl ligand relative to a triphenylphosphine ligand allows the incoming alcohol an approach path to C_α with greatly diminished steric constraints, and, second, the electron-withdrawing nature of the carbonyl increases the electrophilicity of the complex.

Addition of the acetylenic alcohols $HC\equiv C(CH_2)_xOH$ ($x = 3, 4$) to 1 affords a one-pot synthesis of the cyclic carbene complexes (88). The reaction proceeds via initial formation of the vinylidene complexes, followed by an intramolecular attack of the terminal alcohol function on the α carbon [Eq. (84)] (85). Combining the nucleophilicity at the β carbon of

n	Yield (%)
1	91
2	55

acetylide complexes and the electrophilicity of the α carbon of vinylidenes, reaction of 2-iodoethanol and **76** affords the cyclic carbene (**89**) [Eq. (85)] (*78*).

$$\text{(85)}$$

Water also attacks the electrophilic α carbon of the ruthenium vinylidene complex **80**. The reaction does not yield the ruthenium acyl complex, however, as is found for the reaction with the similar iron vinylidene complex $[(\eta^5\text{-}C_5H_5)(CO)_2Fe{=}C{=}CHPh]^+$ (*86*), but rather **91** is the only isolated product (*78*). The mechanism for this transformation most reasonably involves rapid loss of H^+ from the initially formed hydroxycarbene to generate an intermediate acyl complex (**90**). Reversible loss of triphenylphosphine relieves steric strain at the congested ruthenium center, and eventual irreversible migration of the benzyl fragment to the metal leads to formation of the more stable carbonyl complex (**91**) [Eq. (86)].

$$\text{(86)}$$

The reaction of 5-chloropent-1-yne with **92** in refluxing methanol with aqueous hexafluorophosphoric acid affords a one-pot synthesis of the cyclic oxycarbene complex **93**. The mechanism is thought to be similar to that described above, with initial attack by water giving the ruthenium acyl complex. Displacement of the chloride by the acyl oxygen yields the alkoxycarbene complex (**93**) [Eq. (87)] (*78*). A similar reaction in THF–

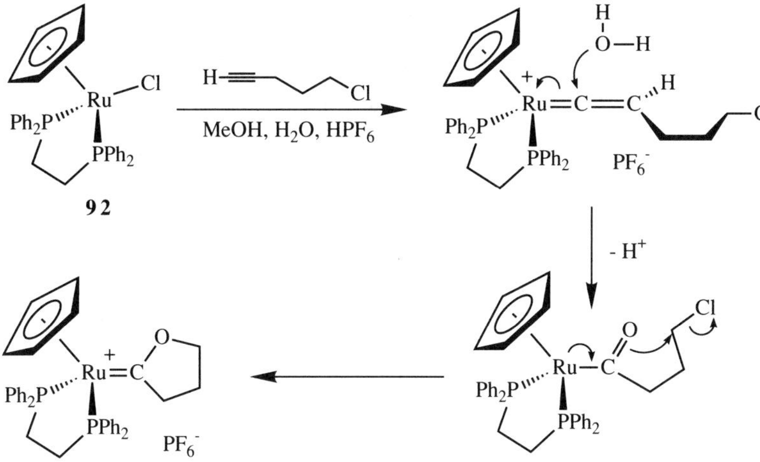

(87)

dichloromethane with ammonia, generated *in situ* from ammonium hexafluorophosphate and Hünigs base, does not form the analogous cyclic aminocarbene complex but the open chain derivative **94** [Eq. (88)] (*78*). Infrared evidence suggests that the complex may be better represented by structure **95**, which explains the reluctance of the nitrogen to act as a nucleophile and displace the chlorine atom.

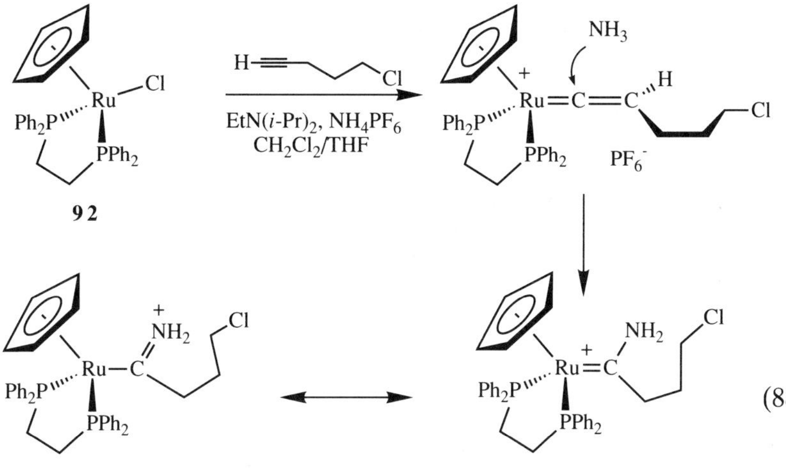

(88)

2. Geometry

As discussed in Section VI,B,3, the maximum π stabilization between the metal fragment $(\eta^5\text{-}C_5H_5)(PPh_3)_2Ru$ and the α carbon of an unsaturated organic fragment arises from interaction of the carbon p-type orbital and a ruthenium-based molecular orbital orthogonal to the symmetry plane of the molecule. Kostic and Fenske (70) have calculated the difference in energy between the resulting vertical geometry of the carbene and the alternate horizontal geometry for $[(\eta^5\text{-}C_5H_5)(PH_3)_2Ru{=}CH_2]^+$ to be a significant 11 kcal/mol. In the absence of severe steric constraints, therefore, the carbene moiety preferentially lies in the symmetry plane of the molecule. The crystal structure of $[(\eta^5\text{-}C_5H_5)(PPh_3)_2Ru{=}C(OMe)Et]PF_6$ (66) clearly shows that the plane of the carbene fragment bisects the P—Ru—P angle (Fig. 13).

As discussed earlier, the chemistry of the vinylidene complexes is influenced by steric constraints imposed by the flanking phosphine groups. The steric congestion about the ruthenium center has an even more pronounced effect on reactivity at C_α in carbene complexes. The crystal structure of complex **96** (61) provides an excellent example of the pro-

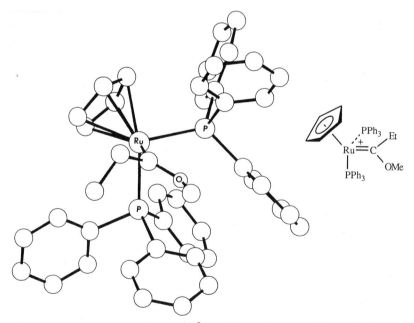

Fig. 13.   X-Ray crystal structure of $[(\eta^5\text{-}C_5H_5)(PPh_3)_2Ru{=}C(OMe)Et]PF_6$ (hydrogens and PF_6^- counterion removed for clarity).

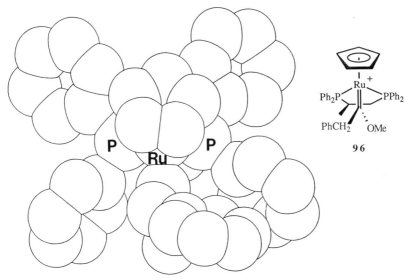

FIG. 14. Space filling representation of (R_C,R_{Ru})-$[(\eta^5\text{-}C_5H_5)(\text{PROPHOS})\text{Ru}{=}$ $C(\text{OMe})CH_2Ph]PF_6$ (**96**) (hydrogens and PF_6^- counterion removed for clarity).

tection of C_α from approaching reactants by two flanking phenyl groups of the (R)-PROPHOS ligand, as best seen in the space filling representation (Fig. 14).

3. Reactions

Most efforts to explore the reactivity of ruthenium carbene complexes have employed the alkoxycarbene species so readily synthesized from the inter- or intramolecular reaction of vinylidene complexes with alcohols. These electrophilic alkoxycarbene complexes exhibit only limited reactivity at C_α, primarily with hydride reagents. For example, treatment of the 2-oxacyclopentylidene complex **97** with $NaAlH_2(OCH_2CH_2OMe)_2$ affords the neutral 2-tetrahydrofuranyl complex (**98**) [Eq. (89)] (*85*), as was anticipated from similar reductions of iron carbene complexes (*87*).

$$\text{97} \quad \xrightarrow[\text{THF}]{NaAlH_2(OCH_2CH_2OMe)_2} \quad \text{98} \qquad (89)$$

yield: 72%

Consiglio *et al.* (*88*) have reported that the reaction of LiAlH$_4$ with the methoxycarbene complex **99** gives the completely reduced 2-phenylethyl complex (**100**) as the exclusive product [Eq. (90)]. The reaction probably involves initial hydride attack at C$_\alpha$, elimination of methoxide, and a second hydride attack to give the final product.

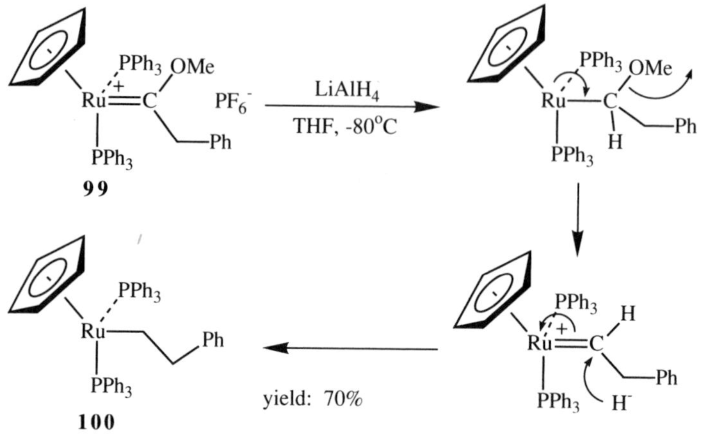

(90)

yield: 70%

Hydrogens on the β carbon of the carbene complexes are fairly acidic, and there are many examples of deprotonation at this position yielding the corresponding vinyl complexes. Two representative examples are shown in Eqs. (91) and (92) (*89*). The deprotonation of the benzylmethoxycarbene

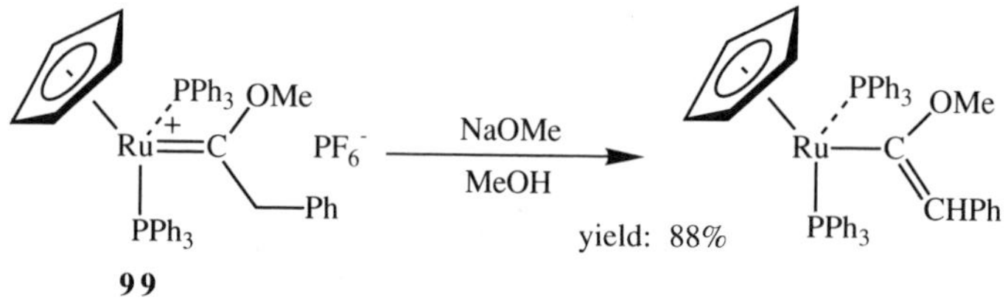

(91)

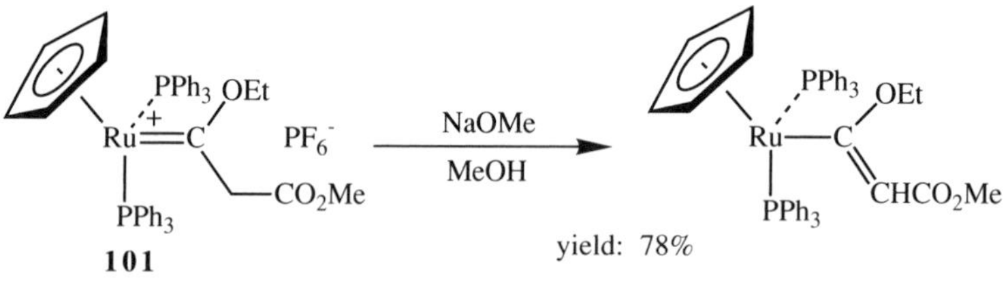

yield: 88%

(92)

yield: 78%

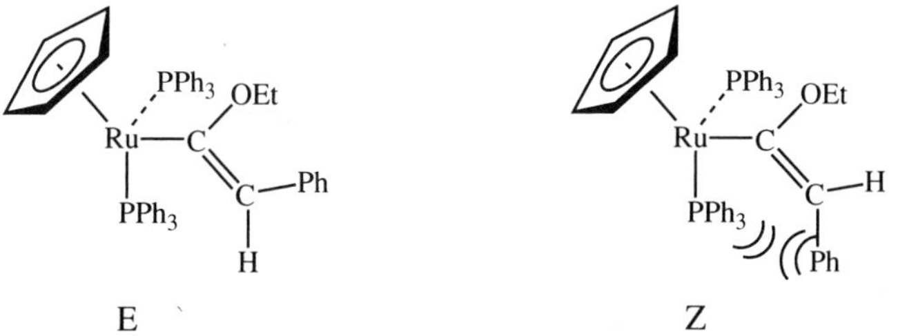

FIG. 15.

complex **99** affords only one isomer as indicated by ^{1}H NMR, although both E and Z isomers could theoretically be formed. In contrast, deprotonation of the methyl ester carbene complex **101** gives a $3:1$ mixture of isomers. The smaller size of the methyl ester group relative to phenyl may be responsible for the observed lack of stereoselectivity; although the stereochemistry has not been assigned, the more abundant isomer should have the E configuration about the double bond for steric reasons (Fig. 15). Alternatively, the methyl ester complex may deprotonate stereoselectively and isomerization occur in the product, as depicted in Scheme 11.

Bruce *et al.* (*85*) have also found that treatment of the 2-oxacyclopentylidene complex **97** with pyridine–D_2O results in initial deuteration at

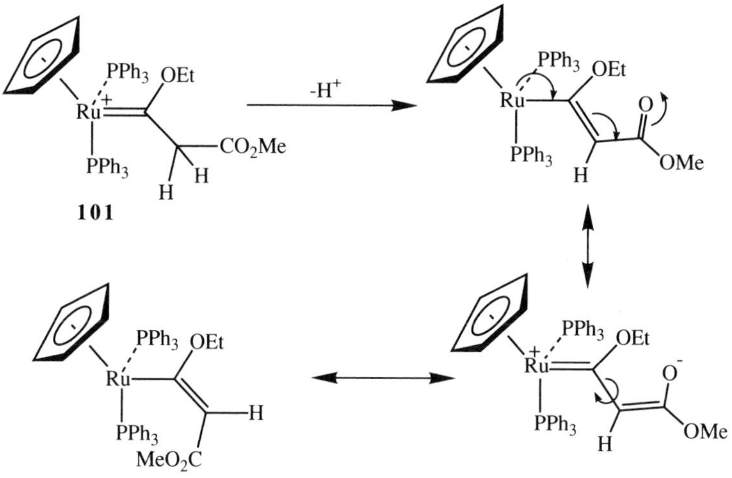

SCHEME 11

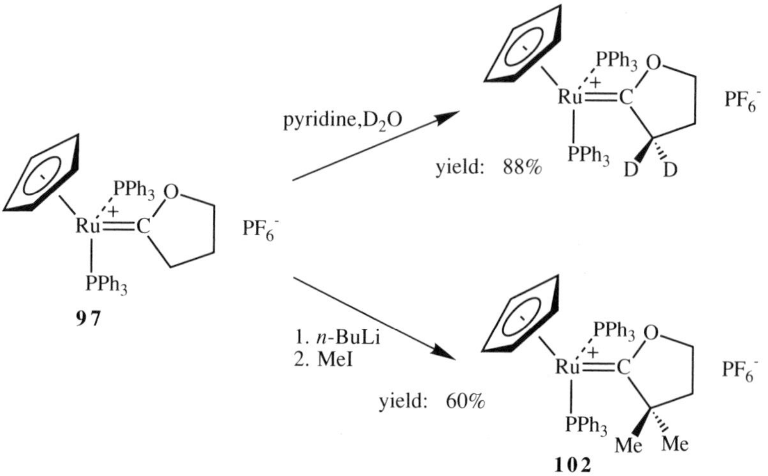

SCHEME 12

the 5 position (Scheme 12). The observed acidity of the two hydrogens in this position suggested that a deprotonation–alkylation sequence could lead to the regiospecific dialkylation of the complex. Indeed, treatment of the complex **97** with *n*-BuLi (or NaOH) followed by MeI affords the 5,5′-dimethylated complex (**102**) in moderately good yield. No evidence for BuLi attack at C_α was observed.

F. Cycloaddition Reactions of Ruthenium Acetylide and Vinylidene Complexes

The addition of substituted alkenes with electron-withdrawing groups to ruthenium acetylide complexes results in the formal [2 + 2] cycloaddition of the olefin to the acetylene moiety. Facile ring opening of the resultant ruthenium cyclobutene complex (**103**) generates the ruthenium butadienyl species (**104**). Subsequent displacement of a phosphine ligand leads to the η^3-allylic product (**105**) [Eq. (93)] (*90–92*). The intermediate cyclobutene complex has been isolated in one instance for the monocarbonyl derivative **106** [Eq. (94)] (*92*).

Another cycloaddition reaction, between molecular oxygen and the electrophilic phenylvinylidene complex **80**, results in the oxidative cleavage of the C_α–C_β bond and affords $[(\eta^5\text{-}C_5H_5)(PPh_3)_2RuCO]^+$ and un-complexed benzoic acid. The initial cleavage product is most likely benzaldehyde, which is then oxidized under the reaction conditions [Eq. (95)] (*83*).

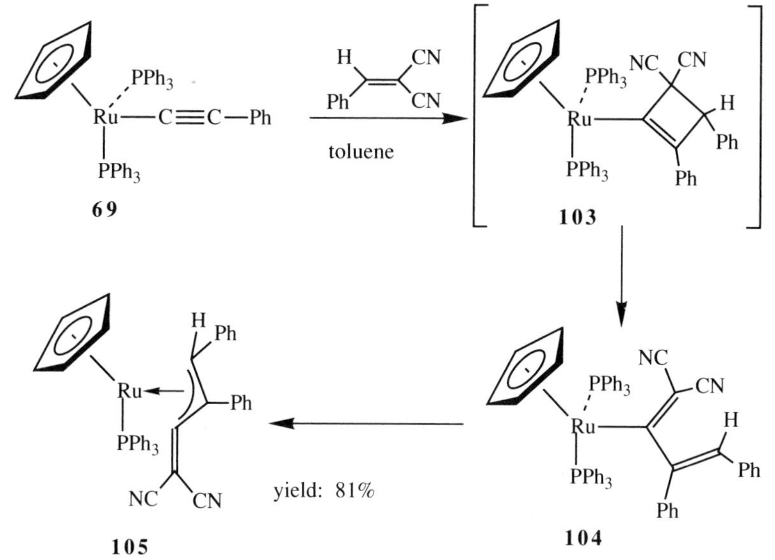

(93)

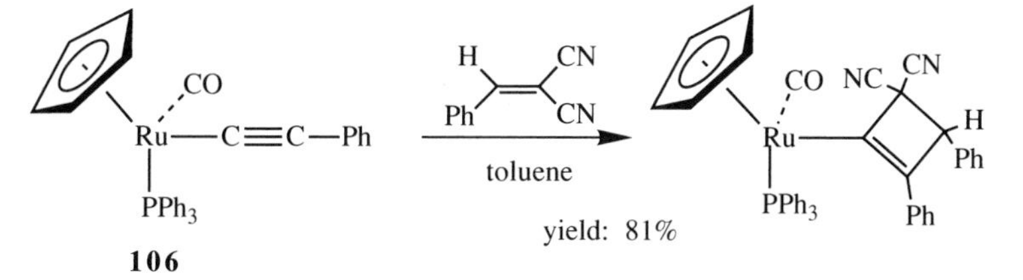

(94)

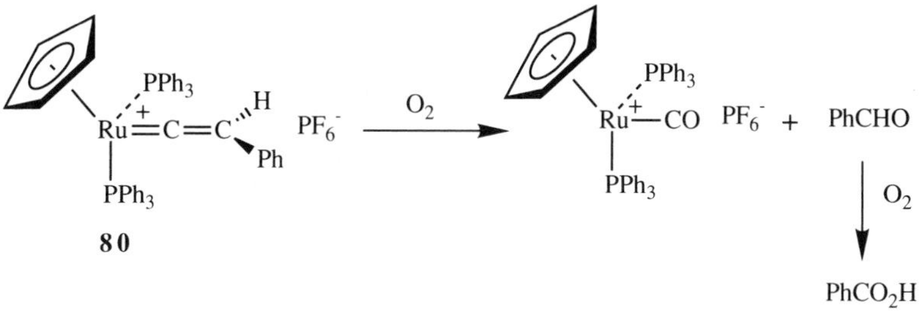

PhCO$_2$H (95)

G. Ruthenium Allenylidene Complexes

1. Preparation

Reaction between the 1,1-diphenylpropargyl alcohol 107 and 11 results in the formation of the diphenylallenylidene complex (109) (93). The proposed mechanism for this reaction involves initial formation of the hydroxyvinylidene complex **108** followed by spontaneous dehydration to form the allenylidene complex [Eq. (96)]. All attempts to isolate the intermediate vinylidene complex have been unsuccessful.

$$(96)$$

Protons on the δ carbon of a ruthenium allenylidene complex are acidic, and deprotonation at this position often occurs to give ene–yne derivatives. Reaction of **1** with cyclic propargyl alcohols of type **110**, for example, did not yield the allenylidene complexes, but rather the ruthenium ene–yne products (**111**) were isolated [Eq. (97)] (78). Reaction of the cor-

$$(97)$$

responding dimethyl propargyl alcohol **112** with **1** initially forms the anticipated dimethylallenylidene complex **113**, which rapidly dimerizes to give
complex **115** as the only observed product (*94*). A plausible mechanism for
the dimerization involves coupling of **113** and the ene–yne derivative **114**,
formed on loss of a proton from **113** (Scheme 13).

If the terminal carbon of the allenylidene complex is unsubstituted, then
attack of nucleophiles at this site is facile. Reaction of the parent propargyl
alcohol (**116**) with **1** does not yield the unsubstituted allenylidene complex,
however, but the triphenylphosphonium salt (**118**) is formed in low yield.

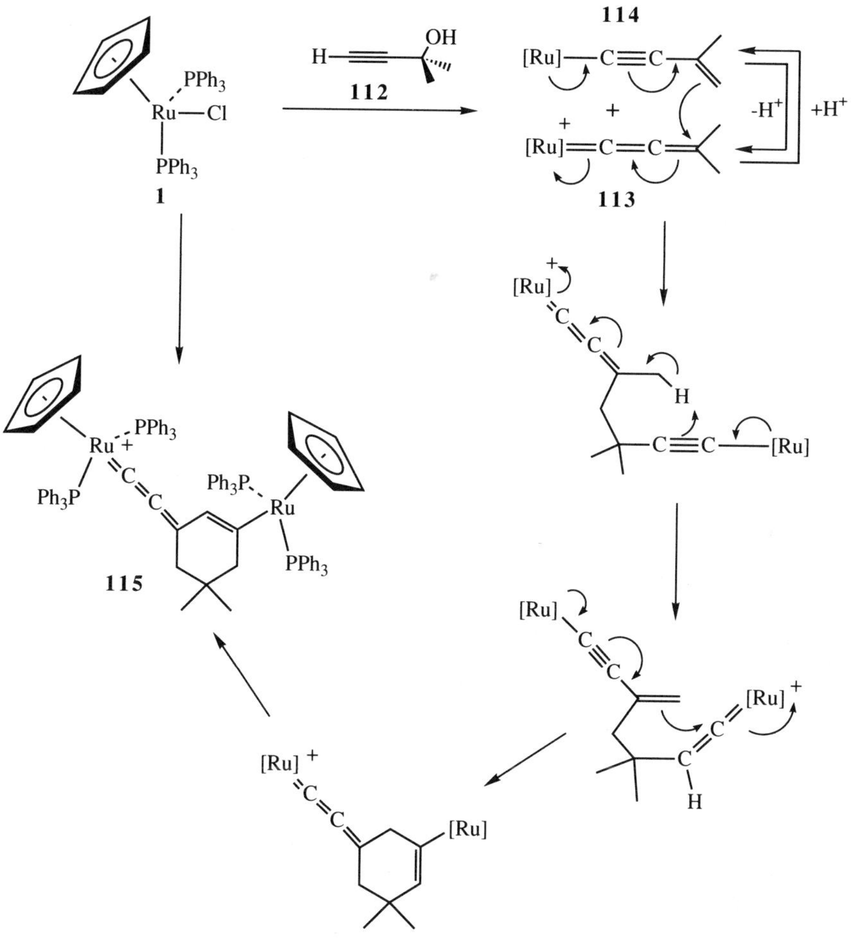

SCHEME 13

The reaction is thought to proceed via initial formation of the unsubstituted allenylidene complex **117**, which then scavenges free triphenylphosphine to form the phosphonium salt (**118**) [Eq. (98)]. Addition of excess triphenylphosphine to the reaction enables the novel acetylide complex to be isolated in good yield (78).

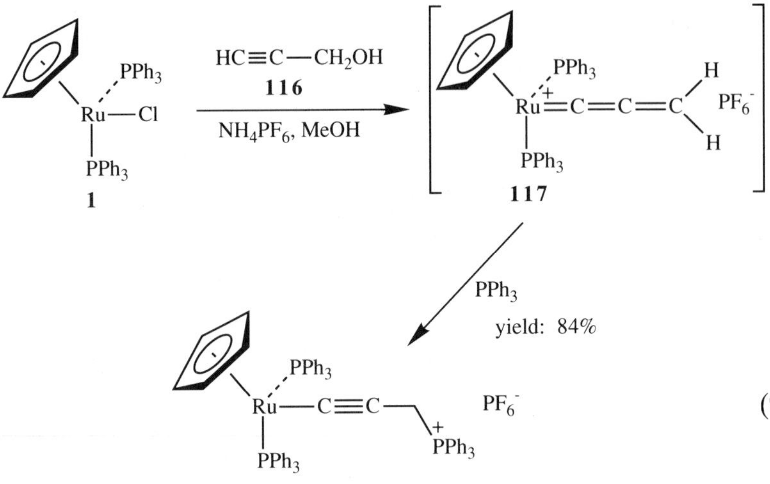

(98)

The dicationic vinylidene complex **119** can be formed under standard conditions (*vide supra*) by the addition of acid to **118** [Eq. (99)] (78).

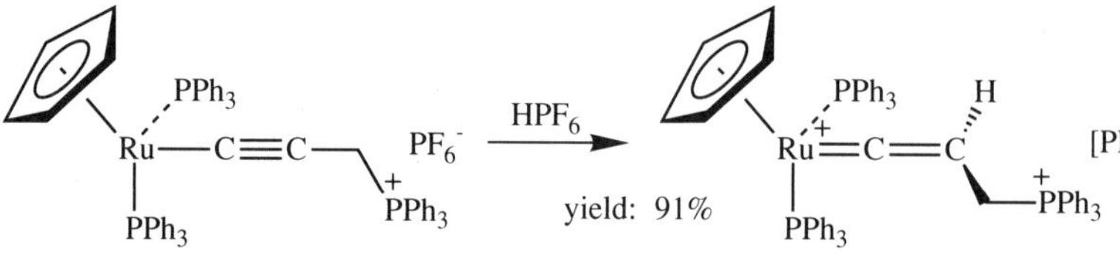

(99)

Addition of butyllithium to **118** forms the ylid, which reacts with a variety of carbonyl compounds to give ruthenium ene–yne complexes [Eq. (100)]. Reaction of aldehydes gave a mixture of *E* and *Z* isomers, dependent on the aldehyde and conditions used, but with the *E* isomer generally the predominant product (78).

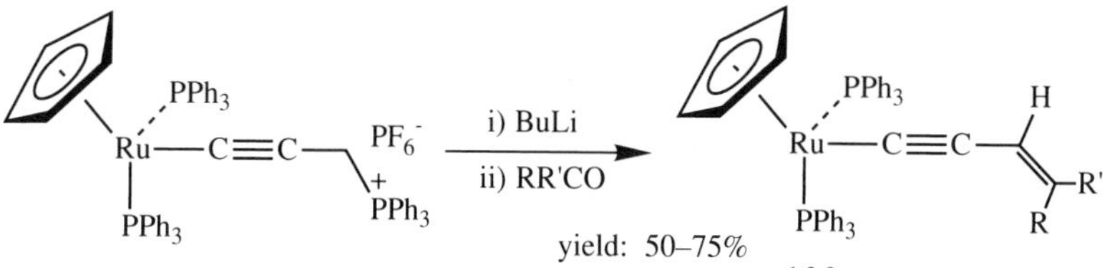

$$\text{(100)}$$

yield: 50–75%

118   **120**

2. Geometry

Using the bonding arguments described in Sections VI,B,3 and VI,E,2, the linear allenylidene fragment is expected to lie in the symmetry plane of the molecule. This is clearly shown in the crystal structure of $[(\eta^5\text{-}C_5H_5)(PMe_3)_2Ru=C=C=CPh_2]PF_6$ (**109**) (93) (Fig. 16). The bond lengths of the allenylidene fragment suggest that there is substantial contribution from two resonance forms which delocalize the charge on the ruthenium and terminal carbon (Fig. 17).

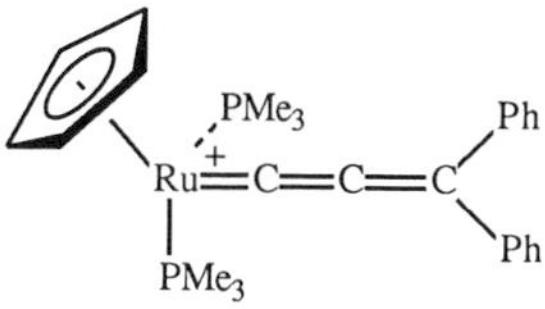

109

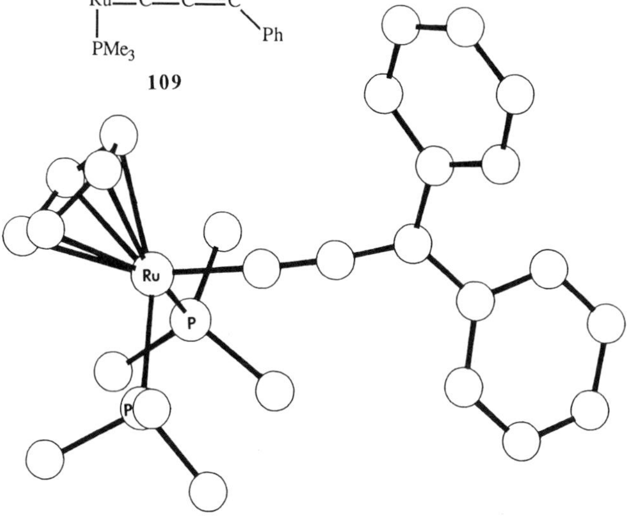

FIG. 16.   X-Ray crystal structure of $[(\eta^5\text{-}C_5H_5)(PMe_3)_2Ru=C=C=CPh_2]PF_6$ (**109**) (hydrogens and PF_6^- counterion removed for clarity).

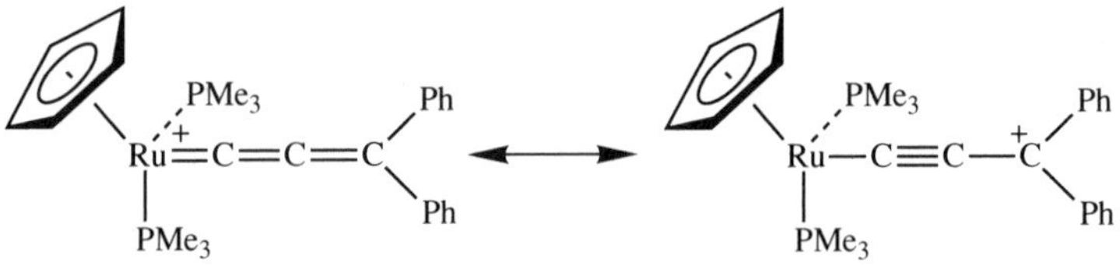

FIG. 17.

VII

DECOMPLEXATION REACTIONS

Although the successful application of organometallic auxiliaries in effecting highly regio- and stereoselective elaborations of bound organic moieties requires a general and selective method for removal of the organic product, this area has so far received little attention in the case of the $(\eta^5\text{-}C_5H_5)(PR_3)_2Ru$ auxiliary. The few successful decomplexation reactions that have been reported are discussed below. Baird and co-workers (95) have investigated the cleavage of simple ruthenium alkyl complexes of the type $(\eta^5\text{-}C_5H_5)(PPh_3)_2RuR$ (R = Me, CH_2Ph). Treatment of these complexes with mercuric bromide or iodide results in almost quantitative production of the organomercury compound **122** and the ruthenium halide complex **121** [Eq. (101)].

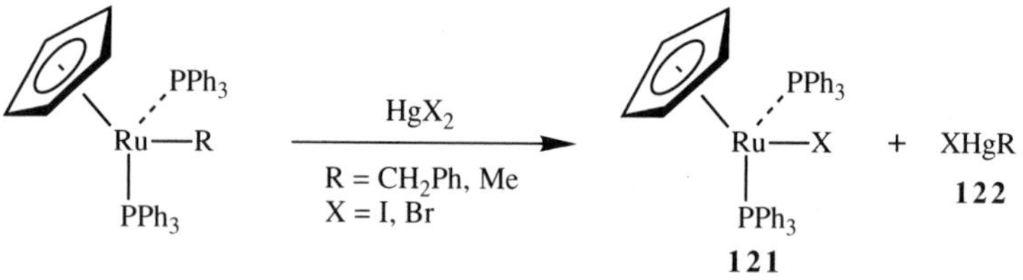

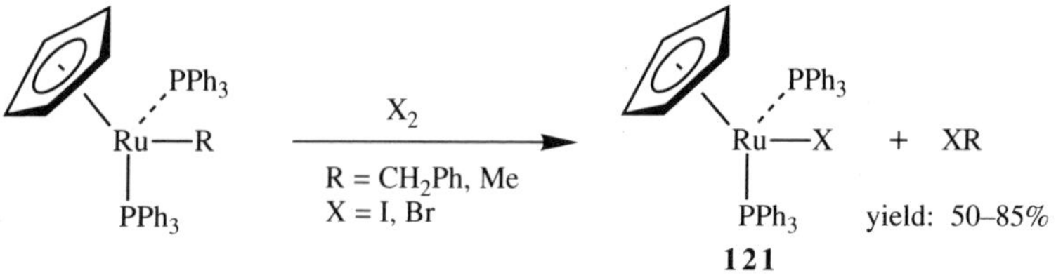

Reaction of the ruthenium alkyl complexes with bromine or iodine also cleaves the ruthenium–carbon bond to form the alkyl halides in reasonable yield [Eq. (102)]. Attempts to decomplex the organic substrate using cupric halides or electrochemical methods were not so successful and gave a mixture of organic products in varying yield. Bruce *et al.* have reported that treatment of the ruthenium phenylacetylide complex **63** or the corresponding vinylidene complex **80** with HCl results in the formation of the ruthenium chloride complex (**1**) [Eqs. (103) and (104)] (*83*). Unfortunately, the fate of the cleaved organic moiety was not determined.

$$
\text{63} \xrightarrow{\text{HCl}} \text{1} \tag{103}
$$

$$
\text{80} \xrightarrow{\text{HCl}} \text{1} \tag{104}
$$

Other attempts to utilize this decomplexation reaction, however, have met with no success (*78*). Treatment of other ruthenium acetylide complexes, such as **123**, with either gaseous HCl in chloroform or concentrated hydrochloric acid lead only to formation of the corresponding vinylidene complexes [Eq. (105)]. No decomplexation products were observed.

$$
\text{123} \xrightarrow{\text{HCl}} \tag{105}
$$

Treatment of the disubstituted vinylidene complex **124** with base in wet methanol cleaves the vinylidene bond to give the cationic ruthenium carbonyl complex **126** and bibenzyl in good yield. This reaction presumably proceeds via initial formation of the acyl complex **125**, which decomposes to give the products [Eq. (106)] (*78*). Other mono- and disubstituted vinylidene complexes, however, do not give identifiable decomplexation

$$(106)$$

products under similar conditions. One other decomplexation reaction, already mentioned in Section VI,F, involves the addition of oxygen to the vinylidene complex **80**, resulting in cleavage of the vinylidene bond to yield benzoic acid [Eq. (107)] (*83*).

$$(107)$$

VIII

PREPARATION OF RUTHENIUM COMPLEXES CONTAINING η^2-UNSATURATED LIGANDS

Addition of a wide range of alkenes, internal alkynes, allenes, and dienes to $(\eta^5\text{-}C_5H_5)(PMe_3)_2RuCl$ (**11**) in the presence of NH_4PF_6 yields the corresponding η^2-unsaturated complexes in high yield (Scheme 14) (*45,46,96*). Reactions with terminal alkynes generally lead to the formation of ruthenium vinylidene complexes by a rapid rearrangement of the η^2-acetylide complex (*vide supra*). Nonetheless, one complex containing an η^2-bound terminal acetylene has recently been isolated (*97*). Bubbling acetylene or propyne through a methanol solution of **11** in the presence of excess NH_4PF_6 results in the rapid precipitation of the η^2 complex before rearrangement to the vinylidene complex **128** can occur. Complex **127** rearranges to the vinylidene complex **128** on dissolution in acetonitrile or methanol [Eq. (108)].

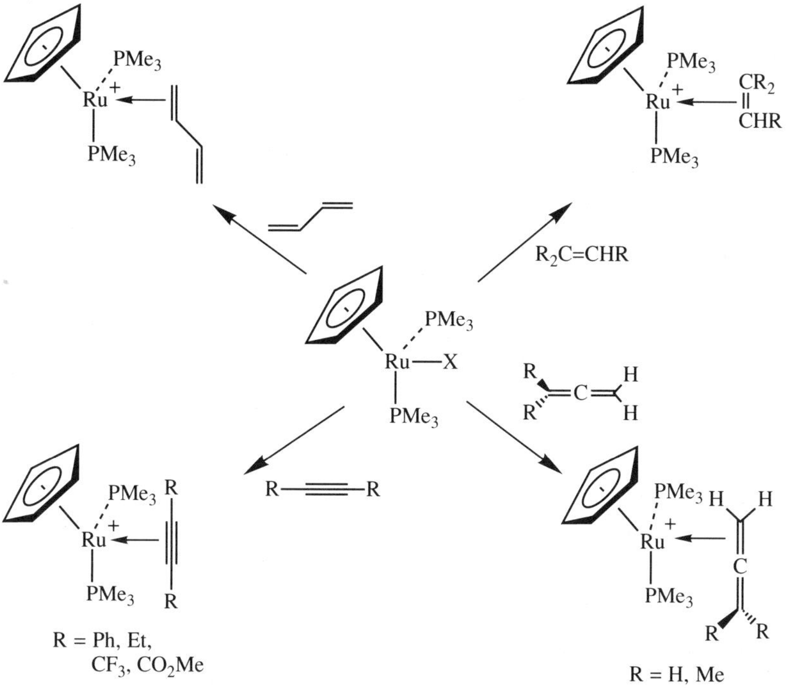

SCHEME 14

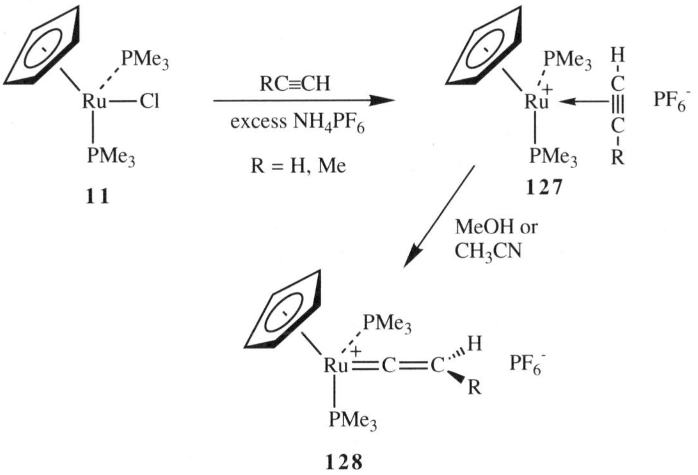

(108)

The interaction between bulky phosphine ligands and the C_2 unit of the unsaturated ligand bound orthogonally to the symmetry plane of the molecule can severely hinder the stability of the η^2 complex. Although reactions

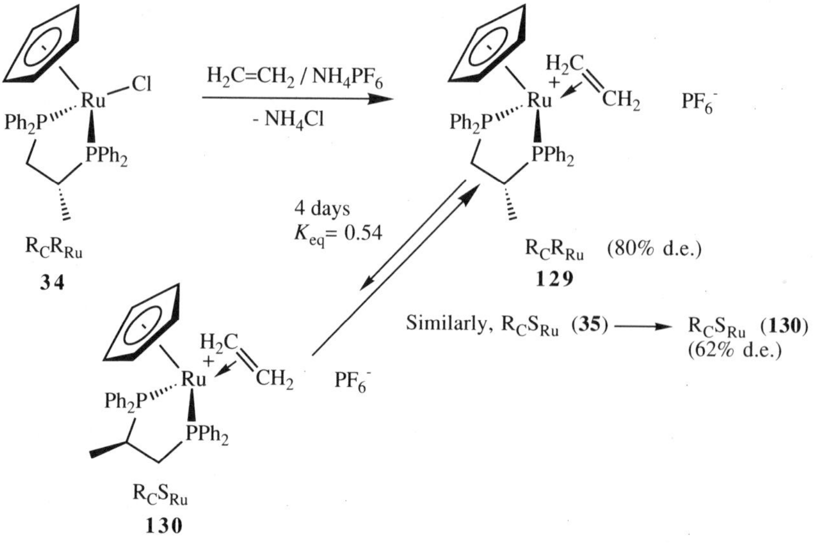

$$\text{SCHEME 15}$$

of both terminal and internal olefins with $(\eta^5\text{-}C_5H_5)(PMe_3)_2RuCl$ lead to the formation of the η^2-alkene complexes, only the sterically less demanding terminal olefins react with the corresponding dppe complex to form stable η^2-alkene complexes (40). Exchanging the chelating dppe ligand for two triphenylphosphine ligands increases the steric constraints at the metal center to the point where the formation of η^2-unsaturated complexes is not observed.

Consiglio and Morandini (98) have determined that the formation of the ethylene complex **129** from the chloro epimer **34** proceeds stereospecifically with retention of configuration at ruthenium (Scheme 15). The alkene complex, however, did epimerize in solution over a period of days to give an equilibrium mixture of the two epimers **129** and **130**. As in the similar reaction of **34** with CH_3CN discussed in Section V,D, the stereochemical integrity of the ruthenium center is maintained in the coordinatively unsaturated intermediate formed on loss of Cl^-.

IX

OLIGOMERIZATION REACTIONS

Reaction between alkynes with electron-withdrawing groups and ruthenium alkyl or hydride complexes results in complexes containing alkenyl or butadienyl ligands formed by insertion or oligomerization of the alkyne (*99*). The addition of hexafluorobut-2-yne to **32**, for example, affords the ruthenium vinyl complex **131** and the ruthenium diene complex **132** in low overall yield [Eq. (109)] (*100,101*). Addition of dimethylacetylenedicar-

$$\text{(109)}$$

boxylate to **32** produces the ruthenium vinyl complex **133** in high yield. On heating this complex in benzene, a bulky triphenylphosphine ligand is displaced by the ester function to form the chelating vinyl complex **134**, which presumably prevents oligomerization with another alkyne [Eq. (110)] (*102*).

$$\text{(110)}$$

X

ORTHOMETALLATION REACTIONS

There is a pronounced tendency for ruthenium cyclopentadienyl complexes with attached phosphorus ligands to undergo a wide variety of intramolecular metallation reactions. Heating $(\eta^5\text{-}C_5H_5)(PPh_3)_2RuMe$ in decalin for 24 hours leads to the orthometallated complex **135** in reasonable yield [Eq. (111)] (6,103). The mechanism presumably involves initial loss of

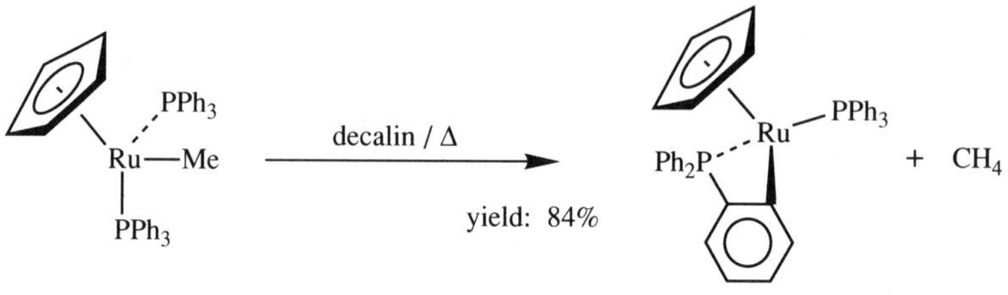

$$\text{decalin / }\Delta \qquad \text{yield: 84\%} \qquad + \ CH_4 \quad (111)$$

135

phosphine, followed by ruthenium insertion into an *ortho* C—H bond of a phenyl group on the phosphine. Elimination of methane and reassociation of phosphine generate the orthometallated product. An X-ray crystal structure of complex **135** is shown in Fig. 18 (*104*).

Thermolysis of ruthenium vinyl complexes also promotes intramolecular orthometallation [Eq. (112)]. However, performing the reaction in the

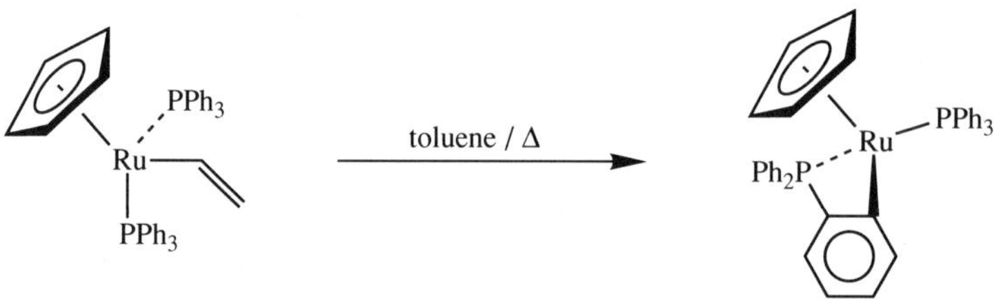

$$\text{toluene / }\Delta \qquad (112)$$

135

presence of ethylene inhibits the orthometallation reaction since the coordinatively unsaturated intermediate is trapped as the ethylene adduct (**136**). Complex **136** subsequently undergoes rearrangement in the presence of phosphine to give a ruthenium allyl complex (**137**) [Eq. (113)] (*105*).

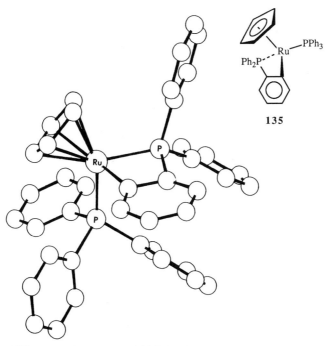

Fɪɢ. 18.   X-Ray crystal structure of $\{(\eta^5\text{-}C_5H_5)(PPh_3)_2Ru[PPh_2(o\text{-phenylene})]\}\cdot CH_2Cl_2$ (**135**) (hydrogens and solvate removed for clarity).

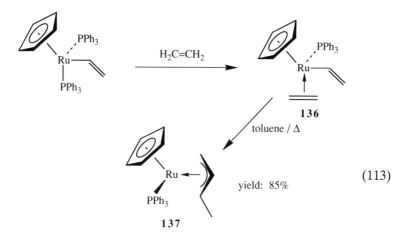

$$(113)$$

Addition of benzene or toluene in the presence of phosphine to the orthometallated species **135** yields ruthenium aryl complexes [Eq. (114)]. The reaction must involve initial dissociation of the nonorthometallated

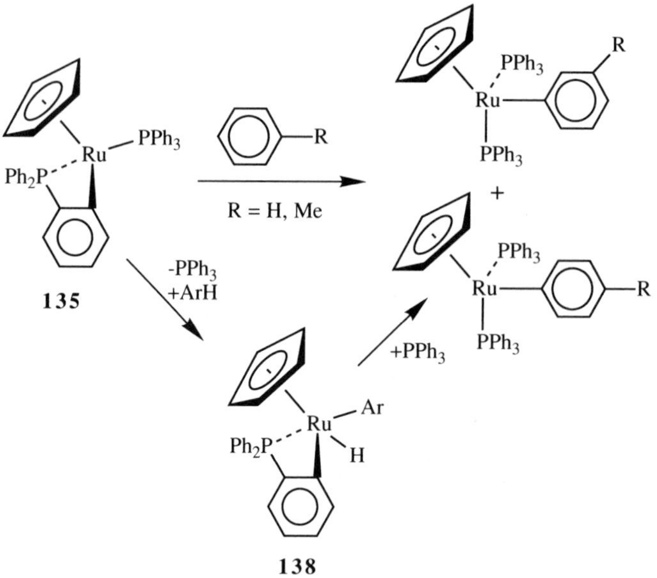

(114)

phosphine followed by insertion of ruthenium into an aryl C—H bond to generate the intermediate hydridoaryl complex **138**. Preferential elimination of the strained orthometallated aryl bond and reassociation of phosphine afford the products. In the toluene reaction, both para and meta adducts are observed (*106*).

Orthometallation also occurs intermolecularly. Addition of azobenzene to $(\eta^5\text{-}C_5H_5)(PPh_3)_2RuMe$ affords the orthometallated species **139** in moderate yield [Eq. (115)] (*104*). The reaction between hexafluorobut-2-yne

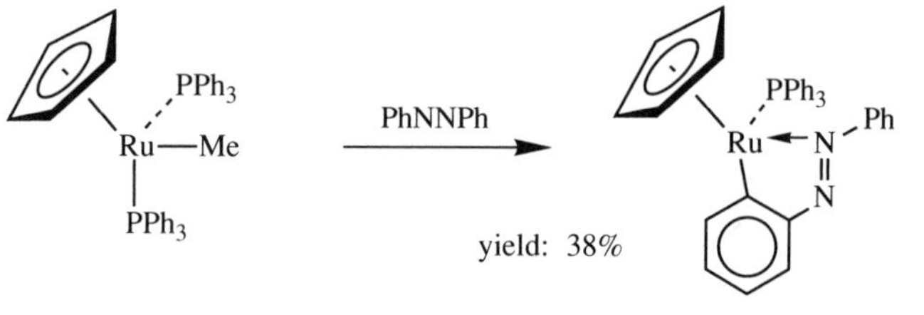

(115)

and $(\eta^5\text{-}C_5H_5)(PPh_3)_2RuCH_2Ph$ presumably proceeds via the addition of alkyne into the ruthenium–hydride bond of an intermediate orthometallated species to form **140** [Eq. (116)] (*107*).

$$(116)$$

140

REFERENCES

1. T. Blackmore, M. I. Bruce, and F. G. A. Stone, *J. Chem. Soc. A*, 2376 (1971).
2. J. D. Gilbert and G. Wilkinson, *J. Chem. Soc. A*, 1749 (1969).
3. M. I. Bruce and N. J. Windsor, *Aust. J. Chem.* **30**, 1601 (1977).
4. M. I. Bruce, C. Hameister, A. G. Swincer, and R. C. Wallis, *Inorg. Synth.* **21**, 78 (1982).
5. P. M. Treichel, D. A. Komar, and P. J. Vincenti, *Synth. React. Inorg. Met. Org. Chem.* **14**, 383 (1984).
6. H. Lehmkuhl, M. Bellenbaum, J. Grundke, H. Mauermann, and C. Kruger, *Chem. Ber.* **121**, 1719 (1988).
7. N. Oshima, H. Suzuki, and Y. Moro-Oka, *Chem. Lett.*, 1161 (1984).
8. L. A. Oro, M. A. Ciriano, and M. Campo, *J. Organomet. Chem.* **289**, 117 (1985).
9. R. Boese, W. B. Tolman, and K. P. C. Vollhardt, *Organometallics* **5**, 582 (1986).
10. K. P. C. Vollhardt and T. W. Weidman, *Organometallics* **3**, 82 (1984).
11. M. A. Huffman, D. A. Newman, M. Tilset, W. B. Tolman, and K. P. C. Vollhardt, *Organometallics* **5**, 1926 (1986).
12. J. Chang and R. G. Bergman, *J. Am. Chem. Soc.* **109**, 4298 (1987).
13. M. I. Bruce, F. S. Wong, B. W. Skelton, and A. H. White, *J. Chem. Soc., Dalton Trans.*, 1398 (1981).
14. F. Morandini, G. Consiglio, B. Straub, G. Ciani, and A. Sironi, *J. Chem. Soc., Dalton Trans.*, 2293 (1983).
15. F. A. Cotton and G. Wilkinson, "Advanced Inorganic Chemistry," 3rd Ed., p. 650. Wiley (Interscience), New York, 1972.
16. G. S. Ashby, M. I. Bruce, I. B. Tomkins, and R. C. Wallis, *Aust. J. Chem.* **32**, 1003 (1979).
17. M. I. Bruce, M. G. Humphrey, J. M. Patrick, and A. H. White, *Aust. J. Chem.* **36**, 2065 (1983).
18. G. Consigli, F. Morandini, and F. Bangerter, *Inorg. Chem.* **21**, 455 (1982).
19. M. I. Bruce, T. W. Hambley, M. R. Snow, and A. G. Swincer, *J. Organomet. Chem.* **273**, 361 (1984).
20. S. G. Davies and S. J. Simpson, *J. Chem. Soc., Dalton Trans.*, 993 (1984).
21. F. M. Conroy-Lewis and S. J. Simpson, *J. Organomet. Chem.* **322**, 221 (1987).
22. M. I. Bruce and R. C. Wallis, *Aust. J. Chem.* **34**, 209 (1981).
23. R. F. N. Ashok, M. Gupta, K. S. Arulsamy, and V. C. Agarwala, *Inorg. Chim. Acta* **98**, 161 (1985).
24. H. K. Gupta, V. Chauhan, and S. K. Dikshit, *Inorg. Chim. Acta* **128**, 175 (1987).
25. L. B. Reventos and A. G. Alonso, *J. Organomet. Chem.* **309**, 179 (1986).
26. K. E. Howard, T. B. Rauchfuss, and S. R. Wilson, *Inorg. Chem.* **27**, 1710 (1988).

27. R. J. Haines and A. L. Dupreez, *J. Organomet. Chem.* **84,** 357 (1975).
28. G. H. Spies and R. J. Angelici, *Organometallics* **6,** 1897 (1987).
29. G. J. Baird and S. G. Davies, *J. Organomet. Chem.* **262,** 215 (1984).
30. A. L. Steinmetz and B. V. Johnson, *Synth. React. Inorg. Met. Org. Chem.* **13,** 367 (1983).
31. W. Weigand, V. Nagel, and W. Beck, *J. Organomet. Chem.* **352,** 191 (1988).
32. G. J. Baird, S. G. Davies, S. D. Moon, S. J. Simpson, and R. H. Jones, *J. Chem. Soc., Dalton Trans.*, 1479 (1985).
33. H. Lehmkuhl, J. Grundke, and R. Mynott, *Chem. Ber.* **116,** 159 (1983).
34. M. S. Kharasch and O. Rheinmuth, "Grignard Reactions of Nonmetallic Substances," pp. 147–159. Prentice-Hall, Englewood Cliffs, New Jersey, 1954.
35. F. Morandini, G. Consiglio, and V. Lucchini, *Organometallics* **4,** 1202 (1985).
36. T. D. Tilley, R. H. Grubbs, and J. E. Bercaw, *Organometallics* **3,** 274 (1984).
37. S. G. Davies, S. D. Moon, and S. J. Simpson, *J. Chem. Soc., Chem. Commun.*, 1278 (1983).
38. T. Wilczewski, M. Bochenska, and J. F. Biernat, *J. Organomet. Chem.* **215,** 87 (1981).
39. M. I. Bruce, M. G. Humphrey, A. G. Swincer, and R. C. Wallis, *Aust. J. Chem.* **37,** 1747 (1984).
40. S. G. Davies and F. Scott, *J. Organomet. Chem.* **188,** C41 (1980).
41. P. M. Treichel and P. J. Vincenti, *Inorg. Chem.* **24,** 228 (1985).
42. M. I. Bruce, R. C. Wallis, M. L. Williams, B. W. Skelton, and A. H. White, *J. Chem. Soc., Dalton Trans.*, 2183 (1983).
43. T. Richardson, G. G. Roberts, M. E. C. Polywka, and S. G. Davies, *Thin Solid Films* **160,** 231 (1988).
44. R. F. N. Ashok, M. Gupta, K. S. Arulsamy, and U. C. Agarwala, *Inorg. Chim. Acta* **98,** 169 (1985).
45. P. M. Treichel and D. A. Komar, *Inorg. Chim. Acta* **42,** 277 (1980).
46. M. I. Bruce and F. S. Wong, *J. Organomet. Chem.* **210,** C5 (1981).
47. P. M. Treichel, D. A. Komar, and P. J. Vincenti, *Inorg. Chim. Acta* **88,** 151 (1984).
48. F. Morandini, G. Consiglio, G. Ciani, and A. Sironi, *Inorg. Chim. Acta* **82,** L27 (1984).
49. M. Draganjac, C. J. Ruffing, and T. B. Rauchfuss, *Organometallics* **4,** 1909 (1985).
50. S. G. Davies, H. Felkin, T. Fillebeen-Khan, F. Tadj, and O. Watts, *J. Chem. Soc., Chem. Commun.*, 341 (1981).
51. S. G. Davies, S. J. Simpson, H. Felkin, and T. Fillebeen-Khan, *Organometallics* **2,** 539 (1983).
52. R. J. Kulawiec and R. H. Crabtree, *Organometallics* **7,** 1891 (1988).
53. M. I. Bruce, I. B. Tomkins, F. S. Wong, B. W. Skelton, and A. H. White, *J. Chem. Soc., Dalton Trans.*, 687 (1982).
54. H. E. Bryndza, L. K. Fong, R. A. Paciello, W. Tam, and J. E. Bercaw, *J. Am. Chem. Soc.* **109,** 1444 (1987).
55. M. S. Chin and D. M. Heinekey, *J. Am. Chem. Soc.* **109,** 5865 (1987).
56. F. M. Conroy-Lewis and S. J. Simpson, *J. Chem. Soc., Chem. Commun.*, 1675 (1987).
57. G. J. Kubas, R. R. Ryan, B. I. Swanson, P. J. Vergamini, and H. J. Wasserman, *J. Am. Chem. Soc.* **106,** 451 (1984).
58. G. J. Kubas, R. R. Ryan, and D. A. Wrobleski, *J. Am. Chem. Soc.* **108,** 1339 (1986); G. J. Kubas, C. J. Unkefer, B. I. Swanson, and E. Fukushima, *J. Am. Chem. Soc.* **108,** 7000 (1986).
59. M. I. Bruce, M. J. Liddell, M. R. Snow, and E. R. T. Tiekink, *J. Organomet. Chem.* **352,** 199 (1988).

60. G. Consiglio, F. Morandini, G. Ciani, A. Sironi, and M. Kretschmer, *J. Am. Chem. Soc.* **105**, 1391 (1983).
61. G. Consiglio, F. Morandini, G. F. Ciani, and A. Sironi, *Organometallics* **5**, 1976 (1986).
62. M. I. Bruce and O. M. Abu Salah, *J. Organomet. Chem* **64**, C48 (1974).
63. O. M. Abu Salah and M. I Bruce, *J. Chem. Soc., Dalton Trans.*, 2311 (1975).
64. K. Sonogashira, T. Yatake, Y. Tohda, and S. Takahashi, *J. Chem. Soc., Chem. Commun.*, 291 (1977).
65. O. M. Abu Salah, M. I. Bruce, M. R. Churchill, and S. A. Bezman, *J. Chem. Soc., Chem. Commun.*, 858 (1972); M. I. Bruce, R. Clark, J. Howard, and P. Woodward, *J. Organomet. Chem.* **42**, C107 (1972).
66. M. I. Bruce, M. G. Humphrey, M. R. Snow, and E. R. T. Tiekink, *J. Organomet. Chem.* **314**, 213 (1986).
67. M. I. Bruce and R. C. Wallis, *J. Organomet. Chem.* **161**, C1 (1978).
68. M. I. Bruce and R. C. Wallis, *Aust. J. Chem.* **32**, 1471 (1979).
69. J. Silvestre and R. Hoffmann, *Helv. Chim. Acta* **68**, 1461 (1985).
70. N. M. Kostic and R. F. Fenske, *Organometallics* **1**, 974 (1982).
71. M. I. Bruce, G. A. Koutsantonis, M. J. Liddell, and B. K. Nicholson, *J. Organomet. Chem.* **320**, 217 (1987).
72. M. I. Bruce, F. S. Wong, B. W. Skelton, and A. H. White, *J. Chem. Soc., Dalton Trans.*, 2203 (1982).
73. M. I. Bruce and A. G. Swincer, *Aust. J. Chem.* **33**, 1471 (1980).
74. M. I. Bruce, M. G. Humphrey, and G. A. Koutsantonis, *J. Organomet. Chem.* **296**, C47 (1985).
75. B. K. Blackburn, S. G. Davies, K. H. Sutton, and M. Whittaker, *Chem. Soc. Rev.* **17**, 147 (1988).
76. G. Consiglio and F. Morandini, *Inorg. Chim. Acta* **127**, 79 (1987).
77. S. Abbott, S. G. Davies, and P. Warner, *J. Organomet. Chem.* **246**, C65 (1983).
78. S. Abbott, S. G. Davies, and P. Warner, manuscripts in preparation; S. Abbott, Ph.D. thesis. Oxford University, Oxford, 1984.
79. M. I. Bruce, M. G. Humphrey, G. A. Koutsantonis, and M. J. Liddell, *J. Organomet. Chem.* **326**, 247 (1987).
80. G. Consiglio, R. Schwab, and F. Morandini, *J. Chem. Soc., Chem. Commun.*, 25 (1988).
81. J. P. Selegue, *J. Am. Chem. Soc.* **104**, 119 (1982).
82. M. I. Bruce, M. G. Humphrey, and M. J. Liddell, *J. Organomet. Chem.* **321**, 91 (1987).
83. M. I. Bruce, A. G. Swincer, and R. C. Wallis, *J. Organomet. Chem.* **171**, C5 (1979).
84. C. A. Tolman, *Chem. Rev.* **77**, 313 (1977).
85. M. I. Bruce, A. G. Swincer, B. J. Thomson, and R. C. Wallis, *Aust. J. Chem.* **33**, 2605 (1980).
86. O. M. Abu Salah and M. I. Bruce, *J. Chem. Soc., Dalton Trans.*, 2302 (1974).
87. M. L. H. Green, L. C. Mitchard, and M. G. Swanwick, *J. Chem. Soc. A*, 794 (1971); T. Bodnar, S. J. LaCroce, and A. R. Cutler, *J. Am. Chem. Soc.* **102**, 3292 (1982); T. Bodnar and A. R. Culter, *J. Organomet. Chem.* **213**, C31 (1981).
88. G. Consiglio, F. Bangerter, and F. Morandini, *J. Organomet. Chem.* **293**, C29 (1985).
89. M. I. Bruce, D. N. Duffy, M. G. Humphrey, and A. G. Swincer, *J. Organomet. Chem.* **282**, 383 (1985).
90. M. I. Bruce, J. R. Rogers, M. R. Snow, and A. G. Swincer, *J. Chem. Soc., Chem. Commun.*, 271 (1981).
91. M. I. Bruce, T. W. Hambley, M. R. Snow, and A. G. Swincer, *Organometallics* **4**, 501 (1985).

92. M. I. Bruce, P. A. Humphrey, M. R. Snow, and E. R. T. Tiekink, *J. Organomet. Chem.* **303**, 417 (1986).
93. J. P. Selegue, *Organometallics* **1**, 217 (1982).
94. J. P. Selegue, *J. Am. Chem. Soc.* **105**, 5921 (1983).
95. M. F. Joseph, J. A. Page, and M. C. Baird, *Organometallics* **3**, 1749 (1984).
96. M. I. Bruce, T. W. Hambley, J. R. Rodgers, M. R. Snow, and F. S. Wong, *Aust. J. Chem.* **35**, 1323 (1982).
97. R. M. Bullock, *J. Chem. Soc., Chem. Commun.*, 165 (1989).
98. G. Consiglio and F. Morandini, *J. Organomet. Chem.* **310**, C66 (1986).
99. M. I. Bruce, R. C. F. Gardiner, F. G. A. Stone, M. Welling, and P. Woodward, *J. Chem. Soc., Dalton Trans.*, 621 (1977).
100. T. Blackmore, M. I. Bruce, and F. G. A. Stone, *J. Chem. Soc., Dalton Trans.*, 106 (1974).
101. M. I. Bruce, R. C. F. Gardiner, and F. G. A. Stone, *J. Chem. Soc., Dalton Trans.*, 906 (1979).
102. T. Blackmore, M. I. Bruce, F. G. A. Stone, R. E. Davis, and N. V. Raghavan, *J. Organomet. Chem.* **49**, C35 (1973).
103. M. I. Bruce, R. C. F. Gardiner, and F. G. A. Stone, *J. Chem. Soc., Dalton Trans.*, 81 (1976).
104. M. I. Bruce, M. P. Cifuentes, M. G. Humphrey, E. Poczman, M. R. Snow, and E. R. T. Tiekink, *J. Organomet. Chem.* **338**, 237 (1988).
105. H. Lehmkuhl, J. Grundke, and R. Mynott, *Chem. Ber.* **116**, 176 (1983).
106. H. Lehmkuhl, M. Bellenbaum, and J. Grundke, *J. Organomet. Chem.* **330**, C23 (1987).
107. M. I. Bruce, R. C. F. Gardiner, and F. G. A. Stone, *J. Organomet. Chem.* **40**, C39 (1972).

ADVANCES IN ORGANOMETALLIC CHEMISTRY, VOL. 30

Thermochromic Distibines and Dibismuthines

ARTHUR J. ASHE III

Department of Chemistry
The University of Michigan
Ann Arbor, Michigan 48109

I
INTRODUCTION

Suitably substituted distibines and related binuclear compounds of the heavier main group elements associate in the solid state. These compounds exhibit short "intermolecular" contacts which indicate a secondary intermolecular bonding. On melting or dissolution in organic solvents the intermolecular bonding is lost simultaneously with dramatic changes in color (thermochromism). However, intermolecular bonding is rather sensitive to substitution since the thermochromic effect is not observed for all distibines. The 1980s have witnessed an intense interest in solid state chemistry, which has produced substantial progress on the distibine problem. Most of this work has been done by the Becker and Breunig groups in Germany and in our own laboratory. Sufficient progress has been made to place this work in overall context, which is the objective of this article.

II
HISTORICAL PERSPECTIVE

The first well-characterized distibine, tetraphenyldistibine (**1**), was reported by Blicke *et al.* of the University of Michigan in 1931 (*1*). Tetraphenyldistibine was originally prepared by reduction of diphenyliodostibine with sodium hypophosphite. The compound is highly reactive toward oxygen and iodine but is otherwise unexceptional in behavior. It is a pale yellow solid which melts sharply at 125°C to a pale yellow liquid. Since no color change is observed on melting, tetraphenyldistibine has been termed a nonthermochromic distibine. The nonthermochromic distibines reported through mid-1988 are collected in Table I (*1–16*). In

TABLE I

Nonthermochromic Distibines

Formula	Compound	Reference(s)
Nonthermochromic distibines reported through mid-1988		
Aryl distibines		
$C_{24}H_{20}Sb_2$	Tetraphenyldistibine (**1**)	*1–3*
$C_{24}H_{16}Br_4Sb_2$	Tetra(p-bromophenyl)distibine (**2**)	*4*
$C_{28}H_{28}Sb_2$	Tetra(p-tolyl)distibine (**3**)	*4*
$C_{36}H_{44}Sb_2$	Tetra(mesityl)distibine (**4**)	*5*
$C_{26}H_{18}Sb_2$	9-Stibaanthracene dimer (**5**)	*6*
Vinyl distibines		
$C_{12}H_{20}Sb_2$	Tetra-(E)-1-propenyldistibine (**6**)	*7*
$C_{12}H_{20}Sb_2$	Tetra-(Z)-1-propenyldistibine (**7**)	*7*
$C_{16}H_{28}Sb_2$	Tetra-(E)-2-but-2-enyldistibine (**8**)	*7*
$C_{16}H_{28}Sb_2$	Tetrakis(2-methyl-1-propenyl)distibine (**9**)	*7*
$C_{16}H_{24}Sb_2$	2,2′,5,5′-Tetraethylbistibole (**10**)	*8*
Alkyl distibines		
$C_4F_{12}Sb_2$	Tetrakis(trifluoromethyl)distibine (**11**)	*9*
$C_{10}H_{20}Sb_2$	1,1′-Bistibacyclohexane (**12**)	*10*
$C_{24}H_{44}Sb_2$	Tetracyclohexyldistibine (**13**)	*11*
Other selected compounds with Sb—Sb bonds		
$C_{16}H_{36}Sb_4$	Tetra(*tert*-butyl)tetrastibine (**14**)	*12–14*
$C_{36}H_{30}Sb_6$	Hexaphenylhexastibine (**15**)	*15,16*
Nonthermochromic distibines not reported as crystalline		
$C_{12}H_{28}Sb_2$	Tetrapropyldistibine (**16**)	*17a,b*
$C_{12}H_{28}Sb_2$	Tetraisopropyldistibine (**16a**)	*17c*
$C_{16}H_{36}Sb_2$	Tetra(*tert*-butyl)distibine (**17**)	*14*
$C_{16}H_{36}Sb_2$	Tetrabutyldistibine (**17a**)	*17c,d*
$C_{14}H_{16}Sb_2$	1,2-Dimethyl-1,2-diphenyldistibine (**18**)	*18*
$C_{16}H_{20}Sb_2$	1,2-Diethyl-1,2-diphenyldistibine (**19**)	*19*
$C_{15}H_{16}Sb_2$	1,2-Diphenyl-1,2-distibacyclopentane (**20**)	*19*
$C_{10}H_{10}Sb_2$	Stibabenzene dimer (**21**)	*20*

addition, there are a number of distibines which have not been reported as solids. They are also listed in Table I (*14,17a–20*).

As a part of his classic study of reactive free radicals in the early 1930s, Paneth observed that methyl radicals react with a heated antimony mirror to produce tetramethyldistibine (**22**) (*21*). This distibine has quite remarkable properties. It forms intensely colored red crystals which melt reversibly to a yellow oil. Similarly, solutions of tetramethyldistibine are pale yellow. Although these color changes are clearly due to changes in phase rather than strictly temperature, they have been termed thermochromic (*22*). It should be noted that Paneth also obtained tetramethyldibismuthine (**23**) in trace quantities from methyl radicals and a heated bismuth mirror.

TABLE II

THERMOCHROMIC DISTIBINES AND DIBISMUTHINES[a]

Compound	E = Sb	Reference(s)	E = Bi	Reference(s)
$(CH_3)_4E_2$	**22:** mp 17.5°C, red solid	*17a,21,23–27*	**23:** mp −12.5°C, violet-blue solid	*21,28–30*
$(C_2H_5)_4E_2$	**24:** mp −61°C, orange	*21,24,31*	**25:** mp not reported, blue	*32*
(ring structure)$_2$	**26:** mp 47°C, orange	*10*	**27:** mp 47°C, purple	*29*
(ring structure)$_2$	**28:** mp 45°C, orange	*33*	**29:** $d > 0$°C, violet	*33*
(ring structure)$_2$	**30:** mp 99°C, purple-blue	*22*	**31:** $d > 95$°C, black	*34*
$(C_2H_3)_4E_2$	**32:** mp −53°C, purple	*7*	**35:** mp 13.5°C, purple	*29*
$[CH_2C(CH_3)]_4E_2$	**34:** mp 1°C, orange	*7*	**37:** mp 148°C, green	*37*
$[(CH_3)_3Si]_4E_2$	**36:** mp 119°C, red	*35,36*		
$[(CH_3)_3Ge]_4E_2$	**38:** mp 125°C, dark red	*38a*		
$[(CH_3)_3Sn]_4E_2$	**40:** mp 143°C, red	*39–41*		

[a] mp, Melting point; *d*, decomposition.

TABLE III

NONTHERMOCHROMIC DIBISMUTHINES

Formula	Compound	Reference(s)
$C_{12}H_{28}Bi_2$	Tetrapropyldibismuthine (42)	32
$C_{12}H_{28}Bi_2$	Tetraisopropyldibismuthine (43)	32
$C_{16}H_{36}Bi_2$	Tetrabutyldibismuthine (44)	32
$C_{16}H_{28}Bi_2$	Tetrakis(2-methyl-1-propenyl)dibismuthine (45)	29
$C_{24}H_{20}Bi_2$	Tetraphenyldibismuthine (46)	29,42
$C_{10}H_{10}Bi_2$	Bismabenzene dimer (47)	20
$C_{12}H_{14}Bi_2$	4-Methylbismabenzene dimer (48)	20

The dibismuthine occurred as violet crystals which melted to a red-yellow liquid prior to decomposition below room temperature.

Nine other thermochromic distibines have been reported. All show a yellow melt, but the solid colors range from deep yellow to violet-blue. Seven thermochromic dibismuthines are known. The liquid colors are all red while solid colors are variable. The thermochromic distibines and dibismuthines are listed in Table II (7,10,17a,21,23–41) along with the melting points and colors of the crystals. Nonthermochromic dibismuthines are collected in Table III (20,29,32,42).

III

SYNTHESIS AND CHEMICAL PROPERTIES

The Blicke and Paneth syntheses are interesting almost exclusively from a historical perspective. The best general method for preparing distibines involves the coupling reactions of the corresponding metal stibides as illustrated in Scheme 1. For example, triisopropenylstibine is available from the reaction of isopropenyllithium with antimony trichloride. Cleavage of triisopropenylstibine with sodium in liquid ammonia followed by oxidation with 1,2-dichloroethane gave a 80% yield of tetraisopropenyldistibine. Alternatively, triisopropenylstibine may be converted to diisopropenylchlorostibine, which on treatment with sodium in liquid ammonia followed by 1,2-dichloroethane gave 81% of tetraisopropenyldistibine (7).

As previously noted, tetraorganodistibines are highly sensitive toward oxygen but may be conveniently handled used Schlenk-ware techniques. Generally, tetraorganodistibines do not react with water or mild acid and base. However, tetramethyldistibine (see Scheme 2) is cleaved by HCl to

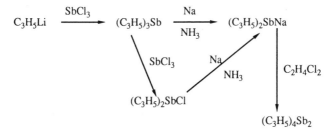

SCHEME 1. Preparation of tetraisopropenyldistibine.

$$(CH_3)_4Sb_2 \xrightarrow{\text{HCl}} (CH_3)_2SbH + (CH_3)_2SbCl$$

$$(CH_3)_4Sb_2 \xrightarrow{C_4H_9Li} (CH_3)_2(C_4H_9)Sb + (CH_3)_2SbLi$$

$$(CH_3)_4Sb_2 \xrightarrow{\Delta} (CH_3)_3Sb + Sb^O$$

SCHEME 2. Reactions of tetramethyldistibine.

give dimethylstibine and dimethylantimony chloride (23), and it reacts with butyllithium to afford butyldimethylstibine and lithium dimethylstibide (25). Distibines are usually stable at room temperature, but on heating above 100°C they produce antimony metal and the corresponding tertiary stibine (7,21).

Tetraorganodibismuthines are prepared in the same manner as the corresponding distibines. Thus, the reaction of trimethylbismuthine with sodium in liquid ammonia gave a red solution of NaBi(CH₃)₂ which on treatment with 1,2-dichloroethane afforded tetramethyldibismuthine (see Scheme 3) in 70% yield. The dibismuthine is conveniently purified by recrystallization from pentane at −20°C. Although 14 dibismuthines have been reported in the literature through mid-1988, only tetramethyldibismuthine has been investigated in detail. All its reported reactions involve cleavage of the Bi—Bi bond (see Scheme 4). It decomposes at 25°C with near quantitative formation of trimethylbismuthine and bismuth metal. Higher alkyl dibismuthines and several tetravinyldibismuthines are less labile, while tetraphenyldibismuthine and 2,2′,5,5′-tetramethylbibismole are stable to 100°C. Tetramethyldibismuthine reacts with oxygen and

$$(CH_3)_3Bi \xrightarrow[NH_3]{Na} (CH_3)_2BiNa \xrightarrow{C_2H_4Cl_2} (CH_3)_4Bi_2$$

SCHEME 3. Preparation of tetramethyldibismuthine.

$(CH_3)_4Bi_2 \xrightarrow{\Delta} Bi^{0} + (CH_3)_3Bi$

$(CH_3)_4Bi_2 + I_2 \longrightarrow Bi^{0} + (CH_3)_2BiI + CH_3BiI_2 + BiI_3 + Bi(CH_3)_3$

$(CH_3)_4Bi_2 + HCl \longrightarrow (CH_3)_2BiCl + [(CH_3)_2BiH] \longrightarrow Bi^{0} + Bi(CH_3)_3 + H_2$

$(CH_3)_4Bi_2 + C_4H_9Li \longrightarrow (C_4H_9)(CH_3)_2Bi + (CH_3)_2BiLi$

SCHEME 4.   Reactions of tetramethyldibismuthine.

with halogenated solvents. Reaction with iodine is rapid and indiscriminate, yielding bismuth metal and all possible iodo- and methylbismuthines. Reaction with HCl gas gives bismuth metal, trimethylbismuthine, dimethylbismuth chloride, and hydrogen gas. Dimethylbismuthine is an intermediate. Reaction of tetramethyldibismuthine with butyllithium apparently gives butyldimethylbismuthine and lithium dimethylbismuthide.

IV

ULTRAVIOLET (UV) AND DIFFUSE REFLECTANCE SPECTRA

Most distibines appear yellow, while dibismuthines are red, either in their liquid phases or in solution in organic solvents. A series of tetravinyldistibines (7) and dibismuthines (24) show absorption maxima in the range of 200–330 nm. 2,2′,5,5′-Tetramethylbistibole (30) shows a maximum at 346 nm (22), while the corresponding bibismole (31) absorbs at 320 nm (34). In all cases, the distibines show a low intensity, featureless absorption tail extending out to about 400 nm, while the tail extends to approximately 800 nm for the dibismuthines. Presumably, this feature corresponds to the observed yellow and red colors. No spectral assignments have been made.

On crystallization, nonthermochromic distibines and dibismuthines show little visual change. The solid colors of two nonthermochromic distibines (8 and 9) and two nonthermochromic dibismuthines (45 and 46) have been characterized by diffuse reflectance. In each case, only very modest changes in the absorption maxima were observed and between solid and solution. On the other hand, the intense colors shown by the solid phases of the thermochromic distibines are red shifted by 200–250 nm from their solution phase maxima (see Table IV) (7,25,29,33,34,37, 38b,40). The dibismuthines are red shifted even further. There is complete correspondence between dibismuthines and the analogous distibines, with the dibismuthines being red shifted by around 100 nm (see Fig. 1).

TABLE IV

DIFFUSE REFLECTANCE ABSORPTION MAXIMA (nm) OF DIBISMUTHINES AND
CORRESPONDING DISTIBINES[a]

Compound	E = Sb	E = Bi	ΔBiSb	Reference(s)
$(CH_3)_4E_2$	22: 530(<215)	23: 665(264)	135	29
(E)₂ cyclopentane	26: 490	27: 580(264)	90	29
(E)₂	28: 475(300)	29: 605(330)	130	33
(E)₂	30: 605(346)	31: 690(320)	85	29,34
$[CH_2C(CH_3)]_4E_2$	34: 475(290)	35: 575(270)	100	7,29
$[(CH_3)_3Si]_4E_2$	36: 520(230)	37: green	—	38b
$[(CH_3)_3Ge]_4E_2$	38: 520(230)	Unknown	—	38b
$[(CH_3)_3Sn]_4E_2$	40: 510(212)	Unknown	—	40

[a] Solution absorption maxima are in parentheses.

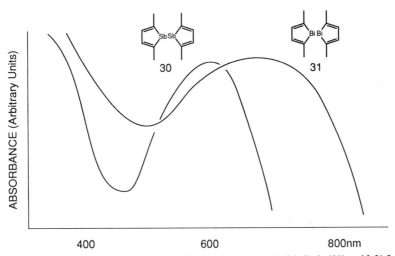

FIG. 1. Diffuse reflectance spectra of solid 2,2′,5,5′-tetramethylbistibole (30) and 2,2′,5,5′-tetramethylbibismole (31). [Reprinted with permission from A. J. Ashe III and F. J. Drone (34), *Organometallics* 3, 495 (1984). Copyright 1984 American Chemical Society.]

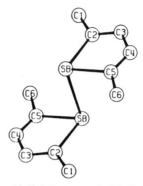

FIG. 2.   Molecular structure of 2,2′,5,5′-tetramethylbistibole (**30**). Selected bond distances
(Å): Sb–C-2, 2.140(5); Sb–C-5, 2.135(4); Sb–Sb, 2.835(1). Selected bond angles (°): C-2–Sb–
C-5, 81.5(2); C-2–Sb–Sb, 91.4(1); C-5–Sb–Sb, 92.2(1). [Reprinted with permission from
A. J. Ashe III, W. Butler, and T. R. Diephouse (*22*), *J. Am. Chem. Soc.* **103**, 207 (1981).
Copyright 1981 American Chemical Society.]

V

STRUCTURE

The five distibines (**1, 22, 30, 36,** and **40**) that have been investigated by
X-ray crystallography show very similar molecular structures (see Fig. 2
for the molecular structure of 2,2′,5,5′-tetramethylbistibole, **30**). In each
case the distibine adopts a staggered trans conformation so that there is an
inversion center through the Sb—Sb bond. The Sb—Sb bond lengths vary
between 2.84 and 2.88 Å, values somewhat shorter than that of elemental
antimony (2.90 Å) (*43*). The bond angles about antimony are close to 90°
and similar to those reported for other trivalent antimony compounds (*44*).
In fact, none of the molecular parameters seems extraordinary. However,
the four thermochromic distibines show unusual crystal packing (see
Fig. 3, Table V) (*22,25,36,40,41,45*). In each case, the antimony atoms are

TABLE V

SELECTED STRUCTURAL DATA FOR DISTIBINES

Compound	Sb—Sb (Å)	Sb---Sb (Å)	∢ Sb—Sb---Sb	Reference
Tetramethyldistibine (**22**)	2.86	3.65	179.2°	*25*
2,2′,5,5′-Tetramethylbistibole (**30**)	2.84	3.63	173.5°	*22*
Tetrakis(trimethylsilyl)distibine (**36**)	2.87	3.99	165.8°	*36*
	2.86	3.89		*41*
Tetrakis(trimethylstannyl)distibine (**40**)	2.88	3.89	173.5°	*40*
	2.87	3.81		*41*
Tetraphenyldistibine (**1**)	2.84	4.29	—	*45*

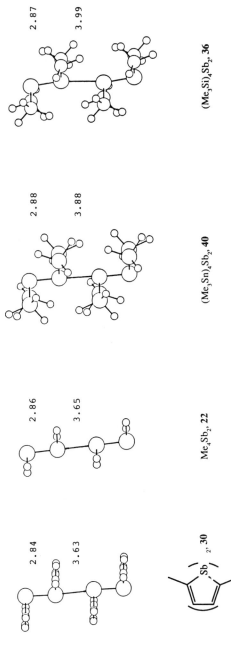

FIG. 3. Crystal packing of the thermochromic distibines 2,2',5,5'-tetramethylbistibibole (**30**) (22), tetramethyldistibine (**22**) (25,26), tetrakis(trimethylstannyl)distibine (**40**) (40,41), and tetrakis(trimethylsilyl)distibine (**36**) (36,41).

aligned in chains with alternate short (intramolecular) Sb—Sb and somewhat longer (intermolecular) Sb---Sb separations. The intermolecular Sb---Sb contacts are quite short, ranging from 3.63 to 3.99 Å. These values are well below the van der Waals radius separation of 4.4 Å and imply that there is an extended bonding along the Sb—Sb---Sb—Sb chain. The dramatic changes in color on melting or dissolution in organic solvents must be associated with disruption of this extended bonding. Thus, the solid phase colors are likely to be due to electronic excitation along the chain. By extrapolation, the thermochromic effect observed for distibines **24, 26, 28, 32, 34**, and **38** implies that they also stack in the solid phase.

Although the Sb chains are qualitatively similar, there is a variation in the linearity of the chain and in the intermolecular distances. Thus, the chains are nearly linear ($\angle$ Sb—Sb---Sb 179.2°) and the Sb---Sb distances are short (3.68 Å) for the sterically undemanding tetramethyldistibine (**22**), while the chains are distinctly zig-zag ($\angle$ Sb—Sb---Sb 165.8°) and the distances (3.99 Å) are longer for the more bulky tetrakis(trimethylsilyl)distibine (**36**). While short intermolecular contacts seem necessary for the thermochromic effect, the UV absorption maxima of the solid distibines do not correlate with the intermolecular Sb---Sb distances. Thus, the red compounds **22, 36**, and **40** have nearly identical absorption maxima despite a variation of more than 0.3 Å in the intermolecular Sb contacts. On the other hand, the intermolecular Sb---Sb separations of **22** and **30** are nearly the same although the absorption maxima differ by 75 nm.

Tetraphenyldistibine (**1**) is the only nonthermochromic distibine for which a crystal structure has been determined (*45*). However, structural data are available for two closely related and also nonthermochromic cyclic polystibines [**14** (*12*) and **15** (*15,16*)]. In no case can a similar chain of antimony atoms be identified. In tetraphenyldistibine the closest intermolecular Sb---Sb contact is 4.29 Å, close to the van der Waals separation (see Fig. 4). Although tetraphenyldistibine (**1**) has a staggered trans conformation, the two phenyl groups attached to each antimony are canted 80° relative to each other. The juxaposition of the two phenyls minimizes intramolecular steric interaction between the ortho hydrogen atoms, but does so at the expense of congesting the back side of the Sb—Sb bond axis. Presumably, this steric hindrance minimizes the possibility of close intermolecular Sb---Sb interaction. It has been suggested that similar steric problems prevent association of the other nonthermochromic distibines (*22*).

Fewer structural data are available for dibismuthines. Published structures include only the thermochromic tetrakis(trimethylsilyl)dibismuthine (**37**) (*37*) and the nonthermochromic tetraphenyldibismuthine (**46**) (*42*) although some data from an unpublished structure of tetramethyl-

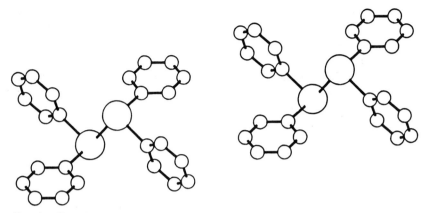

FIG. 4.   Crystal structure of tetraphenyldistibine (1) showing two neighboring molecules (45).

dibismuthine (23) have been summarized by Becker and colleagues (46). Both thermochromic dibismuthines are isostructural with the corresponding distibines (see Fig. 5 for the structure of 37), and there is a Bi—Bi---Bi—Bi chain. Intramolecular Bi—Bi bonds (3.12 Å for 23 and 3.04 Å for 37) are somewhat longer than for the nonthermochromic

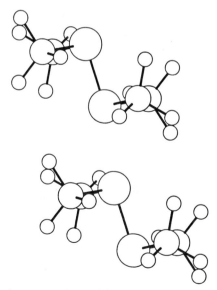

FIG. 5.   Crystal structure of tetrakis(trimethylsilyl)dibismuthine (37) (37).

tetraphenyldibismuthine (**23**) (2.99 Å). Surprisingly, the intermolecular Bi---Bi separations (3.80 Å for **23** and 3.58 Å for **37**) are shorter than the Sb---Sb separations of the corresponding distibines. Thus, the metal–metal chains are more symmetrical.

While tetraphenyldibismuthine (**46**) is not isostructural with tetraphenyl-distibine (**1**), it does not show any close intramolecular Bi---Bi contacts. Although no other structural data are available, it might be noted that there is a complete match of the thermochromic properties for the dibismuthines with the corresponding distibines. Thus, in all cases if a distibine is thermochromic the analogous dibismuthine is as well. Conversely, nonthermochromic distibines correspond to nonthermochromic dibismuthines. A structural correspondence also seems likely.

VI
THEORY AND BONDING

The thermochromic effect of distibines has been treated in three papers by Hoffmann and colleagues using a tight bonding model based on extended Hückel calculations (*33,47,48*). These calculations treated only unsaturated distibines, and major attention was focused on bistibole (*47*). The important orbitals of the stacked bistibole are derived from molecular orbitals of the SbC_4H_4 unit (see Fig. 6). The HOMO results from the in-phase mixing of the $Sb(p_z)$ and $Sb(n_x)$ orbitals and is largely localized on the Sb atoms. At the zone center ($k = 0$) this band has primarily lone pair

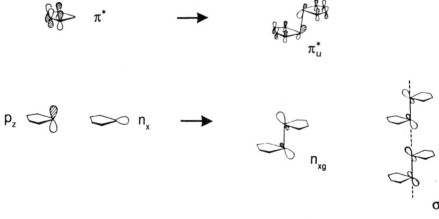

FIG. 6.   Derivation of the HOMO and LUMO of stacked bistibole from C_4H_4Sb orbitals.

character (n_{xg}) but has σ character at the zone edge (σ_g) (see Fig. 7). At the zone edge this valence band is destabilized since it is σ antiboding between neighboring cells, although it remains σ bonding within each unit cell. The LUMO is made from the out-of-phase combination of the π^* orbitals stabilized by the σ^*_{SbSb}. This conduction band is mostly localized on the diene and changes little on moving across the zone.

Since the HOMO–LUMO gap of the stacked bistibole decreases on moving to the zone edge, crystallization is predicted to result in a red shift, as is observed. In this model the conduction band is sensitive mainly to the energy of the π_u^* orbital of the distibine. The unconjugated distibine **28** is calculated to have a higher π^* energy, resulting in a larger band gap (*33*). As predicted, the solid phase color of **28** is blue shifted relative to **30**. By extension the as yet unknown more conjugated distibine **49** is predicted to

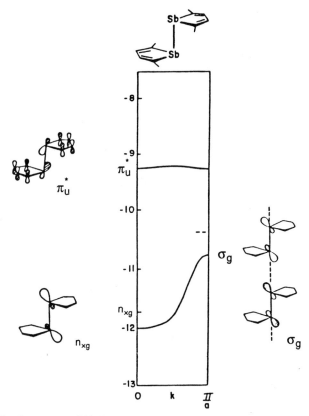

Fig. 7. Band structure of bistibole. The higher and lower bands have been omitted for clarity. [Reprinted with permission from A. J. Ashe III, C. M. Kausch, and O. Eisenstein (*33*), *Organometallics* **6**, 1185 (1987). Copyright 1987 American Chemical Society.]

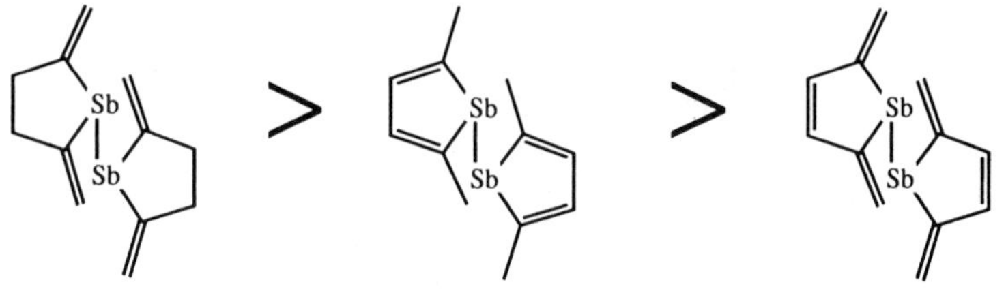

SCHEME 5.   Decreasing energy of the HOMO–LUMO gap of solid distibines with increasing conjugation.

be red shifted relative to **30** (see Scheme 5). The major shortcoming of this model is that the alkyl, silyl, germyl, and stannyl thermochromic distibines seem to have no low-lying π^* orbitals to serve as the LUMO. Hoffmann has suggested that the vacant Sb $(5d)$ orbitals may mix with the σ^*_{SbSb} orbital to fulfill this function. Thus, the color of these distibines may be due to a $\sigma \rightarrow \sigma^*$ transition where the σ^* orbital has appreciable d orbital contributions. Certainly, a more detailed treatment of these compounds is desirable.

No theoretical treatment is available for dibismuthines. However, the qualitative features of the band structures of the thermochromic dibismuthines can be anticipated from the Hoffmann model for the corresponding distibines. Thus, the substitution of Bi for Sb is unlikely to affect greatly the ligand π^* orbitals. But the σ_g orbital will likely be raised to reflect the weaker metal–metal bond (28). A smaller band gap is predicted, as is observed.

VII

RAMAN SPECTRA, ELECTRON DIFFRACTION, AND MÖSSBAUER SPECTRA

Raman spectroscopy is extremely useful for characterization of distibines and dibismuthines, since the metal–metal stretching vibrations give rise to intense, easily identifiable peaks. Several thermochromic and nonthermochromic distibines have been examined as both solids and liquids (see Table VI) (7,40,49–51). With the exception of tetramethyldistibine, there are only small shifts in ν_{SbSb} between solid and liquid. Nor are there systematic differences between thermochromic and nonthermochromic distibines. The observed range of ν_{SbSb} appears to be predominantly

TABLE VI

ν_{SbSb} FROM RAMAN SPECTRA OF DISTIBINES IN SOLID AND LIQUID PHASES

Distibine	ν_{SbSb} (cm^{-1})		Reference(s)
	Liquid	Solid	
Tetravinyldistibine (**32**)	163	—	7
Tetraisopropenyldistibine (**34**)	158	152	7
Tetra-(Z)-1-propenyldistibine (**7**)	154	165	7
Tetra-(E)-1-propenyldistibine (**6**)	152	155	7
Tetrakis(2-methyl-1-propenyl)distibine (**9**)	144	151	7
Tetra-(E)-2-but-2-enyldistibine (**8**)	148	154	7
Tetramethyldistibine (**22**)	143, 175	180	*49,50*
Tetraethyldistibine (**24**)	163	—	*49*
Tetraphenyldistibine (**1**)	—	141	*49*
Tetrakis(trimethylsilyl)distibine (**36**)	140	137	*49*
Tetrakis(trimethylstannyl)distibine (**40**)	—	162	*40*
2,2′,5,5′-Tetramethylbistibole (**30**)	—	170	*51*

a mass effect, which suggest the vibrations are due largely to localized Sb—Sb bonds (7).

Tetramethyldistibine (**22**) is unique in that new bands appear for the liquid which were not present for the solid. It has been suggested that the new bands are due to a population of gauche conformation in the liquid (49). Although this view has been challenged (50), it seems particularly plausible since $(CH_3)_4P_2$ (52), $(CH_3)_4As_2$ (53), $(CH_3)_2PAs(CH_3)_2$, $(CH_3)_2AsSb(CH_3)_2$, and $(CH_3)_2SbBi(CH_3)_2$ (51) show the same effect. Furthermore, tetramethyldistibine has a dipole moment of 0.95 D in benzene solution (25). Since the symmetrical trans conformation can have no dipole moment, a significant fraction of the molecules must be in another conformation.

The Raman spectra of the thermochromic tetramethyldistibine (**22**) and tetrakis(trimethylsilyl)distibine (**36**) show a great enhancement of the intensity of the ν_{SbSb} band in the solid over the liquid. In addition, new low-frequency bands near 50 cm^{-1} are found for the solid which are not observed for the liquid (50). These bands have been assigned to the intermolecular Sb_2---Sb_2 stretch. Neither effect is observed for nonthermochromic distibines.

Raman spectra have also been reported for several dibismuthines. The ν_{BiBi} stretch is close to 110 cm^{-1} in all cases for both thermochromic and nonthermochromic dibismuthines in either solid or liquid (29). This insensitivity to substitution suggests that these bands are largely due to localized Bi–Bi vibrations.

The structure of tetramethyldistibine (**22**) has been investigated in the gas phase by electron diffraction (*27*). Although gauche and trans conformations are probably present, it was not possible to determine the conformational preference. Of major interest was the finding that the Sb—Sb bond length of 2.818(4) Å was shorter than found in the crystal (2.862 and 2.831 Å) (*25,26*). Presumably, the extended bonding in the crystal along the Sb—Sb---Sb—Sb chain reduces the intramolecular Sb—Sb bond order so that it is less than in the monomeric gas. A similar effect has been observed for I_2 (*vide infra*) (*54,55*).

Finally, [121]Sb-Mössbauer spectra have been reported for two thermochromic distibines (**22** and **26**) and two nonthermochromic distibines (**1** and **12**) (*56*). There are no meaningful variations in the isomer shift, quadruple coupling constants, or anisotropy in the series.

VIII

COMPARISON WITH SIMILAR COMPOUNDS

No diarsines or diphosphines show thermochromic behavior similar to that of the distibines and dibismuthines. For example, in the series of dipnictogen compounds A, B, and C illustrated in Scheme 6, the thermochromic distibines and dibismuthines correspond to nonthermochromic diarsines and diphosphines. Structural data are available to compare the three diarsines **51** (*46*), **52** (*57*), and **53** (*58*) with the corresponding distibines **22** (*25,26*), **36** (*37*), and **30** (*22*). Diarsines **52** and **53** crystallize in gauche conformations as opposed to the trans-staggered conformation of the distibines. In neither case are there intermolecular As---As contacts shorter than 4 Å. However, the lack of conformational correspondence makes any comparison tenuous.

A $(CH_3)_4E_2$	B $[(CH_3)_3Si]_4 E_2$	C
E = P, **50**		
E = As, **51**	E = As, **52**	E = As, **53**
E = Sb, **22**	E = Sb, **36**	E = Sb, **30**
E = Bi, **23**	E = Bi, **37**	E = Bi, **31**

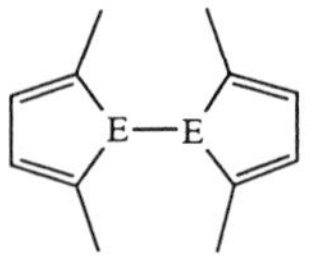

SCHEME 6. Analogous series of dipnictogen compounds.

TABLE VII

INTRAMOLECULAR E—E AND INTERMOLECULAR E---E BOND LENGTHS OF THE
TETRAMETHYLDIPNICTOGENS $(CH_3)_4E_2$

Compound	E—E (Å)	E---E (Å)	van der Waals separation (Å)	E---E/E—E	Reference
$(CH_3)_4P_2$ (50)	2.21	3.81	3.80	1.72	46
$(CH_3)_4As_2$ (51)	2.43	3.70	4.00	1.52	46
$(CH_3)_4Sb_2$ (22)	2.84	3.68	4.40	1.30	26
	2.86	3.65		1.28	25
$(CH_3)_4Bi_2$ (23)	3.12	3.58	4.60	1.15	46

A more informative comparison may be made for the four tetramethyl-dipnictogens **50, 51, 22,** and **23.** Although tetramethyldistibine (**22**) is not isostructural with tetramethyldiarsine (**51**) and tetramethyldiphosphine (**50**), all four tetramethyldipnictogens crystallize in similar trans conformations. Crystal packing shows a linear array of pnictogen atoms E—E---E—E. Although the intermolecular contacts (P---P, 3.81 Å, and As---As, 3.70 Å) are only slightly longer than the Sb---Sb contact (3.65 Å), the ratio of intramolecular and intermolecular E distances (E---E/E—E) falls regularly in the series from 1.72 and 1.52 for **50** and **51** to 1.30 and 1.15 for **22** and **23** (see Table VII) (*25,26,46*). This suggests that the brake in appreciable intermolecular interaction occurs in the range of 1.50 to 1.30.

An examination of the interpnictogen compound dimethylarsino-(dimethylstibine) [$(CH_3)_2AsSb(CH_3)_2$, **54**] would be particularly interesting in order to test the limits of this brake. Unfortunately, **54** could not be obtained in pure form owing to the mobility of the equilibrium between tetramethyldistibine (**22**) and tetramethyldiarsine (**51**) (*51*). However, the thermochromic arsinostibine 1-(2,5-dimethylarsolo)-2,5-dimethylstibole (**55**) was obtained from sublimation of mixtures of the corresponding biarsole **53** and bistibole **30** (see Scheme 7). The observed blue shift from the bistibole is an expected consequence of a stronger As—Sb than Sb—Sb σ bond.

Comparison of the tetramethyldipnictogens with the isoelectronic dihalogens is particularly informative. Cl_2, Br_2, and I_2 crystallize in isostructural molecular lattices with increasing intermolecular interaction (*54,59*). For iodine, the atoms are connected intramolecularly at 2.72 Å and intermolecularly in a two-dimensional rectangular net at 3.50 and 3.97 Å. The ratio E---E/E—E drops from 1.68 for Cl_2 to 1.29 for I_2. Only for I_2 is there an appreciable intermolecular interaction. Again the brake occurs between the fourth and fifth periods of elements (see Table VIII) (*54,59*). The

$(CH_3)_4As_2$ + $(CH_3)_4Sb_2$ ⇌ $2 (CH_3)_2AsSb(CH_3)_2$

51   **22**   **54**

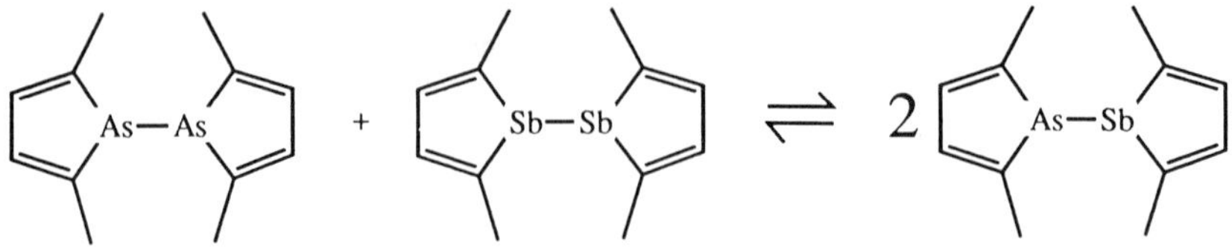

53   **30**   **55**

λ_{max} = 360 nm λ_{max} = 605 nm λ_{max} = 410 nm

SCHEME 7.   Preparation of arsinostibines.

iodine association has been subject to theoretical treatment that may be applicable to the dipictnogen compounds (60).

Ditellurides, also in the fifth period, seem quite analogous to distibines. Like tetraphenyldistibine (1) the red diphenylditelluride (56) does not associate in the solid state. The closest intermolecular Te---Te contact is 4.255 Å, near the van der Waals separation of 4.40 Å (61). On the other hand, di(p-methoxyphenyl)ditelluride (57), which has a brown-green metallic luster in the solid, has close intermolecular Te---Te contacts of 3.57 and 3.98 Å (62). The ratio Te---Te/Te—Te is 1.32. Just as in the distibines the intermolecular bonding in ditellurides is sensitive to substitution. It is also interesting to note that the intermolecular interaction in ditellurides and dihalogens occurs normal to the metal–metal axis, as well as colinear as in distibines (63). Thus, it is clear that the intermolecular association shown by distibines is a general property of many of the diatomic like compounds of the heavier main group elements.

TABLE VIII

COMPARISON OF SHORTENED INTERMOLECULAR E---E AND INTRAMOLECULAR E—E
DISTANCES OF DIHALOGENS

Compound	E—E (Å)	E---E (Å)	van der Waals separation (Å)	E---E/E—E	Reference
Cl_2	1.98	3.32	3.60	1.68	59
Br_2	2.27	3.31	3.90	1.46	59
I_2	2.72	3.50	4.30	1.29	54

IX

CONCLUDING REMARKS

In summary, distibines and dibismuthines show a propensity to stack in the solid state. The search for new organic conductors and superconductors has focused on compounds with stacked structures in the solid state (64). The metal atom chains of thermochromic distibines and dibismuthines appear to have sufficient overlap to create channels for conduction. However, distibines do not have an average non integral charge which is essential for high conductivity. Indeed, the thermochromic tetrakis-(trimethylstannyl)distibine (40) and 2,2',5,5'-tetramethylbistibole (65) have been found to be insulators. A recent theoretical study by Canadell and Shaik (48) however, has suggested that incorporation of antimony into conjugated organic networks such as phenalenyl will lead to distibinelike stacked molecules with an open band structure. Synthetic exploration along these lines should be very interesting.

ACKNOWLEDGMENTS

I would like to thank Dr. W. Butler for formatting figures of crystal structures and Drs. O. Mundt and H. J. Breunig for providing preprints of papers prior to publication. I also thank the donors of the Petroleum Research Fund administered by the American Chemical Society and the AFOSR (Grant #81-094) for partial support of our research.

REFERENCES

1. F. F. Blicke, U. O. Oakdale, F. D. Smith, *J. Am. Chem. Soc.* **53**, 1025 (1931).
2. W. Hewertson and H. R. Watson, *J. Chem. Soc.,* 1490 (1962).
3. K. Issleib and B. Hamann, *Z. Anorg. Allg. Chem.* **343**, 196 (1966).
4. F. F. Blicke and U. O. Oakdale, *J. Am. Chem. Soc.* **55**, 1198 (1933).
5. M. Ates, H. J. Breunig, A. Soltani-Neshan, and M. Tegeler, *Z. Naturforsch. B: Anorg. Chem. Org. Chem.* **41b**, 321 (1986).
6. F. Bickelhaupt, R. Lourens, H. Vermeer, and R. J. M. Weustink, *Recent Trav. Chim. Pays-Bas* **98**, 3 (1979).
7. A.J. Ashe III, E. G. Ludwig, Jr., and H. Pommerening, *Organometallics* **2**, 1573 (1983).
8. F. J. Drone, *Diss. Abstr. Int. B* **47**, 1059 (1986).
9. J. W. Dale, H. J. Eméleus, R. N. Haszeldine, and J. H. Moss, *J. Chem. Soc.*, 3708 (1957).
10. H. A. Meimema, H. F. Martens, J. G. Noltes, N. Bertazzi, and R. Barbieri, *J. Organomet. Chem.* **136**, 173 (1977).
11. K. Issleib and B. Hamann, *Z. Anorg. Allg. Chem.* **332**, 179 (1964).
12. O. Mundt, G. Becker, H.-J. Wessely, H. J. Breunig, and H. Kischkel, *Z. Anorg. Allg. Chem.* **486**, 70 (1982).
13. H. J. Breunig, *Z. Naturforsch. B: Anorg. Chem. Org. Chem.* **33B**, 242 (1978).
14. K. Issleib, B. Hamann, and L. Schmidt, *Z. Anorg. Allg. Chem.* **339**, 298 (1965).
15. H. J. Breunig, K. Häberle, M. Dräger, and T. Severengiz, *Angew. Chem.* **97**, 62 (1985);

H. J. Breunig, K. Häberle, M. Dräger, and T. Severengiz, *Angew. Chem., Int. Ed. Engl.* **24,** 72 (1985).

16. H. J. Breunig, A. Soltani-Neshan, K. Häberle, and M. Dräger, *Z. Naturforsch. B: Anorg. Chem. Org. Chem.* **41B,** 327 (1986).

17a. H. A. Meinema, H. F. Martens, and J. G. Noltes, *J. Organomet. Chem.* **51,** 223 (1973).

17b. H. J. Breunig and H. Jawad, *Z. Naturforsch. B: Anorg. Chem. Org. Chem.* **37b,** 1104 (1982).

17c. H. J. Breunig and W. Kanig, *J. Organomet. Chem.* **186,** C5 (1980).

17d. H. J. Breunig and T. Severengiz, *Z. Naturforsch. B: Anorg. Chem. Org. Chem.* **37b,** 395 (1982).

18. E. G. Ludwig, Jr., *Diss. Abstr. Int. B* **45,** 563 (1984).

19. K. Issleib and A. Balszuwert, *Z. Anorg. Allg. Chem.* **419,** 87 (1978).

20. A. J. Ashe III, T. R. Diephouse, and M. Y. El-Sheikh, *J. Am. Chem. Soc.* **104,** 5693 (1982).

21. F. A. Paneth, *Trans. Faraday Soc.* **30,** 179 (1934); F. A. Paneth and H. Loleit, *J. Chem. Soc.,* 366 (1935).

22. A. J. Ashe III, W. Butler, and T. R. Diephouse, *J. Am. Chem. Soc.* **103,** 207 (1981).

23. A. B. Burg and L. R. Grant, *J. Am. Chem. Soc.* **81,** 1 (1959).

24. H. J. Breunig, V. Breunig-Lyriti, and T. P. Knobloch, *Chem-Ztg.* **101,** 399 (1977).

25. A. J. Ashe III, E. G. Ludwig, Jr., J. Oleksyszyn, and J. C. Huffman, *Organometallics* **3,** 337 (1984).

26. O. Mundt, H. Riffel, G. Becker, and A. Simon, *Z. Naturforsch. B: Anorg. Chem. Org. Chem.* **39B,** 317 (1984).

27. A. G. Császár, L. Hedberg, K. Hedberg, E. G. Ludwig, Jr., and A. J. Ashe III, *Organometallics* **5,** 2257 (1986).

28. A. J. Ashe III and E. G. Ludwig, Jr., *Organometallics* **1,** 1408 (1982).

29. A. J. Ashe III, E. G. Ludwig, Jr., and J. Oleksyszyn, *Organometallics* **2,** 1859 (1983).

30. M. Wieber and I. Sauer, *Z. Naturforsch. B: Anorg. Chem. Org. Chem.* **39B,** 887 (1984).

31. K. Issleib and B. Hamann, *Z. Anorg. Allg. Chem.* **339,** 289 (1965).

32. H. J. Breunig and D. Müller, *Angew. Chem.* **94,** 448 (1982); H. J. Breunig and D. Müller, *Angew. Chem. Int. Ed. Engl.* **21,** 439 (1982); H. J. Breunig and D. Müller, *Z. Naturforsch B: Anorg. Chem. Org. Chem.* **38B,** 125 (1983).

33. A. J. Ashe III, C. M. Kausch, and O. Eisenstein, *Organometallics* **6,** 1185 (1987).

34. A. J. Ashe III and F. J. Drone, *Organometallics* **3,** 495 (1984).

35. H. J. Breunig and V. Breunig-Lyriti, *Z. Naturforsch. B: Anorg. Chem. Org. Chem.* **34B,** 926 (1979).

36. G. Becker, H. Freudenblum, and C. Witthauer, *Z. Anorg. Allg. Chem.* **492,** 37 (1982).

37. G. Becker and M. Rössler, *Z. Naturforsch. B: Anorg. Chem. Org. Chem.* **37B,** 91 (1982); O. Mundt, G. Becker, M. Rössler, and C. Witthauer, *Z. Anorg. Allg. Chem.* **506,** 42 (1983).

38a. H. J. Breunig, *Z. Naturforsch. B: Anorg. Chem. Org. Chem.* **33B,** 244 (1978).

38b. H. J. Breunig, private communication.

39. H. J. Breunig, *Z. Naturforsch. B: Anorg. Chem. Org. Chem.* **33B,** 990 (1978); H. J. Breunig, *Z. Naturforsch. B: Anorg. Chem. Org. Chem.* **39B,** 111 (1984).

40. S. Roller, M. Dräger, H. J. Breunig, M. Ates, and S. Gülec, *J. Organomet. Chem.* **329,** 319 (1987).

41. G. Becker, M. Meiser, O. Mundt, and J. Weidlein, *Z. Anorg. Allg. Chem.* **569,** 62 (1989).

42. F. Calderazzo, A. Morvillo, G. Pelizzi, and R. Poli, *J. Chem. Soc., Chem. Commun.,* 507 (1983).

43. P. Fisher, I. Sosnowska, and M. Syzmanski, *J. Phys. C* **11**, 1043 (1978).
44. H. J. M. Bowen, *Trans. Faraday Soc.* **50**, 463 (1954); B. Beagley and A. R. Medwid, *J. Mol. Struct.* **38**, 229 (1977).
45. K. von Deuten and D. Rehder, *Cryst. Struct. Commun.* **9**, 167 (1980).
46. O. Mundt, H. Riffel, G. Becker, and A. Simon, *Z. Naturforsch. B: Anorg. Chem. Org. Chem.* **43B**, 952 (1988).
47. T. Hughbanks, R. Hoffmann, M.-H. Whangbo, K. R. Stewart, O. Eisenstein, and E. Canadell, *J. Am. Chem. Soc.* **104**, 3876 (1982).
48. E. Canadell and S. S. Shaik, *Inorg. Chem.* **26**, 3797 (1987).
49. H. J. Breunig, V. Breunig-Lyriti, and W. Fichtner, *Z. Anorg. Allg. Chem.* **487**, 111 (1982).
50. H. Bürger, R. Eujen, G. Becker, O. Mundt, M. Westerhausen, and C. Witthauer, *J. Mol. Struct.* **98**, 265 (1983).
51. A. J. Ashe III and E. G. Ludwig, Jr., *J. Organomet. Chem.* **303**, 197 (1986).
52. J. R. Durig and J. S. DiYorio, *Inorg. Chem.* **8**, 2796 (1969).
53. J. R. Durig and J. M. Casper, *J. Chem. Phys.* **55**, 198 (1971).
54. F. van Bolhuis, P. B. Koster, and T. Michelson, *Acta Crystallogr.* **23**, 90 (1967).
55. T. Ukaji and K. Kuchitsa, *Bull. Chem. Soc. Jpn.* **39**, 2153 (1966).
56. J. G. Stevens, J. M. Trooster, H. F. Martens, and H. A. Meinema, *Inorg. Chim. Acta* **115**, 197 (1986).
57. G. Becker, G. Gutekunst, and C. Witthauer, *Z. Anorg. Allg. Chem.* **486**, 90 (1982).
58. A. J. Ashe III, W. M. Butler, and T. R. Diephouse, *Organometallics* **2**, 1005 (1983).
59. J. Donohue and S. H. Goodman, *Acta Crystallogr.* **23**, 90 (1965).
60. R. Bersohn, *J. Chem. Phys.* **36**, 3445 (1962).
61. G. Llabres, O. Dideberg, and L. Dupont, *J. Acta. Crystallogr., Sect. B* **28B**, 2438 (1972).
62. S. Ludlow and A. E. McCarthy, *J. Organomet. Chem.* **219**, 169 (1981).
63. D. J. Sandman, J. C. Stark, and B. M. Foxman, *Organometallics* **1**, 739 (1982).
64. J. Y. Becker, J. Bernstein, and S. Bittner, *Isr. J. Chem.* **27**(4) (1986).
65. F. Wudl, private communication.

Syntheses, Structures, Bonding, and Reactivity of Main Group Heterocarboranes

NARAYAN S. HOSMANE and
JOHN A. MAGUIRE

Department of Chemistry
Southern Methodist University
Dallas, Texas 75275

I
INTRODUCTION

This review covers the research on the main group (groups 2 and 13–16) heterocarboranes published since 1982.[1] Earlier work is discussed only as background to current results or for purposes of comparison. There are several monographs (*1–3*) and a number of reviews (*4–6b*) that adequately cover the earlier literature. Recent advances in the research of main group metallacarboranes containing groups 13 and 14 metals and/or metalloids have also been reviewed (*7*).

Carboranes, or carbaboranes, are mixed hydrides of carbon and boron in which atoms of both elements are incorporated in an electron-deficient molecular skeleton. These compounds can react with other main group elements or moieties to form heterocarboranes. Our discussion is restricted to heterocarboranes in which the main group element is incorporated as an integral part of the polyhedral framework. No attempt is made to cover the inclusion of transition metals or those compounds where the main group heteroatom is in a bridging group linking several carborane polyhedra together or when it is involved solely as a member of a substituent group. This restriction is dictated both by space and by the fact that the heteroatoms incorporated into carborane cages have unique properties that are not typical of their usual chemical behavior. Since there are no known examples of atoms in groups 1 or 17 being incorporated into carborane cages, these groups are not specifically discussed. A great deal of structural information has recently become available on main group

---

[1] The former groups 2A, 3A, 4A, 5A, and 6A have been redesignated, respectively, as groups 2, 13, 14, 15, and 16 in accordance with recent IUPAC nomenclature rules.

heterocarboranes. Therefore, this article concentrates as much on the structural and bonding features of these compounds as on their reaction chemistry.

II
GENERAL VIEW OF BONDING AND STRUCTURAL RELATIONSHIPS

The main geometries encountered in this review are shown in Fig. 1. In the closed (closo) structure (Fig. 1b) the skeletal atoms occupy all corners of a polyhedron, while in the open (nido) structure (Fig. 1a) one corner of the polyhedron is vacant. This open face is usually the location of attachment when the carborane bonds to a metal group. In some cases, the metal can occupy common vertices of two polyhedra to give a commo structure (Fig. 1c).

There is a simple relationship between the number of skeletal electron pairs and the geometry (8). For a carborane containing n skeletal atoms with formula $(CH)_a(BH)_{n-a}H_b{}^{c-}$, the number of electron pairs (P) involved in cage bonding is $n + \frac{1}{2}(a + b + c)$. That is, each CH unit can furnish three electrons, each BH unit furnishes two electrons, while the bridged hydrogens (H_b) each contribute one. The electrons are contained in $n + 1$ bonding molecular orbitals formed by interaction of three orbitals (one radially oriented p or sp hybrid and two tangentially oriented p orbitals) from each of the framework atoms (Cs and Bs). The general rule states that a closo structure is preferred if $P = n + 1$; a nido structure, if $P = n + 2$; and an arachno structure, if $P = n + 3$ (8). Thus, while C_2B_{10}-

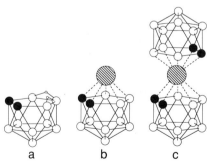

FIG. 1.   (a) nido (open), (b) closo (closed), and (c) commo (sandwich) geometries of main group heterocarboranes. (O) BH, (●) CH, (o) H, and (◍) heteroatom.

H_{12} and $[C_2B_9H_{11}]^{2-}$ (or $C_2B_9H_{13}$) both have 13 electron pairs, $C_2B_{10}H_{12}$ will have a closo structure ($13 = 12 + 1$) and $C_2B_9H_{13}$ will have a nido structure ($13 = 11 + 2$).

These ideas can be extended to heterocarboranes by noting that the replacement of a BH unit in a carborane by some other group which can furnish two electrons and three similarly oriented orbitals (or isolobal with BH) to the cage will not change the general geometry. Groups such as AlR, GaR, and Sn all have these characteristics (9). Since $C_2B_{10}H_{12}$ is known to have a closo geometry (10), $RAlC_2B_9H_{11}$, $RGaC_2B_9H_{11}$, and $SnC_2B_9H_{11}$ should also have the same geometry. As will be seen in later discussions, these geometries have been confirmed experimentally. The same isolobal arguments can be extended to transition metal heterocarboranes (8). It should be noted that the replacement of a terminal hydrogen with some other group does not change the above arguments; the electron counting rules can be applied to derivatized carboranes as well. The geometries of closo polyhedra having 4 to 14 vertices with the standard numbering system are shown in Fig. 2 (6).

The most cited heterocarboranes are those derived from the dianions of the nido-carboranes 7,8-$R_2C_2B_9H_{11}$ and 2,3-$R_2C_2B_4H_6$ (R = substituents on the cage carbons) in which the two cage carbons occupy adjacent positions in the open pentagonal faces of the polyhedra. The dianions have six delocalized electrons in π-type orbitals on the open pentagonal face of the C_2B_3 rings, the same as found in the cyclopentadienyl anions, $[C_5R_5]^-$. This similarity was first recognized by Hawthorne et al. (11) and has been profitably exploited by many investigators (1-6). However, there are a number of important differences. First, instead of being orthogonal to the open pentagonal faces, the orbitals of the carborane anions are tilted inward (Fig. 3) (3,6). The metal orbitals will thus tend to overlap more effectively with the carborane orbitals than with those of cyclopentadienyl anion. Second, the higher polarizability of boron versus carbon would encourage stronger metal–carborane bonding. Third, the C_2B_3 face of the carborane is not necessarily flat but can be slightly folded away from the capping metal group. This folding, coupled with the heteronuclear nature of the carborane bonding face, encourages a slippage of the capping metal atom away from the centroidal position above the pentagonal face. Slip distortion seems to be a common feature of the main group heterocarboranes, and slippage is invariably toward the boron side of the C_2B_3 face. The last difference to be noted is that of charge; the carborane ligands bear a 2– charge while the cyclopentadienide ligands are monoanions. This increased charge should help stabilize bonding with positively charged metal or metal groups and can effect the metal oxidation state most stabilized by bonding to the carborane.

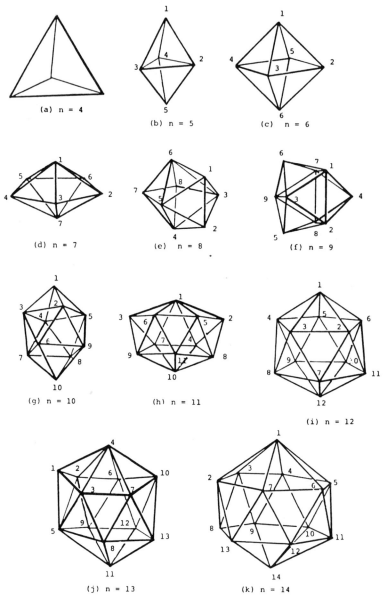

FIG. 2.   Geometries of closo polyhedra having 4 to 14 vertices, showing standard numbering system: (a) tetrahedron, (b) trigonal bipyramid, (c) octahedron, (d) pentagonal bipyramid, (e) dodecahedron, (f) tricapped trigonal prism, (g) bicapped square antiprism,
(h) octadecahedron, (i) icosahedron, (j) docosahedron, and (k) bicapped hexagonal antiprism. [Reprinted with permission from R. N. Grimes (6a), *in* "Comprehensive Organometallic Chemistry" (G. Wilkinson, F. G. A. Stone, and E. W. Abel, eds.), Copyright
1982, Pergamon Press PLC.]

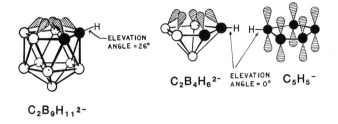

$C_2B_9H_{11}^{2-}$

π overlap with metal $C_2B_4H_6^{2-} > C_2B_9H_{11}^{2-} > C_5H_5^-$

FIG. 3. Comparison of η^5-bonding capabilities of carborane ligands and cyclopenta-dienide anion. [Reprinted with permission from R. N. Grimes (6b), in "Molecular Structure and Energetics" (J. F. Liebman, A. Greenberg, and R. E. Williams, eds.), Copyright 1988, VCH, New York.]

III

STABILITY

The *closo*-metallacarboranes containing transition metals possess high thermal stability, many being stable at temperatures of 500°C or more for short periods of time. On the other hand, many of the main group metalla-carboranes decompose above 300°C. *C*-Trimethylsilyl-substituted main group heterocarboranes are often more air stable than the *C*-H- and/or alkyl-substituted ones. The hydrolytic stability of the *closo*-metalla-carboranes varies greatly. For example, $\{Ni^{III}[\eta^5\text{-}(CH)_2B_9H_9]_2\}^-$ can be prepared in aqueous media, while $C_2H_5Al[\eta^5\text{-}(CH)_2B_9H_9]$ must be pre-pared in high vacuum systems (3). Since it is difficult to predict the sta-bility of a particular main group heterocarborane, each new compound should be handled in an inert atmosphere or in a high vacuum line until its stability is determined.

IV

STRUCTURAL NOMENCLATURE

The systematic naming of the metallacarboranes is exceedingly awk-ward and is generally avoided, with heavy reliance being placed on the use of figures and line formulas to convey molecular structures. For exam-ple, *closo*-1-Sn-2,3-[(CH$_3$)$_3$Si]$_2$-2,3-C$_2$B$_4$ and *closo*-1-Sn(2,2'-C$_{10}$H$_8$N$_2$)-2,3-[Si(CH$_3$)$_3$]$_2$-2,3-C$_2$B$_4$H$_4$ have the IUPAC names "[2,3,4,5,6-η]-1,4,5,6-tetrahydro-2,3-bis(trimethylsilyl)-2,3-dicarbahexaborato(2−)]tin"   and

"(2,2'-bipyridine-N,N')[(4,5,6-η)-1,4,5,6-tetrahydro-2,3-bis(trimethyl-silyl)-2,3-dicarbahexaborato(2−)]tin," respectively (12a,b). The commo germanium compound 1,1'-Ge{η^5-2,3-[(CH$_3$)$_3$Si]$_2$-2,3,-C$_2$B$_4$H$_4$}$_2$ could be named as either "2,2',3,3'-tetrakis(trimethylsilyl)[1,1'-commo-bis-(2,3-dicarba-1-germa-closo-heptaborane)]" (12a,b) or "bis[(2,3,4,5,6-η)-1,4,5,6-tetrahydro-2,3-bis(trimethylsilyl)-2,3-dicarbahexaborato(2−)] germanium."

The IUPAC rules are also ambiguous with respect to the order in which hetero (nonboron) atoms are to be numbered in a given polyhedral cage, when more than one type of heteroatom is present. The situation led to confusion, with different numbering conventions being used by different investigators. Hawthorne and co-workers formerly assigned lowest framework numbers to the metal atoms but more recently have given priority to the carbon atoms; thus, the compound closo-1-C$_2$H$_5$-1,2,3-AlC$_2$B$_9$H$_{11}$ was later designated as closo-3-C$_2$H$_5$-3,1,2-AlC$_2$B$_9$H$_{11}$. It has been a general practice that ligand groups attached to the metal atom(s) are not designated by number, and ligands not specified by a number may be assumed to be bonded to the metal. Because of these ambiguities, the reader is cautioned to refer to the relevant figures when compounds are mentioned in the text.

V

HETEROCARBORANES OF GROUP 2 ELEMENTS

Among group 2 elements, only beryllium is known to have been inserted into a carborane polyhedron. Even though cyclopentadienyl π complexes of the heavier alkaline earth meals are known, analogous carborane complexes have not been reported.

The reaction of nido-1,2-C$_2$B$_9$H$_{13}$ wiith dimethyl- or diethylberyllium in ether produces an extremely air-sensitive white solid etherate of the icosahedral beryllacarborane as shown in Eq. (1) (13,14). The reaction of

$$1,2\text{-C}_2\text{B}_9\text{H}_{13} + \text{BeR}_2 \cdot 2[\text{O}(\text{C}_2\text{H}_5)_2] \xrightarrow[\text{C}_6\text{H}_6]{(\text{C}_2\text{H}_5)_2\text{O}} closo\text{-}3\text{-}[\text{O}(\text{C}_2\text{H}_5)_2]\text{-}3,1,2\text{-BeC}_2\text{B}_9\text{H}_{11}$$

$$+ (\text{C}_2\text{H}_5)_2\text{O} + 2\,\text{RH} \qquad \text{R} = \text{CH}_3, \text{C}_2\text{H}_5 \tag{1}$$

1,2-C$_2$B$_9$H$_{13}$ with diethylberyllium containing only one-third of a mole equivalent of complexed diethyl ether proceeds differently, yielding a polymer which is proposed to have repeating BeC$_2$B$_9$H$_{11}$ units linked by B—H—B bridges (13). The treatment of either the diethyl etherate monomer or the polymer with triethylamine produces the solid amine adduct (CH$_3$)$_3$NBeC$_2$B$_9$H$_{11}$, which is considerably less air sensitive but is

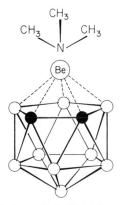

FIG. 4.   Proposed structure of *closo*-1-Be[N(CH$_3$)$_3$]-2,3-C$_2$B$_9$H$_{11}$. (○) BH and (●) CH. [Reprinted with permission from G. Popp and M. F. Hawthorne (*13*), *Inorg. Chem.* **10**, 391 (1971). Copyright 1971 by the American Chemical Society.]

degraded to 1,2-C$_2$B$_9$H$_{12}$$^-$ by ethanolic KOH. The (CH$_3$)$_3$NBe group, a two-electron donor, is isolobal with BH. Thus, the closo icosahedral structure of (CH$_3$)$_3$NBeC$_2$B$_9$H$_{11}$ (Fig. 4) can be inferred by noting that it can be formally derived by replacing a BH group in *closo*-1,2-C$_2$B$_{10}$H$_{12}$ by (CH$_3$)$_3$NBe. Presumably, the etherate produced in Eq. (1) also has a closo geometry. Attempts to produce BeC$_2$B$_9$H$_{11}$ and other base-stabilized beryllacarboranes have not been successful (*13*). There have been no reports of smaller cage beryllacarboranes.

VI

HETEROCARBORANES OF GROUP 13 ELEMENTS

Since this review involves a discussion of the insertion of heteroatoms in carborane cages, the group 13 elements discussed are aluminum, gallium, indium, and thallium. Aluminum, gallium, and indium have been successfully inserted into both C$_2$B$_4$ and C$_2$B$_9$ carborane systems. Thallium(I) acetate was found to react with [7,8-B$_9$C$_2$H$_{12}$]$^-$ in aqueous alkaline solution to yield (Tl)$_2$B$_9$C$_2$H$_{11}$ (*15*). The crystal structure of the (C$_6$H$_5$)$_3$PCH$_3$$^+$ salt of [TlB$_9$C$_2$H$_{11}$]$^-$ showed the Tl to occupy the apical position above the B$_9$C$_2$H$_{11}$ open face with a very slight slippage away from the carbon atoms. The fact that thallium–cage bond distances are greater than expected from the covalent radii of the atoms involved has been taken to indicate that their interaction is essentially ionic (*16*). However, the pale yellow color of the [TlB$_9$C$_2$H$_{11}$]$^-$ derived from the colorless Tl$^+$ and C$_2$B$_9$H$_{11}$$^{2-}$ implies some covalent interactions. The thallium is loosely

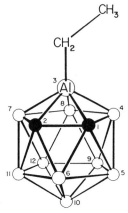

FIG. 5. Crystal structure of *closo*-1-C$_2$H$_5$-1,2,3-AlC$_2$B$_9$H$_{11}$. (O) BH and (●) CH. [Reprinted with permission from M. R. Churchill and A. H. Reis, Jr. (*20*), *J. Chem. Soc., Dalton Trans.*, 1317 (1972). Copyright 1972 by the Royal Society of Chemistry.]

ligated by the carborane cage; hence, the thallacarboranes are useful precursors for the syntheses of other metallacarboranes.

In 1968, Mikhailov and Potapova (*17*) reported the synthesis of the first *closo*-aluminacarborane, 1-C$_2$H$_5$-1,2,3-AlC$_2$B$_9$H$_{11}$, as its bis(tetrahydrofuran) adduct, by the reaction of [C$_2$B$_9$H$_{11}$]$^{2-}$ with C$_2$H$_5$AlCl$_2$ in THF at −50°C. Later, Hawthorne and co-workers (*18,19*) reported the syntheses of *closo*-1-C$_2$H$_5$-AlC$_2$B$_9$H$_{11}$, *closo*-1-CH$_3$-AlC$_2$B$_9$H$_{11}$, and *closo*-1-C$_2$H$_5$-GaC$_2$B$_9$H$_{11}$ as shown in Eq. (2). The syntheses involve the initial forma-

$$7,8\text{-B}_9\text{C}_2\text{H}_{13} + \text{MR}_3 \rightarrow \text{RH} + nido\text{-7,8-B}_9\text{C}_2\text{MH}_{12}\text{R}_2 \xrightarrow{\text{heat}} closo\text{-1-R-MC}_2\text{B}_9\text{H}_{11} + \text{RH}$$

$$\text{M = Al; R = CH}_3, \text{C}_2\text{H}_5; \text{ and M = Ga; R = C}_2\text{H}_5 \qquad (2)$$

tion of the corresponding *nido*-B$_9$C$_2$MH$_{12}$R$_2$ derivatives which undergo cage closure with the elimination of 1 mol of RH to form the closo complexes. The closo geometry of 1-C$_2$H$_5$-1,2,3-AlC$_2$B$_9$H$_{11}$ was confirmed by Churchill and Reis (*20*) by X-ray diffraction (Fig. 5). The (C$_2$H$_5$)Al is situated above the C$_2$B$_3$ open face with the CH$_2$ carbon lying approximately along the aluminum–apical boron axis. The aluminum–cage bond distances [Al–C-1,2 = 2.173(7); Al–B-4,7 = 2.138(8); and Al–B(unique) = 2.136(9) Å] indicate that the Al(C$_2$H$_5$) is situated almost symmetrically above the carborane open face with only a slight slippage away from the cage carbons (*20*).

Jutzi and Galow (*21*) found the aluminacarborane 2,3-(CH$_3$)$_2$-1-C$_2$H$_5$-1,2,3-AlC$_2$B$_9$H$_9$ to be a convenient carbollyl transfer agent when reacted with a number of main group halides as outlined in Scheme 1. The structures of the products were inferred from NMR data. These investigators

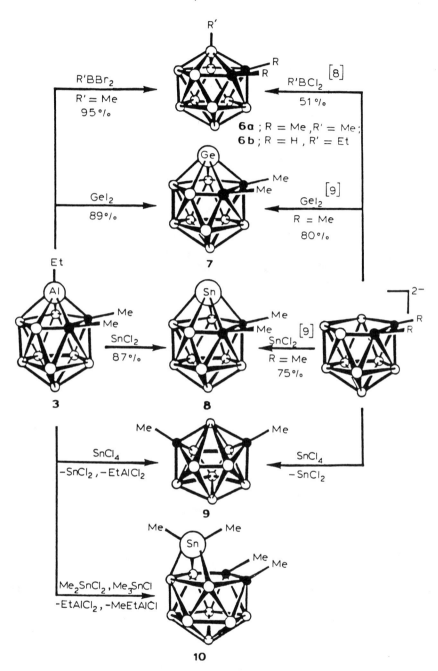

SCHEME 1.   (○) BH and (●) C. [Reprinted with permission from P. Jutzi and P. Galow (21), *J. Organomet. Chem.* **319**, 139 (1987). Copyright 1987 by Elsevier Sequoia S.A.]

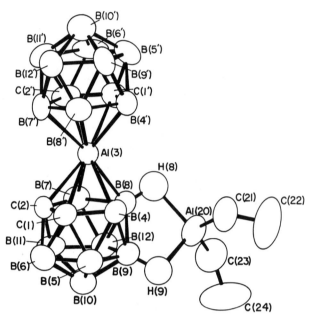

FIG. 6. Crystal structure of *commo*-3,3'-Al{[*exo*-8,9-(μ-H)$_2$Al(C$_2$H$_5$)$_2$-3,1,2,AlC$_2$B$_9$H$_9$]-(3',1',2'-AlC$_2$B$_9$H$_{11}$]}. [Reprinted with permission from W. S. Rees, Jr., *et al.* (22), *J. Am. Chem. Soc.* **108**, 5367 (1986). Copyright 1986 by the American Chemical Society.]

also report that the aluminacarborane forms adducts with Lewis bases, such as diethyl ether and tetrahydrofuran, where the aluminum acts as a Lewis acid site. The use of stronger bases evidently removes the apical aluminum group, yielding the [(CH$_3$)$_2$C$_2$B$_9$H$_9$]$^{2-}$ which was detected spectroscopically (*21*). No X-ray structural data were given for the alumincarborane–Lewis base adducts. This is unfortunate, since the electron counting rules would predict that these adducts should have nido structures. As discussed in Section VII, electron counting rules for such acid–base adducts of metalla-carboranes are of limited use in many cases.

The syntheses of the *commo*-aluminacarboranes have been reported extensively by Hawthorne and co-workers. These investigators found that *commo*-3,3'-Al{[*exo*-8,9-(μ-H)$_2$Al(C$_2$H$_5$)$_2$-3,1,2-AlC$_2$B$_9$H$_9$](3',1',2'-AlC$_2$B$_9$H$_{11}$)} was formed in 93% yield on stirring a benzene solution of *closo*-3-C$_2$H$_5$-3,1,2-AlC$_2$B$_9$H$_{11}$[2] under an atmosphere of CO at ambient temperature (*22*). The structure of this unusual compound (Fig. 6) shows

---

[2] This compound was designated as *closo*-1-C$_2$H$_5$-1,2,3-AlC$_2$B$_9$H$_{11}$ when its synthesis was first reported (*18,19*). The present designation is due to a change in the numbering system since its first report.

one aluminum to be η^5 bonded to the faces of two C_2B_9 carborane cages. The other aluminum is involved in an exopolyhedral diethylaluminum group which is attached to one of the carborane cages via two B—H—Al bridges. Thus, the compound may be regarded as a "zwitterion" of the $[Al(C_2H_5)_2]^+$ cation and the $[Al(C_2B_9H_{11})_2]^-$ anion. The two carborane cages are essentially planar, and the *commo*-aluminum atom is approximately in the center of each carborane face but is dislocated toward the carborane cage having the exopolyhedral aluminum.

The Tl(I) salt of the isolated $[commo\text{-}3,3'\text{-}Al(3,1,2\text{-}AlC_2B_9H_{11})_2]^-$ ion was reported most recently (*23*). This compound was synthesized from the reaction of $Tl_2B_9C_2H_{11}$ with a variety of alkylaluminum chlorides or trialkylaluminum compounds as outlined in Eq. (3). This commo complex has

$$2\ Tl_2[nido\text{-}7,8\text{-}C_2B_9H_{11}] + 4\ AlR_2Cl \xrightarrow[\text{reflux}]{\text{toluene}} Tl[commo\text{-}3,3'\text{-}Al(3,1,2\text{-}AlC_2B_9H_{11})_2]$$

$$+\ 2\ Tl[AlR_3Cl] + Tl[AlR_2Cl_2] \qquad R = C_2H_5 \qquad\qquad (3)$$

been characterized by ^{1}H and ^{11}B NMR, IR, and X-ray crystallography (*23*). The unit cell consists of three crystallographically unique anionic *commo*-aluminacarboranes, one of which is shown in Fig. 7. In all three, the aluminum is equidistant from the two planar, parallel faces of the carborane ligands, but slipped toward the three borons on the C_2B_3 faces [the Al–B-8 distances are 2.14(2) Å]. The three anions differ in that in one form the aluminum resides in the mirror plane containing B-8, B-10, and

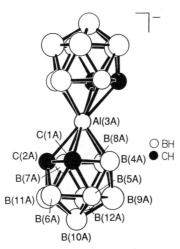

FIG. 7. One of the three crystallographically unique structures of Tl[*commo*-3,3′-Al(3,1,2-AlC$_2$B$_9$H$_{11}$)$_2$]. [Reprinted with permission from M. A. Bandman *et al.* (*23*), *Inorg. Chem.* **27**, 2399 (1988). Copyright 1988 by the American Chemical Society.]

the midpoint of the C-1–C-2 bond, while in the other two forms the aluminum seems to be slightly displaced out of the plane, toward B-7 in one case and B-4 in the other (23). It is of interest that the removal of the exopolyhedral aluminum in the zwitterionic dialumina compound causes a centering of the remaining aluminum between the two carborane ligands and an increased slippage of the metal across the carborane faces.

In a series of papers, Hawthorne and co-workers have also reported the syntheses and structures of the smaller cage aluminacarborane clusters, *nido*-{μ-6,9-Al(C$_2$H$_5$)[O(C$_2$H$_5$)$_2$]-6,9-C$_2$B$_8$H$_{10}$}, [Al(η^2-6,9-C$_2$B$_8$H$_{10}$)$_2$]$^-$, and [Al(η^2-2,7-C$_2$B$_6$H$_8$)$_2$]$^-$ as shown in Figs. 8–10 (24–27). In these complexes the carborane dianions act as η^2 ligands that donate four electrons, via two carbon-based orbitals, to a tetrahedrally coordinated aluminum. The rather long aluminum–boron distances suggest an absence of Al—B bonding. The aluminum can best be described as a bridging, exopolyhedral group that does not participate in cage framework bonding.

Although the icosahedral *commo*- and *closo*-aluminacarboranes have been extensively characterized and their reaction chemistry explored, the corresponding aluminacarboranes in the pentagonal bipyramidal system

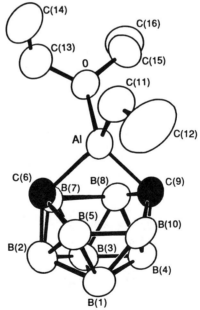

FIG. 8.   Crystal structure of *nido*-{μ-6,9-Al(C$_2$H$_5$)[O(C$_2$H$_5$)$_2$]-6,9-C$_2$B$_8$H$_{10}$}, with hydrogen atoms omitted for clarity. [Reprinted with permission from D. M. Schubert *et al.* (24), *Organometallics* **6**, 201 (1987). Copyright 1987 by the American Chemical Society.]

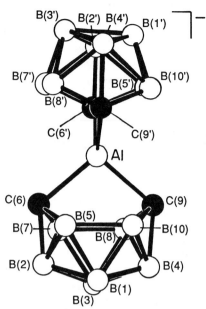

FIG. 9. Crystal structure of $[Al(\eta^2\text{-}6,9\text{-}C_2B_8H_{10})_2]^-$ anion, with hydrogen atoms omitted for clarity. [Reprinted with permission from D. M. Schubert *et al.* (*25*), *Organometallics* **6,** 203 (1987). Copyright 1987 by the American Chemical Society.]

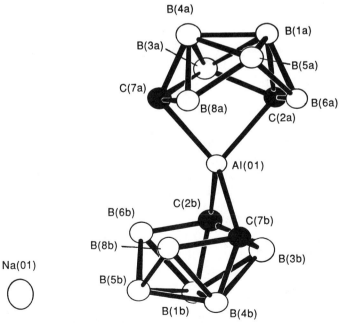

FIG. 10. Crystal structure of one enantiomer of $Na[Al(\eta^2\text{-}2,7\text{-}C_2B_6H_8)_2]$, with hydrogen atoms omitted for clarity. [Reprinted with permission from D. M. Schubert *et al.* (*26*), *Organometallics* **6,** 1353 (1987). Copyright 1987 by the American Chemical Society.]

have proved to be illusive. This is somewhat surprising since there is a striking parallelism in the two systems of main group metallacarboranes. Grimes and co-workers (28) have reported the syntheses of a number of bridged main group and transition metal complexes as described in Eq. (4). The μ-(CH$_3$)$_2$AlC$_2$B$_4$H$_7$ was extremely unstable, decomposing even in carefully dried solvents or in neat liquid form. Its existence was inferred by mass spectrometry. The compound was much more stable in the gas phase; the pyrolysis at 100°C yielded a volatile material whose mass and IR spectra were consistent with a closo-(CH$_3$)AlC$_2$B$_4$H$_6$ species (28).

$$Na^+[2,3\text{-}C_2B_4H_7]^- + M(CH_3)_2X \rightarrow \mu\text{-}(CH)_3MC_2B_4H_7 + MX$$

$$M = Ga, Al; X = Cl, Br \qquad (4)$$

Beck and Sneddon (29) have recently described the reaction of nido-2,3-(C$_2$H$_5$)$_2$C$_2$B$_4$H$_6$ with triethylaminealane [(C$_2$H$_5$)$_3$N·AlH$_3$] under various conditions to produce three new small cage aluminacarboranes as outlined in Scheme 2. At 0°C, the reaction proceeds by the elimination of 1 equiv of H$_2$ to produce the bridged nido-4,5-μ-AlH$_2$-N(C$_2$H$_5$)$_3$-2,3-(C$_2$H$_5$)$_2$C$_2$B$_4$H$_5$ in 91% yield. Subsequent heating of this compound to 50°C results in the loss of another equivalent of H$_2$ to produce 6-AlH[N(C$_2$H$_5$)$_3$]-3,4-(C$_2$H$_5$)$_2$C$_2$B$_4$H$_4$. This compound was described as a seven-vertex nido-aluminacarborane in which the [(C$_2$H$_5$)$_3$N]AlH group is η^3 bonded to the carborane cage through the aluminum. This is interesting since seven-vertex nido cage systems are rare. Further heating of this compound at 70°C in the presence of 2,3-(C$_2$H$_5$)$_2$C$_2$B$_4$H$_6$ produces a novel compound, commo-(AlN(C$_2$H$_5$)$_3$-[6-AlN-(C$_2$H$_5$)$_3$-3,4-(C$_2$H$_5$)$_2$C$_2$B$_4$H$_4$][4′,5′-μ-AlN-(C$_2$H$_5$)$_3$-2′,3′-(C$_2$H$_5$)$_2$C$_2$B$_4$H$_5$], in which the aluminum in the seven-vertex nido cage also occupies a bridging position in an adjacent C$_2$B$_4$ carborane. All three of the aluminacarboranes could be prepared directly from (C$_2$H$_5$)$_3$N·AlH$_3$ and 2,3-(C$_2$H$_5$)$_2$C$_2$B$_4$H$_6$ under the appropriate temperature and stoichiometric conditions. The structures of the compounds were inferred from the ^{1}H- and ^{11}B-NMR and IR spectra (29). The commo complex is a solid at room temperature, and it may be possible to obtain an X-ray crystal structure of this compound that would verify all three proposed molecular geometries.

In contrast to the rather rich and varied chemistry reported for aluminacarboranes, only limited information is available on the heavier group 13 metallacarboranes. The preparation of closo-1-C$_2$H$_5$-GaC$_2$B$_9$H$_{11}$ by Hawthorne and co-workers (18,19) was described earlier [see Eq. (2)]. Grimes et al. (30) have reported the synthesis of closo-1-(CH$_3$)-1,2,3-GaC$_2$B$_4$H$_6$ and 1-(CH$_3$)-1,2,3-InC$_2$B$_4$H$_6$ by the gas-phase reaction of Ga(CH$_3$)$_3$ and In(CH$_3$)$_3$ with C$_2$B$_4$H$_8$. Reaction conditions and yields are

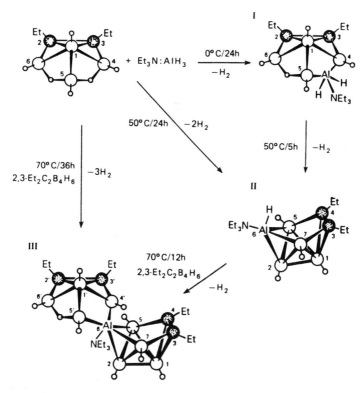

SCHEME 2. [Reprinted with permission from J. S. Beck and L. G. Sneddon (*29*), *J. Am. Chem. Soc.* **110**, 3467 (1988). Copyright 1988 by the American Chemical Society.]

given in Eqs. (5) and (6). This preparative route is similar to that given in

$$C_2B_4H_8 + Ga(CH_3)_3 \xrightarrow{215^\circ C} (CH_3)GaC_2B_4H_6 + B(CH_3)_3 + \text{solids} \qquad (5)$$
$$\underset{20-30\%}{}$$

$$C_2B_4H_8 + In(CH_3)_3 \xrightarrow{95-110^\circ C} (CH_3)InC_2B_4H_6 + B(CH_3)_3 + \text{solids} \qquad (6)$$
$$\underset{50-60\%}{}$$

Eq. (2) for the C_2B_9 system. Evidently, this type of reaction cannot be used to prepare $CH_3AlC_2B_4H_6$. As one goes from indium to gallium, the production of the metallacarborane requires higher temperatures and the yield decreases. It may be that the temperatures necessary to cause the reaction of $Al(CH_3)_3$ would result in vanishingly small yields.

The mechanisms of the reactions in Eqs. (2), (5), and (6) have not been fully elucidated, and the similarities between the preparative routes may be superficial. X-Ray crystal structures (*19,20*) have confirmed the bridging

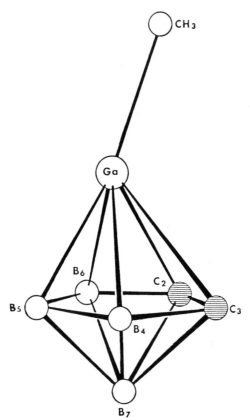

FIG. 11.   Crystal structure of *closo*-1-CH$_3$-1-GaC$_2$B$_4$H$_6$. [Reprinted with permission from R. N. Grimes *et al. (30), J. Am. Chem. Soc.* **94**, 1865 (1972). Copyright 1972 by the American Chemical Society.]

nature of the Al(CH$_3$)$_2$ group in *nido*-7,8-B$_9$C$_2$H$_{12}$Al(CH$_3$)$_2$ [in Eq. (2)], and cage closure on heating to form *closo*-1-CH$_3$-AlC$_2$B$_9$H$_{11}$ is quite reasonable. In the case of the smaller cage systems [Eqs. (5) and (6)], however, the interpretation of the reaction sequence is not so straightforward. Grimes and co-workers prepared the bridged μ-(CH$_3$)$_2$GaC$_2$B$_4$H$_7$ complex [Eq. (4)] but could not find suitable experimental conditions under which it would rearrange to the *closo*-gallacarborane (*30*). It may be that the chemistry of C$_2$B$_4$ system containing group 13 metals is different from that of the C$_2$B$_9$ system.

The structure of *closo*-1-CH$_3$-1-GaC$_2$B$_4$H$_6$ was determined by X-ray crystallography (*30*) and is shown in Fig. 11. The structure is that of a distorted pentagonal bipyramid in which the position of the Ga is shifted slightly so that the primay Ga—C distances are longer than the Ga—B distances by about 0.1 Å. Also the Ga—CH_3 bond is tilted by about 20° away from the Ga–B-7 axis. This can be compared to the structure of *closo*-1-C_2H_5-$AlC_2B_9H_{11}$ (Fig. 5) in which a slight distortion is also found (*20*) but the C_2H_5 group is not tilted. Grimes *et al.* speculated that the distortion may be due to the participation of filled *d*-orbitals on the Ga in back π bonding to the vacant e_2 orbitals of the C_2B_4 cage (*30*). Although the structure of 1-CH_3-$InC_2B_4H_6$ was not determined, the similarities in the ^{11}B- and ^{1}H-NMR, mass, and IR spectra of this compound and 1-CH_3-$GaC_2B_4H_6$ would lead one to expect a similar distortion in the indacarboranes.

Recently, Canadell *et al.* (*31*) have reinvestigated the reason for the tilt in 1-CH_3-$GaC_2B_4H_6$. Extended Hückel calculations on this system indicate that *d* orbital back π bonding is not needed to explain this distortion. Walsh diagrams show that the energy of two occupied molecular orbitals (Fig. 12) change as the Ga—CH_3 bond is tilted away from the Ga–B-7 axis. As the tilt angle increases, the energy of 4*s* drops fairly rapidly while that of 2*s* increases, but less sharply. The 4*s* orbital is concentrated mainly on the Ga and the B-5 atom (see Fig. 12). Bending would increase the overlap between the $GaCH_3$ fragment and the carborane ligand, thereby stabilizing the orbital. The 2*s* orbital, which can be described as a $\pi_{C=C} + \sigma_{GaCH_3}$ interaction, will become less stabilized with bending. Since the $GaCH_3$ fragment interacts much less with the carborane in 2*s* than in 4*s*, there is a net increase in bonding with distortion. Although these authors did not mention slippage of the Ga away from a centroidal position above the C_2B_3 face, the same types of interactions could be used to rationalize this distortion. As the Ga moves toward the B-5 atom 4*s* should decrease in energy because of increased overlap between the $GaCH_3$ and carborane fragments, while the energy of 2*s* should increase. As was the case for tilting, the difference between the extent of fragment interaction in the two molecular orbitals would lead to an overall stabilization on slippage. It would be of great interest to see if the icosahedral *closo*-alkylgallacarboranes show this type of distortion.

It is unfortunate that there is not more work appearing on the chemistry and structures of group 13 heterocarboranes. The results that are available raise intriguing questions and tend to pique one's curiosity. From the recent work of Hawthorne *et al.* on the aluminum sandwich complexes and of Sneddon *et al.* on the smaller cage systems, it is apparent that there is a wealth of fascinating chemistry yet to be explored in the group 13 carborane systems.

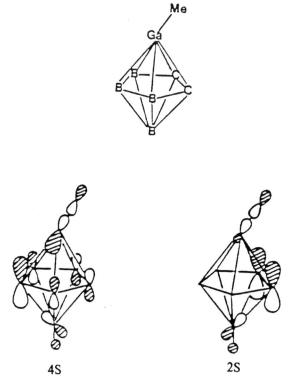

FIG. 12.   4s and 2s molecular orbitals in 1-CH$_3$-GaC$_2$B$_4$H$_6$. [Reprinted with permission from E. Canadell *et al.* *(31)*, *Organometallics* **3**, 759 (1984). Copyright 1984 by the American Chemical Society.]

VII

HETEROCARBORANES OF GROUP 14 ELEMENTS

The compounds discussed in this section are those involving the reactions of carboranes with compounds of silicon, germanium, tin, and lead. Reactions leading solely to the expansion of the carborane cage to produce larger cage systems with increased carbon content are not considered. Research in the reaction of group 14 compounds with carboranes has been quite active. All elements in group 14 have been inserted into carborane cages. There are many examples in which insertion has been accomplished for the group 14 elements in both their +2 and +4 oxidation states. In contrast to group 13 heterocarboranes, a great deal of structural information is available on these systems.

The first group 14 insertions into carborane cages were reported by Rudolph and co-workers (*32a–c*), who synthesized the series 1,2,3-$MC_2B_9H_{11}$ (M = Ge, Sn, Pb). In these compounds the group 14 heteroatom, in its +2 oxidation state, can formally be thought of as replacing an isolobal BH group in the $C_2B_{10}H_{12}$ icosahedron. The +2 oxidation state of the tin in $SnC_2B_9H_{11}$ was confirmed by ^{119m}Sn Mössbauer effect spectroscopy (*33*). The closo structures were consistent with IR, mass, and NMR spectroscopy, but characterization of the compounds by single-crystal X-ray diffraction was frustrated by the pronounced tendency of these almost spherically symmetric compounds to form disordered lattices. These investigators also reported the synthesis of the 3-Ge-1,7-$B_9C_2H_{11}$ isomer by the reaction of GeI_2 with $[7,9-B_9C_2H_{11}]^{2-}$ and by the thermal isomerization of 1,2,3-$GeB_9C_2H_{11}$ at 600°C (*32c*). A similar reaction using $SnCl_2$ and $[7,9-B_9C_2H_{11}]^{2-}$ did not produce the expected product but rather led to an oxidative closure of the carborane cage. At about the same time, Todd and co-workers (*34a,b*) reported the synthesis of the germamonocarbaborane $CH_3GeCB_{10}H_{11}$. The CH_3 group could be removed by reaction with piperidine to give the $[GeCB_{10}H_{11}]^-$ ion. This ion was found to undergo photocatalyzed reactions with $M(CO)_6$ (M = Cr, Mo, W) to give the corresponding $[(OC)_5MGeCB_{10}H_{11}]^-$. Although no X-ray crystal structures were determined for these compounds, it was thought that the $[GeCB_{10}H_{11}]^-$ bonds to the transition metal, presumably through the germanium, as a two-electron σ donor (*34a,b*). Since then, a number of closo compounds, such as $GePCB_9H_{10}$ and $GeAsCB_9H_{10}$ (*35*), have been reported. Although structural data were not available, other evidence was consistent with structures derived by replacing one of the cage CH groups in the 1,2,3-$MC_2B_9H_{11}$ compounds reported by Rudolph (see above) with the group 15 atom.

Wong and Grimes (*36*) first reported the pentagonal bipyramidal analogs $MC_2B_4H_6$ (M = Sn, Pb, Ge). From IR, mass, and NMR spectral data, the group 14 metal in its +2 oxidation state was proposed to occupy one of the apices of seven-vertex closo structure. The $GeC_2B_4H_6$ was too unstable to be isolated or characterized, and its existence was inferred by mass spectral data. The *C,C'*-dimethyl substituted smaller cage stannacarborane $Sn(CH_3)_2C_2B_4H_4$ was found to react with $(\eta^5-C_5H_5)Co(CO)_2$ to yield $(\eta^5-C_5H_5)CoSn(CH_3)_2C_2B_4H_4$ (*36*) in which the Sn and Co atoms are thought to occupy adjacent positions in an eight-vertex *closo*-$CoSnC_2B_4$ cage, with the $[C_5H_5]^-$ η^5 bonded to the Co. The reaction of MCl_2 (M = Ge, Sn) with $[(CH_3)_2C_2B_4H_4]_2FeH^-$ produces $MFe(CH_3)_4C_4B_8H_8$ (*37*). Although the structure of this compound has not been determined, it is proposed to consist of a $[(CH_3)_2C_2B_4H_4]_2Fe$ sandwich complex with the M bridging the two carborane cages and forming an Fe—M bond (*37*).

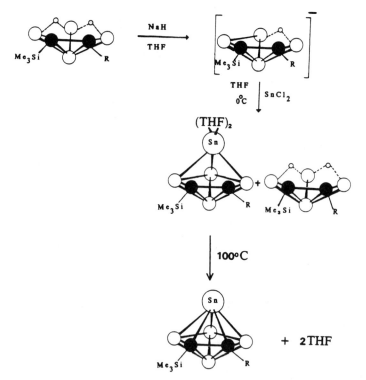

SCHEME 3. (○) BH, (●) C, and (o) H. R = Me₃Si, Me, or H. [Reprinted with permission from N. S. Hosmane *et al.* (*39*), *Organometallics* **5,** 772 (1986). Copyright 1986 by the American Chemical Society.]

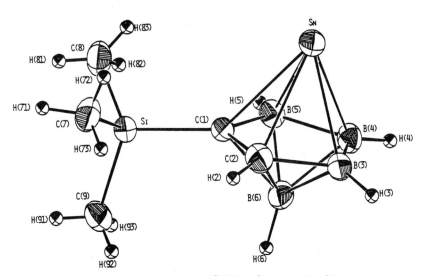

FIG. 13. Crystal structure of *closo*-1-Sn-2-[Si(CH₃)₃]-2,3-C₂B₄H₅. [Reprinted with permission from N. S. Hosmane *et al.* (*39*), *Organometallics* **5,** 772 (1986). Copyright 1986 by the American Chemical Society.]

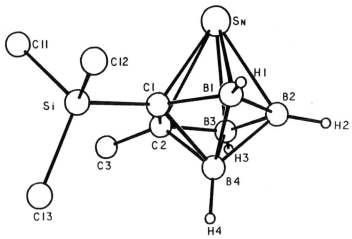

Fig. 14. Crystal structure of *closo*-1-Sn-2-[Si(CH$_3$)$_3$]-3-(CH$_3$)-2,3-C$_2$B$_4$H$_4$. [Reprinted with permission from A. H. Cowley *et al.* (*40*), *J. Chem. Soc., Chem. Commun.*, 1564 (1984). Copyright 1984 by the Royal Society of Chemistry.]

The first definitive crystal structures of the group 14 *closo*-metalla-carboranes were those reported by Hosmane and co-workers (*38–41*) from the reactions involving the monoanion of *nido*-2-[(CH$_3$)$_3$Si]-3-(R)-2,3-C$_2$B$_4$H$_6$ [R = (CH$_3$)$_3$Si, CH$_3$, H] with SnCl$_2$ in THF (C$_4$H$_8$O) to produce the corresponding stannacarboranes (see Scheme 3). The structures of the complexes (Figs. 13–15) show Sn^{2+} to occupy an apical position in

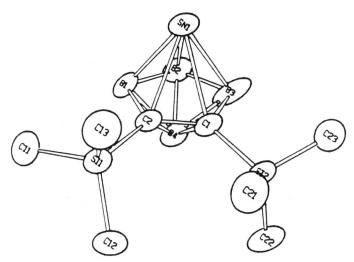

Fig. 15. Crystal structure of *closo*-1-Sn-2,3-[Si(CH$_3$)$_3$]$_2$-2,3-C$_2$B$_4$H$_4$.

TABLE I

Tin–Cage Bond Distances[a] in 1-Sn-2-[Si(CH$_3$)$_3$]-3-(R)-2,3-C$_2$B$_4$H$_4$

R	Sn–C-1	Sn–C-2	Sn–B-3	Sn–B-4	Sn–B-5
H	2.518(5)	2.475(6)	2.432(7)	2.397(8)	2.431(7)
CH$_3$	2.476(3)	2.489(4)	2.426(6)	2.378(6)	2.402(5)
Si(CH$_3$)$_3$	2.503(3)	2.492(3)	2.425(4)	2.425(5)	2.434(4)

[a] Distances in angstroms; numbering system is for R = H (Fig. 13).

the SnC$_2$B$_4$ pentagonal bipyramidal cage. The ^{11}B- and ^{1}H-NMR spectra of these compounds are quite similar to those of the MC$_2$B$_4$H$_6$ compounds of Wong and Grimes (36), thus confirming the assumption of a closo geometry. It seems that the large (CH$_3$)$_3$Si group introduces enough asymmetry that lattice disorder is minimized. The tin atoms are not symmetrically situated above the pentagonal open faces of the carboranes but are slightly slipped toward the boron side of the faces. Table I lists the tin–cage bond distances for these compounds. In addition to slippage, the open C$_2$B$_3$ face is not planar but is slightly folded away from the capping heteroatom. This slip distortion was found in the group 13 heterocarboranes and is a common feature of the closo and commo complexes of d^8 and d^9 transition metals (42).

MNDO–SCF calculations on the model compound closo-1,2,3-SnC$_2$B$_4$H$_6$ have been carried out by Maguire et al. (43). Figure 16 shows some of the tin-containing molecular orbitals calculated for this compound. Orbitals 10a' and 5a″, which arise from the interaction of the two HOMOs of C$_2$B$_4$H$_6^{2-}$, are polarized toward the boron side of the C$_2$B$_3$ face of the carborane and are stabilized by a slight slippage of the tin in that direction. This stabilization is offset by an accompanying destabilization in orbitals 4a″, 9a', and 1a', which are polarized toward the cage carbons. The net effect is that tin–carborane interactions would seem to favor slippage toward the cage carbons; an increase in intracage bonding favors slippage of the tin toward the boron side of the cage.

Although the capping group 14 heteroatoms in the closo complexes all have a lone pair of electrons exo to the cage, there is very little evidence that the lone pairs are chemically active. With the exception of Todd's compound, [GeCB$_{10}$H$_{11}$]$^-$, the group 14 heterocarboranes show no tendency to react with Lewis acids (32a–c,36,38). In fact, the heterocarboranes form donor–acceptor complexes with Lewis bases, with the group 14 atoms acting as acid sites. The most studied are the stannacarboranes. Structures have been determined for the 2,2'-bipyridine complexes (C$_{10}$H$_8$N$_2$)Sn[(CH$_3$)$_3$Si](R)C$_2$B$_4$H$_4$ [R = (CH$_3$)$_3$Si, CH$_3$] and

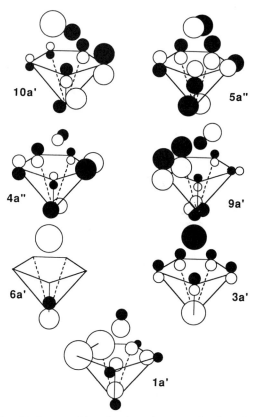

FIG. 16. Atomic orbital composition sketches of some molecular orbitals of *closo*-1,2,3-$SnC_2B_4H_6$. [Reprinted with permission from J. A. Maguire *et al.* (*43*), *Inorg. Chem.* **27**, 3354 (1988). Copyright 1988 by the American Chemical Society.]

$(C_{10}H_8N_2)Sn(CH_3)_2C_2B_9H_9$ (*39,40,44*), which are shown in Figs. 17–19. All show an extreme slippage of the tin toward the boron side of the C_2B_3 face in such a way that the tin could be considered to be η^3 bonded to the carborane. The bipyridines are opposite the cage carbons and make rather severe bond angles with the cage. For example, in $(C_{10}H_8N_2)Sn$-$[(CH_3)_3Si]_2C_2B_4H_4$ (Fig. 17), the dihedral angle between the plane of the bipyridine and the C_2B_3 face is 18.4°. The tin–carbon (cage) bond distances are about 0.38 Å longer than the tin–B-4 distance (2.75 versus 2.37 Å).

The reaction of *closo*-$Sn[(CH_3)_3Si]_2C_2B_4H_4$ with 2,2'-bipyrimidine in a 2:1 molar ratio yielded the bridged donor–acceptor complex 1,1'-(2,2'-$C_8H_6N_4$)-{*closo*-1-Sn-2,3-$[(CH_3)_3Si]_2$-2,3-$C_2B_4H_4$}$_2$ in good yield (*45*). The

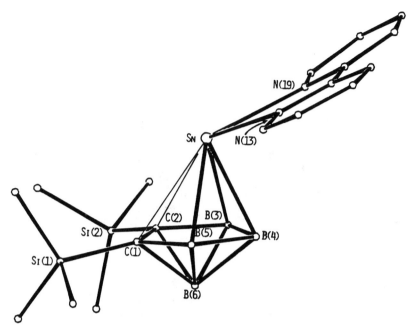

FIG. 17. Crystal structure of 1-Sn(C$_{10}$H$_8$N$_2$)-2,3-[Si(CH$_3$)$_3$]$_2$-2,3-C$_2$B$_4$H$_4$. [Reprinted with permission from N. S. Hosmane *et al.* (*39*), *Organometallics* **5,** 772 (1986). Copyright 1986 by the American Chemical Society.]

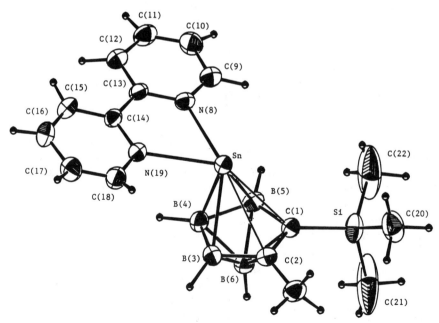

FIG. 18. Crystal structure of 1-Sn(C$_{10}$H$_8$N$_2$)-2-[Si(CH$_3$)$_3$]-3-(CH$_3$)-2,3-C$_2$B$_4$H$_4$. [Reprinted with permission from U. Siriwardane *et al.* (*44*), *Acta Crystallogr., Sect. C* **C43,** 1067 (1987). Copyright 1987 by the International Union of Crystallography.]

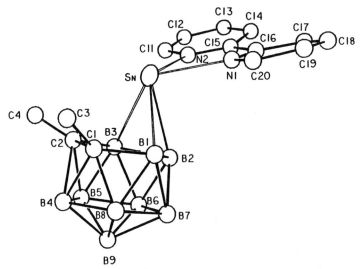

FIG. 19. Crystal structure of 1-Sn($C_{10}H_8N_2$)-2,3-$(CH_3)_2$-2,3-$C_2B_9H_9$. [Reprinted with permission from A. H. Cowley *et al.* (*40*), *J. Chem. Soc., Chem. Commun.*, 1564 (1984). Copyright 1984 by the Royal Society of Chemistry.]

structure (Fig. 20) shows a similar base–stannacarborane orientation, except that the tin slippage is slightly less than that found for the analogous bipyridine complex. The trans configuration of the two stannacarborane cages is probably dictated by steric factors. The structures of several *closo*-stannacarborane monodentate Lewis base adducts have also been

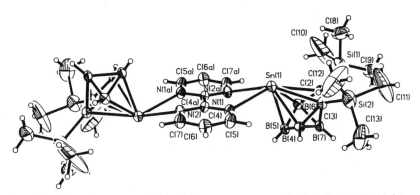

FIG. 20. Crystal structure of bridged donor–acceptor complex 1,1'-(2,2'-$C_8H_6N_4$)-{*closo*-1-Sn-2,3-[Si$(CH_3)_3$]$_2$-2,3-$C_2B_4H_4$}$_2$. [Reprinted with permission from N. S. Hosmane *et al.* (*45*), *Organometallics* **6**, 2447 (1987). Copyright 1987 by the American Chemical Society.]

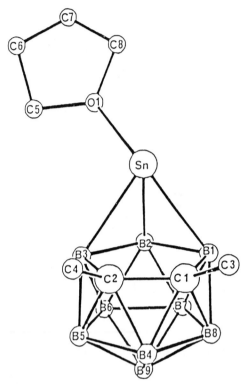

FIG. 21.   Crystal structure of 1-Sn(C$_4$H$_8$O)-2,3-(CH$_3$)$_2$-2,3-C$_2$B$_9$H$_9$. [Reprinted with permission from P. Jutzi et al. (46), *Organometallics* **6**, 1024 (1987). Copyright 1987 by the American Chemical Society.]

determined. Figures 21 and 22 show structures of the THF complex of 2,3-(CH$_3$)$_2$-1-Sn-2,3-C$_2$B$_9$H$_9$ (46) and the ferrocenylmethyl-N,N-dimethylamine {(η^5-C$_5$H$_5$)Fe[η^5-C$_5$H$_4$CH$_2$N(CH$_3$)$_2$]} complex of 2,3-[(CH$_3$)$_3$Si]$_2$-1-Sn-2,3-C$_2$B$_4$H$_4$ (47). Both show less slip distortion than found in the corresponding bipyridine complexes.

The studies have been extended to include at least one tridentate Lewis base adduct of the stannacarboranes. Figure 23 shows the structure of the terpyridine complex of 1-Sn-2-[(CH$_3$)$_3$Si]-3-(CH$_3$)-2,3-C$_2$B$_4$H$_4$ (48). This compound is of interest in that one of the end hexagonal rings of the terpyridine is oriented at an angle of about 13.9° from the plane of the other two rings. As a consequence of this nonplanarity, one Sn—N bond is significantly longer than the others (bond distances: Sn–N-11 = 2.687, Sn–N-12 = 2.573, Sn–N-13 = 2.560 Å). The SnC$_2$B$_4$ cage geometry of the terpyridine complex is essentially the same as its bipyridine analog. It is an

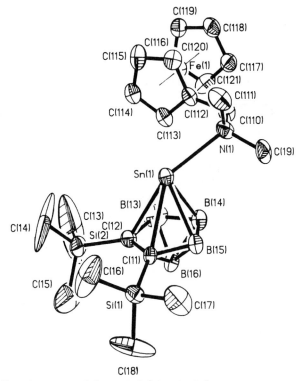

FIG. 22.   Crystal structure of *closo*-1-Sn(η^5-C$_5$H$_5$)Fe[η^5-C$_5$H$_4$CH$_2$N(CH$_3$)$_2$]-2-[Si(CH$_3$)$_3$]-3-(R)-2,3-C$_2$B$_4$H$_4$.

open question as to whether the terpyridine is acting as a tridentate or a bidentate ligand such as bipyridine.

There have been several theoretical investigations of the geometries of model compounds in the bipyridine–stannacarborane system. Fehlner and co-workers (49) conducted Fenske–Hall calculations on (C$_{10}$H$_8$N$_2$)-Sn(CH$_3$)$_2$C$_2$B$_4$H$_4$, and Maguire *et al.* (43) reported MNDO studies on (C$_{10}$H$_8$N$_2$)SnC$_2$B$_4$H$_6$. Both treatments show that complexation with bi-pyridine gives rise to orbitals with antibonding tin–carbon (cage) interac-tions, which would encourage slippage. MNDO calculations indicate that the major bonding interactions are between the bipyridine mole-cule and the tin orbitals oriented parallel to the C$_2$B$_3$ face of the carborane. Hence, maximum tin–bipyridine bonding would be expected when the rings of the bipyridine molecule and the C$_2$B$_3$ face are essentially parallel. Repulsion between the two coordinating groups would prevent such an ideal alignment. An increased slip distortion of tin would tend to decrease

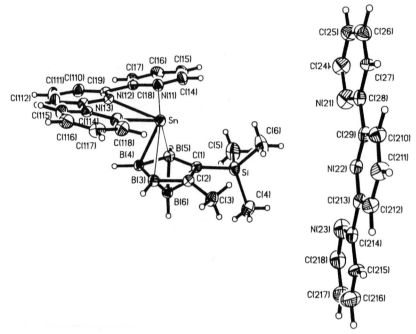

FIG. 23.   Crystal structure of *closo*-1-Sn(2,2′:6′,2″-$C_{15}H_{11}N_3$)-2-[Si$(CH_3)_3$]-3-(CH_3)-2,3-$C_2B_4H_4$ · 2,2′:6′,2″-$C_{15}H_{11}N_3$. [Reprinted with permission from U. Siriwardane and N. S. Hosmane (*48*), *Acta Crystallogr., Sect. C* **C44,** 1572 (1988). Copyright 1988 by the International Union of Crystallography.]

ligand–ligand repulsion and yield a more favorable bipyridine orientation and stronger tin–bipyridine bonding. In general, one would expect a decrease in the base–carborane dihedral angle and an increase in slip distortion on forming stronger tin–base adducts. This has been generally borne out by experiment. Adducts with monodentate bases (Figs. 21 and 22) show less slip distortion than found in the bipyridine complexes, as does the complex with the weaker bipyrimidine base (Fig. 20).

The C-Si$(CH_3)_3$-substituted stannacarboranes have been found to be useful starting materials for the synthesis of the other metallacarboranes. The apical tin is loosely ligated and can be replaced by other metal groups. For example, the reaction of Os$_3$(CO)$_{12}$ with *closo*-Sn[$(CH_3)_3$Si]$_2$-$C_2B_4H_4$ in the absence of solvent yields 1-Os(CO)$_3$-2,3-[$(CH_3)_3$Si]$_2C_2B_4H_4$ (*50*) in essentially quantitative yield [see Eq. (7)], while direct reaction of

$$3 \text{ Sn}[(CH_3)_3Si]_2C_2B_4H_4 + Os_3(CO)_{12} \xrightarrow{150°C} 3 \text{ Os(CO)}_3[(CH_3)_3Si]_2C_2B_4H_4$$
$$+ 3 \text{ Sn}^0 + 3 \text{ CO} \tag{7}$$

$Os_3(CO)_{12}$ with *nido*-$[(CH_3)_3Si]_2C_2B_4H_6$ gave the *closo*-osmacarborane in much lower yield (<3%). The stannacarboranes have also been used extensively in the preparation of the corresponding germacarboranes.

The only metallacarboranes of tin are the closo complexes, or their adducts, in which the tin is present in a formal +2 oxidation state. In contrast, germacarboranes are reported in which the germanium is present in formal +2 and +4 oxidation states. The Ge(II) inserted complexes have closo structures, while Ge(IV) yields *commo*-germacarboranes. Several *closo*-germacarboranes have already been discussed. As in the case of the stannacarboranes, what limited structural information available on the *closo*-germacarboranes is on those derived from *nido*-$\{[(CH_3)_3Si](R)$-$C_2B_4H_6\}^{2-}$. Tetrachlorogermane was found to react with the lithium or the sodium/lithium salt of $\{[(CH_3)_3Si](R)C_2B_4H_4\}^{2-}$ [R = $(CH_3)_3Si$, CH_3, H] in THF to produce a mixture of the Ge(IV) inserted sandwich compound, *commo*-germacarborane, and the corresponding Ge(II) inserted closo complex, *closo*-germacarborane, as shown in Scheme 4 (*51,52*). Although the yields varied with the nature of R group, the closo complexes formed by the reductive insertion of Ge^{II} were always produced in larger amounts than the commo products. The *commo*-germacarboranes could be obtained in good yield from the direct reaction of $GeCl_4$ with the corresponding stannacarborane [Eq. (8)].

$$2\ closo\text{-}Sn[(CH_3)_3Si](R)C_2B_4H_4 + GeCl_4 \xrightarrow[\text{no solvent}]{150°C}$$

$$commo\text{-}\{[(CH_3)_3Si](R)C_2B_4H_4\}_2Ge^{IV} + 2\ SnCl_2 \qquad R = (CH_3)_3Si, CH_3, H \qquad (8)$$

The mechanism for the reductive insertion is not known; since the corresponding neutral *nido*-carborane is also produced, it would seem to involve the THF solvent.

The closo geometry of Ge(II)-carboranes was assigned on the basis of mass, IR, and NMR spectroscopy. Since the compounds were all liquids, X-ray crystal structures could not be determined. When the reaction shown in Eq. (8) was carried out with different stoichiometric ratios, the product was quite different. The reaction of *closo*-1-Sn-2,3-$[(CH_3)_3Si]_2C_2B_4H_4$ with only a slight excess of $GeCl_4$ above a 1:1 molar ratio at 135°C in the absence of solvent produced the mixed-valence germacarborane, *closo*-1-Ge^{II}-2,3-$[(CH_3)_3Si]_2$-5-$(Ge^{IV}Cl_3)$-2,3-$C_2B_4H_3$ in 44% yield (*53*). The X-ray crystal structure shows that the Ge^{II} is η^5 bonded to the open pentagonal face of the carborane and symmetrically situated above the face [bond distances: Ge–C(cage) = 2.251(4) and 2.244(4); Ge–B = 2.265(6), 2.243(6), and 2.250(6) Å]. The other germanium, in a +4 oxidation state, is involved in an exopolyhedral $GeCl_3$ group bonded to the unique boron of the cage via a Ge—B sigma bond. This is one of the few cases of a main

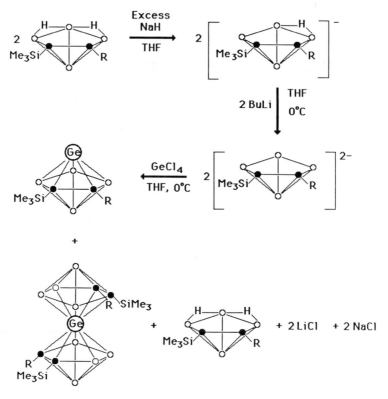

SCHEME 4. (O) BH and (●) C. R = SiMe$_3$, Me, or H. [Reprinted with permission from N. S. Hosmane *et al.* (*52*), *Organometallics* **7**, 2340 (1988). Copyright 1988 by the American Chemical Society.]

group *closo*-metallacarborane that is not slip distorted (see Fig. 24). Theoretical studies on the stannacarboranes indicate that electron-withdrawing groups on the unique boron should favor a more centroidal location of the capping heteroatom (*43*), which may be the case for the mixed-valence germacarborane.

As for the case of the stannacarboranes, the *closo*-germacarboranes, synthesized in Scheme 4, form adducts with Lewis bases such as 2,2'-bipyridine (*52,54*). The structure of 1-Ge^{II}(C$_{10}$H$_8$N$_2$)-2,3-[(CH$_3$)$_3$Si]$_2$-2,3-C$_2$B$_4$H$_4$ (Fig. 25) is similar to those of the bipyridine–stannacarborane adducts in that the germanium is slipped away from the cage carbons and the bipyridine is situated above the ring borons. However, the structure differs in that the apical germanium is twisted away from the carborane mirror plane so that the Ge–B-3 and Ge–B-5 bond distances are unequal (see Fig. 25). The germanium can be considered to be η^2 bonded to the

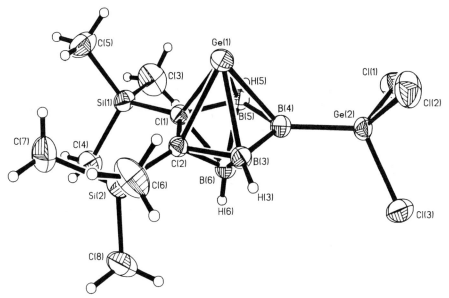

FIG. 24.   Crystal structure of *closo*-1-Ge^{II}-2,3-[Si(CH$_3$)$_3$]$_2$-5-(Ge^{IV}Cl$_3$)-2,3-C$_2$B$_4$H$_4$. [Reprinted with permission from U. Siriwardane *et al.* (*53*), *Organometallics* **7**, 1893 (1988). Copyright 1988 by the American Chemical Society.]

unique boron and one basal boron. The two Ge—N bonds are also non-equivalent, with one bond distance being 0.153 Å longer than the other. Since the 2,2'-bipyridine nitrogens are equivalent and the *closo*-germacarborane is, presumably, symmetric, there is no ready explanation for these distortions. The solution behavior of the bipyridine–germacarborane complexes is also unusual in that the room temperature proton-decoupled ^{11}B-NMR spectrum shows a single boron resonance, indicating fluxional behavior (*52*). It may be that the structure shown in Fig. 25 represents only one of several that exist in solution.

The Ge(IV) sandwiched carboranes produced in Eq. 8 have been characterized by the usual spectroscopic techniques and by X-ray crystallography (*51,55*). The structure of *commo*-{[(CH$_3$)$_3$Si]$_2$C$_2$B$_4$H$_4$}$_2$Ge^{IV} (Fig. 26) is that of two distorted pentagonal bipyramids joined by the germanium atom. The carbon atoms of the opposing cages reside on opposite sides of the germanium, which is slipped toward the boron sides of the cages [bond distances: Ge–C(cage) = 2.38, Ge–B-4 = 2.08, Ge–B-5 and Ge–B-3 = 2.15 Å]. The trans configuration of the cage carbons may be dictated in part by the steric repulsion of the bulky Si(CH$_3$)$_3$ groups. However, the fact that the trans orientations of the cage carbons have been found for a number of transition metal sandwiched carborane complexes (*42*) and in

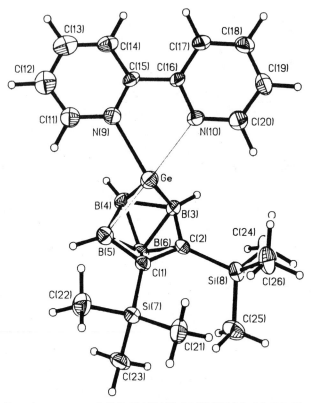

FIG. 25.   Crystal structure of 1-Ge($C_{10}H_8N_2$)-2,3-[Si(CH_3)$_3$]$_2$-2,3-$C_2B_4H_4$. [Reprinted with permission from N. S. Hosmane *et al.* (*52*), *Organometallics* **7**, 2340 (1988). Copyright 1988 by the American Chemical Society.]

the *commo*-aluminacarboranes (Figs. 6 and 7) in which hydrogens are bound to the carbons indicates that other factors are also involved.

    There have been several reports of the incorporation of silicon into both icosahedral and pentagonal bipyramidal cages. Hawthorne and co-workers (*56*) have described the synthesis of *commo*-3,3′-Si(3,1,2-Si$C_2B_9H_{11}$)$_2$ as summarized in Eq. (9). The X-ray crystal structure (Fig. 27) shows that the

$$2 \; \text{Li}_2[\textit{nido-}7,8\text{-}C_2B_9H_{11}] + \text{SiCl}_4 \;\xrightarrow[\text{reflux}]{\text{benzene}}\; \textit{commo-}3,3'\text{-Si}(3,1,2\text{-SiC}_2B_9H_{11})_2 + 4 \; \text{LiCl} \quad (9)$$

silicon is η^5 bonded to the C_2B_3 faces of two [$C_2B_9H_{11}$]$^{2-}$ ligands. The silicon atom is slightly slipped away from the cage carbons. The relevant bond distances are as follows: Si–C-1,2 = 2.22(1); Si–B-4,7 = 2.14; Si–B-8 = 2.05 Å. The compound is sufficiently stable to undergo conventional

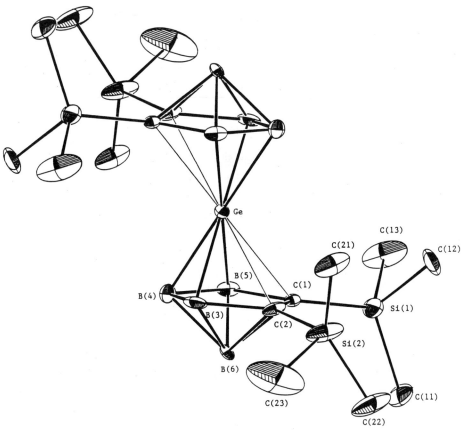

FIG. 26. Crystal structure of 2,2',3,3'-[Si(CH₃)₃]₄-*commo*-1,1'-Ge(1,2,3-C₂B₄H₄)₂. [Reprinted with permission from N. S. Hosmane *et al.* (*51*), *J. Am. Chem. Soc.* **108**, 6050 (1986), and from M. S. Islam *et al.* (*55*), *Organometallics* **6**, 1936 (1987). Copyright 1986 and 1987 by the American Chemical Society.]

carborane cage nucleophilic derivatization reactions at the cage carbons [see Eq. (10)].

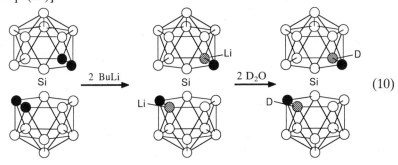

(10)

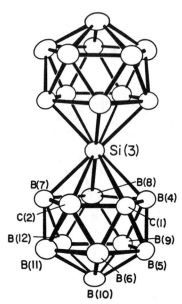

FIG. 27.   Crystal structure of *commo*-3,3'-Si(3,1,2-SiC$_2$B$_9$H$_{11}$)$_2$. [Reprinted with permission from W. S. Rees, Jr., *et al.* (*56*), *J. Am. Chem. Soc.* **108**, 5369 (1986). Copyright 1986 by the American Chemical Society.]

The smaller cage carborane analogs have also been prepared (*57,58*). Their synthesis and yields are given in Eq. (11). This reaction is similar to

$$2 \text{ NaLi}\{[\text{Si}(\text{CH}_3)_3](\text{R})\text{C}_2\text{B}_4\text{H}_4\} + \text{SiCl}_4 \xrightarrow[0°C]{\text{THF}} commo\text{-}\{[\text{Si}(\text{CH}_3)_3](\text{R})\text{C}_2\text{B}_4\text{H}_4\}_2\text{Si}$$

$$18\text{--}57\%$$

$$+ 2 \text{ NaCl} + 2 \text{ LiCl} \quad \text{R} = (\text{CH}_3)_3\text{Si}, \text{CH}_3, \text{H} \tag{11}$$

the preparation of the corresponding germacarboranes (see above). In addition to the commo product, a small amount (trace to 1%) of extremely air-sensitive compounds whose ^{1}H-, ^{11}B-, ^{13}C-, and ^{29}Si-NMR, mass, and IR spectra are consistent with the formula 1-Si^{II}-2-[(CH$_3$)$_3$Si]-3-(R)-2,3-C$_2$B$_4$H$_4$ and a closo structure (see Scheme 5). Since these Si(II) compounds are extremely unstable, decomposing slowly at room temperature even in high vacuum, complete characterization was not possible. On the other hand, the *commo*-silacarborane products [Eq. (11)] are quite stable and can be handled in air for short periods of time (*58*). The crystal structures of *commo*-1,1'-Si^{IV}-{2-[(CH$_3$)$_3$Si]-3-(R)-2,3-C$_2$B$_4$H$_4$}$_2$ [R = (CH$_3$)$_3$Si, CH$_3$] (Figs. 28 and 29) are quite similar to that of *commo*-[C$_2$B$_9$H$_{11}$]$_2$Si (Fig. 27) in that they show the silicon η^5 bonded to each of the carborane cages with a slight slippage away from the cage carbons.

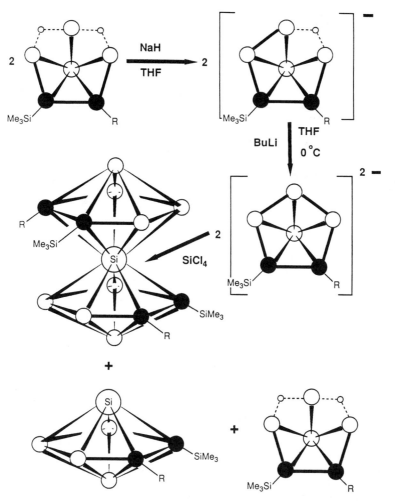

SCHEME 5. (○) BH, (●) C, and (o) H. R = SiMe$_3$, Me, or H. [Reprinted with permission from U. Siriwardane *et al.* (*58*), *J. Am. Chem. Soc.* **109**, 4600 (1987). Copyright 1987 by the American Chemical Society.]

The Na/Li salt of *nido*-{[Si(CH$_3$)$_3$](CH$_3$)C$_2$B$_4$H$_4$}$^{2-}$ was also found to react with SiH$_2$Cl$_2$ to give Cl(H)Si{[Si(CH$_3$)$_3$](CH$_3$)C$_2$B$_4$H$_4$} as a moderately air-sensitive, colorless liquid of low volatility [see Eq. (12)] (*58*).

$$\textit{nido-}\text{NaLi}\{[\text{Si}(\text{CH}_3)_3](\text{CH}_3)\text{C}_2\text{B}_4\text{H}_4\} + \text{SiH}_2\text{Cl}_2 \text{ (excess)} \xrightarrow[0°\text{C}]{\text{THF}}$$

$$\underset{58\%}{\text{Cl(H)Si}\{[\text{Si}(\text{CH}_3)_3](\text{CH}_3)\text{C}_2\text{B}_4\text{H}_4\}} + \textit{nido-}\{[\text{Si}(\text{CH}_3)_3](\text{CH}_3)\text{C}_2\text{B}_4\text{H}_6\} + \text{NaCl} + \text{LiCl} \quad (12)$$

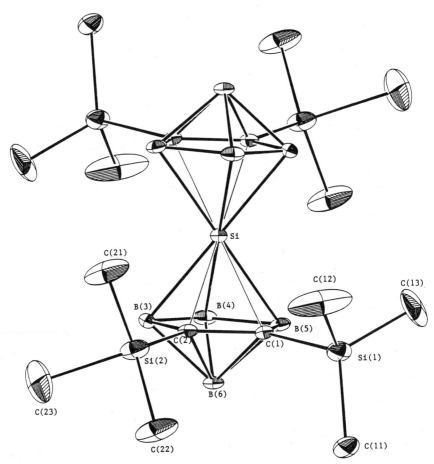

FIG. 28.   Crystal structure of $2,2',3,3'$-$[Si(CH_3)_3]_4$-$commo$-$1,1'$-$Si(1,2,3$-$SiC_2B_4H_4)_2$. [Reprinted with permission from N. S. Hosmane *et al.* (*57*), *J. Chem. Soc., Chem. Commun.*, 1421 (1986). Copyright 1986 by the Royal Society of Chemistry.]

This compound could be quantitatively converted to $H_2Si\{[Si(CH_3)_3]$-$(CH_3)C_2B_4H_4\}$ by reaction with NaH (*58*). These compounds have been characterized by mass, IR, ^{1}H-, ^{11}B-, ^{13}C-, and ^{29}Si-NMR spectroscopy but not by X-ray diffraction. Their ^{11}B-NMR spectra bear a striking resemblance to those of the stannacarboranes and do not exhibit resonances indicative of a bridging SiH(X) group. Thus, the SiH(X) group seems to be incorporated into the cage, giving a seven-vertex silacarborane. It is of interest to note that Tabereaux and Grimes (*59*) found that the reaction of *nido*-Na[C$_2$B$_4$H$_7$] with SiH$_2$Cl$_2$ yielded only μ,μ'-SiH$_2$(C$_2$B$_4$H$_7$)$_2$ in

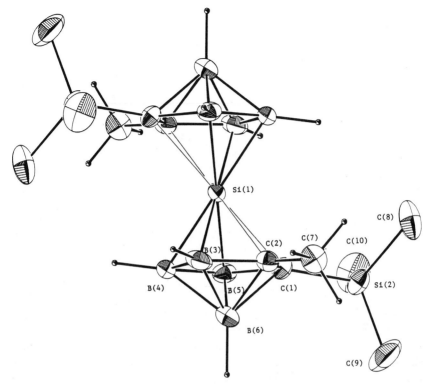

FIG. 29. Crystal structure of 2,2′-[Si(CH₃)₃]₂-3,3′-(CH₃)₂-*commo*-1,1′-Si(1,2,3-C₂-B₄H₄)₂. [Reprinted with permission from U. Siriwardane *et al.* (58), *J. Am. Chem. Soc.* **109,** 4600 (1987). Copyright 1987 by the American Chemical Society.]

which the silicon is doubly bridged by two three-center, two-electron B—Si—B bonds. It is not apparent what factors dictate the different products in this reaction and the one described in Eq. (12). Since attempts to prepare doubly bridged compounds with groups other than hydrogen on the silicon have not been successful (5,60), steric factors could be important.

The ^{29}Si-NMR spectra of $H_2Si\{[Si(CH_3)_3](CH_3)C_2B_4H_4\}$ indicate that structural changes from its H(Cl)Si precursor may be more than a simple hydrogen–chlorine exchange. The proton-coupled ^{29}Si-NMR spectrum of the SiH₂-containing silacarborane did not show the expected 1:2:1 triplet due to ^{1}H coupling but consisted of a broad major doublet with a large coupling constant ($^1J = 362$ Hz). Each line of this doublet was further split into doublets with a much smaller coupling constant ($^1J = 42$ Hz). The large splitting could arise from ^{29}Si–H$_{terminal}$ spin coupling and the secondary splitting from a much weaker coupling of the second hydrogen. This

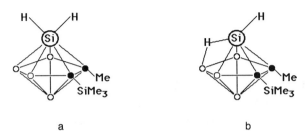

FIG. 30.   (a and b) Possible structures for $1\text{-Si(H)}_2\text{-}2\text{-[Si(CH}_3)_3]\text{-}3\text{-(CH}_3)\text{-}2,3\text{-}C_2B_4H_4$. (O) BH and (●) C.

suggests a structure, such as that shown in Fig. 30b, in which one hydrogen is involved in a Si—H—B three-center, two-electron bond. If this structure is correct the silicon would formally be in a +2 oxidation state. It is of interest to note that if, as the ^{29}Si-NMR spectrum indicates, one of the Si—H bonds lies in the polyhedral structure, the SiH$_2$ group could be considered as furnishing four electrons and three atomic orbitals to cage bonding, yielding a seven-vertex, 18-skeletal electron system which, according to Wade's rules (61), would be a nido structure.

Some caution must be exercised in applying simple electron counting rules to heterocarboranes containing heavily ligated metal or metalloid groups. This is true not only for metallacarboranes of transition metals (62) but also for some main group carborane compounds. The stannacarboranes are good examples. The n vertex structures of the simple stannacarboranes $SnC_2(R)_2B_4H_4$ (and presumably $SnC_2B_9H_{11}$) have $n + 1$ skeletal electron pairs and are predicted by Wade's rule to have closo structures, which was confirmed experimentally (see Figs. 13–15). According to these same rules, coordination of the tin by a monodentate ligand would give $n + 2$ skeletal electron pairs, and a nido structure would be expected. The structures of two such complexes are known (Figs. 21 and 22) and show a slippage of the tin away from the centroidal position above the C_2B_3 faces of the carboranes. Although it is an open question as to when a slip-distorted closo structure could be better described as an open, nido one, the slip distortions in the compounds shown as Figs. 21 and 22 are small enough that a distorted closo geometry is a better structural description. The problem is even worse in the case of bidentate ligands, such as 2,2′-bipyridine, where Wade's rules would predict arachno structures, which are clearly not the case (see Figs. 17–20). The convention of treating the coordinating electron pairs of the Lewis base as skeletal electrons may lead to erroneous predictions. In situations involving strong metal–base bonding, such pairs should better be considered as exopolyhedral, in which the base coordination would remove orbitals from cage bonding and

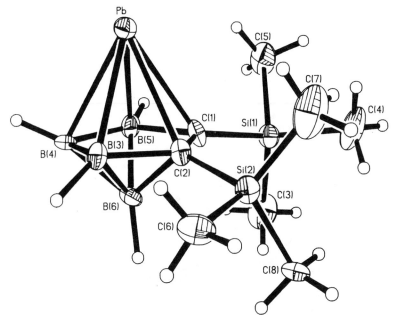

Fig. 31. Crystal structure of *closo*-1-Pb-2,3-[Si(CH$_3$)$_3$]$_2$-2,3-C$_2$B$_4$H$_4$. [Reprinted with permission from N. S. Hosmane *et al.* (*63*), *Organometallics* **8**, 566 (1989). Copyright 1989 by the American Chemical Society.]

the metal would no longer be isolobal with BH (or CH). In such cases, Wade's rules would not be expected to apply.

Of all the group 14 carboranes, the least studied have been the plumba-carboranes. The first example of such compounds was reported by Rudolph *et al.* (*32b*) in 1970 as a member of the series MC$_2$B$_9$H$_{11}$ (M = group 14 metal). The smaller cage pentagonal bipyramidal plumba-carboranes of the type *closo*-1-Pb-2,3-(R)$_2$-2,3-C$_2$B$_4$H$_4$ (R = CH$_3$, H) were reported somewhat later by Wong and Grimes (*36*). Structural data were not reported for either of these compounds, but ^{11}B-NMR spectra indicated closo structures. However, it was not clear whether the plumba-carboranes exhibit a slip distortion as has been found in many metallacar-boranes. The splitting of the basal BH resonances in the ^{11}B-NMR spectrum of PbC$_2$B$_9$H$_{11}$ was taken by Rudolph *et al.* as indicating a possible slippage (*32b*), but such an interpretation has been questioned by Wong and Grimes (*36*). The X-ray crystal structure of 1-Pb-2,3-[Si(CH$_3$)$_3$]$_2$-2,3-C$_2$B$_4$H$_4$ has recently been determined and is shown in Fig. 31 (*63*). The structure confirms the closo geometry in that it shows the Pb to be essentially symmetrically bonded above the C$_2$B$_3$ face; the Pb–cage atom distances are 2.582(17), 2.634(14), 2.601(16), 2.579(17), and 2.520(2) Å for

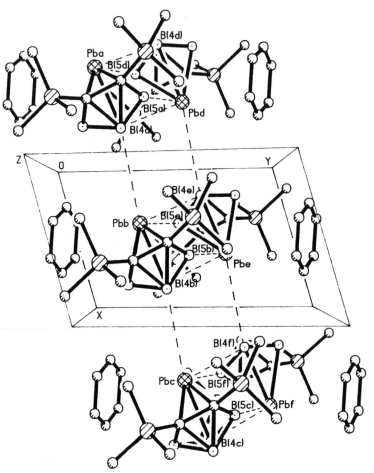

FIG. 32.   Crystal packing diagram showing the extended chain network of {*closo*-1-Pb-2,3-[Si(CH$_3$)$_3$]$_2$-2,3-C$_2$B$_4$H$_4$}$_2$ molecular dimers with benzene molecules of crystallization.

atoms C-1 through B-5, respectively. The bond distances indicate that, if any distortion exists, it is one in which the Pb is slipped toward B-5, a basal boron that is bonded to a cage carbon. The crystal packing diagram (Fig. 32) shows that the solid consists of closely associated {Pb[Si-(CH$_3$)$_3$]$_2$C$_2$B$_4$H$_4$}$_2$ dimers with a crystallographic center of symmetry halfway between the two Pb atoms (*63*). This dimeric structure could give rise to the distortions found in the plumbacarborane cage and, if the dimer exists in solution, to the splitting of the basal BH resonances in its ^{11}B-NMR spectrum.

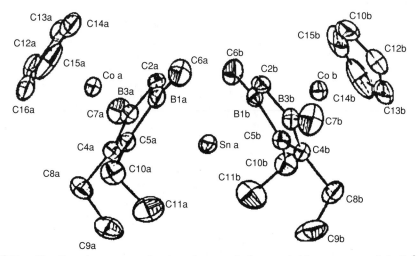

FIG. 33. Crystal structure showing the tetradecker sandwich geometry of $Sn\{(\eta^5\text{-}C_5H_5)Co[\eta^5\text{-}C_2(C_2H_5)_2B_2(CH_3)_2CH]\}_2$. [Reprinted with permission from H. Wadepohl *et al.* (*64*), *Organometallics* **2**, 1899 (1983). Copyright 1983 by the American Chemical Society.]

In addition to reactions with polyhedral carborane clusters, some recent work has been published on the interaction of main group metals with planar carboranes of the diborolene system. Siebert and co-workers have reported that the reaction of $\{(C_5H_5)Co[C_2(C_2H_5)_2B_2(CH_3)CH]\}^-$ with $SnCl_2$ produced a dark orange crystalline product which was found to be highly soluble in both polar and nonpolar solvents (*64*). The crystal structure of this compound (Fig. 33) is that of a tetradecker complex in which two $(C_5H_5)Co(C_2B_2C)$ units are bound to a central tin. There are two crystallographically independent molecules in which the angles between the normals to the diborolenyl rings are 112 and 114°, respectively. The same type of bending was found in the stannocene system (*65*). The analogy between the stannabis(cobaltadiborolene) and stannocene was further demonstrated in that, like stannocene, $[(C_5H_5)Co(C_2B_2C)]_2Sn$ can be cleaved by HBF_4 to give $[(C_5H_5)Co(C_2B_2C)Sn]^+BF_4^-$ (*64*).

In related work, bis(1-*tert*-butyl-2,3-dimethyl-1,2-azaborolinyl)tin was synthesized by the reaction of $SnCl_2$ with azaborolinyllithium to give the sandwich compound (*66*). The structure (Fig. 34) is a bent one that is similar to stannocene and the stannadiborolene (Fig. 33). It is of interest in that the tin atom is significantly slipped away from the nitrogen atom toward the BCC side of the ring so as to be considered as η^3 bonded to the azaborolines (*66*).

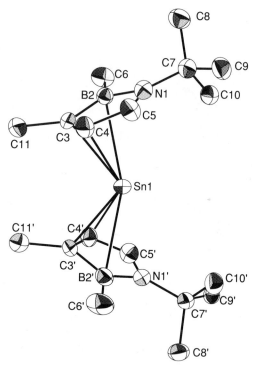

Fig. 34. Crystal structure of bis(1-*tert*-butyl-2,3-dimethyl-1,2-azaborolinyl)tin. [Reprinted with permission from G. Schmid *et al.* (*66*), *Angew. Chem., Int. Ed. Engl.* **24,** 602 (1985). Copyright 1985 by VCH Verlagsgesellschaft, Weinheim, Federal Republic of Germany.]

VIII

HETEROCARBORANES OF GROUP 15 ELEMENTS

Todd has extensively reviewed the preparative and reaction chemistry of carboranes containing group 15 elements, covering the literature up to 1982 (*4*). The chemistry of phospha- and arsametallacarboranes containing transition metals has been reviewed by Grimes (*67*). An inspection of the literature since 1982 reveals very little additional work that falls under the purview of this article. Therefore, for completeness, some of the earlier chemistry of the group 15 carboranes is summarized here.

A group 15 atom is isoelectronic and isolobal with a CH group. Therefore, one would expect that substitution of a group 15 element for a carbon group in a dicarbaborane would yield compounds with similar structures and comparable reactivities. Generally, this has been found to be true, and, except for nitrogen, all the group 15 elements have been inserted into monocarbaborane cages. Todd and co-workers reported that the reaction of $Na_3B_{10}H_{10}CH \cdot (THF)_n$ with PCl_3, $AsCl_3$, or SbI_3 produced 1,2-$PCHB_{10}H_{10}$, 1,2-$AsCHB_{10}H_{10}$, or 1,2-$SbCHB_{10}H_{10}$ in respectable yields (45, 25, and 41%, respectively) (34b,68–70). Although no crystal structures were given, icosahedral structures, with the carbon and group 15 elements occupying adjacent positions, were inferred from ^1H- and ^{11}B-NMR spectra. These icosahedral heteromonocarbaboranes were found to undergo thermal rearrangement to produce both 1,7 and 1,12 isomers, which is similar to the behavior exhibited by *ortho*-dicarbaboranes (34b). In procedures identical to those employed for the *ortho-*, *meta-*, *para*-carboranes, the C—H bonds in 1,7- and 1,12-$CHPB_{10}H_{10}$ can be lithiated without cage rearrangement, and a number of C-substituted derivatives have been synthesized (71–73).

Todd and co-workers have reported the synthesis of *nido*-$C[N(CH_3)_3]P(R)B_{10}H_{10}$ (R = C_6H_5, C_2H_5, CH_3) from the reaction of $C[N(CH_3)_3]B_{10}H_{12}$ with $P(R)Cl_2$ (74). The X-ray crystal structure of the compound when R is C_6H_5 was determined (Fig. 35). The P-C_6H_5 group bridges B-9 and B-10 on the open face of the carborane [bond lengths: P–B-9 = 1.998(5), P–B-10 = 2.018(5), P–B-11 = 2.381, P–B-8 = 2.347, P–C-7 > 2.4 Å]. This structure is very similar to the 12-atom *nido*-$[(C_6H_5)CHC(C_6H_5)B_{10}H_{10}]^-$ reported by Tolpin and Lipscomb (75). It was also reported that sealed tube pyrolysis of $C[N(CH_3)_3]P(C_6H_5)B_{10}H_{10}$ at 475°C produced *closo*-$C[N(CH_3)_3]PB_{10}H_{10}$, but no details were given (74).

The $AlCl_3$-catalyzed halogenation reactions of all three isomers of phospha- and arsamonocarbaboranes have been studied by Zakharkin and Kyskin (76). It was found that electrophilic monosubstitution at a B—H bond occurred more readily with arsamonocarbaboranes, and, in general, the 1,2 isomers are more reactive than the 1,12 isomers. Wong and Lipscomb (77) have studied isomer distributions in the thermal rearrangements of the chlorophosphacarboranes and reported the X-ray structure of 9,10-Cl_2-1,7-$CHPB_{10}H_8$. The structure of this compound (Fig. 36) is that of a distorted icosahedron having C_s symmetry with the C, P, B-5, and B-12 atoms constituting the mirror plane. The presence of the phosphorus causes a distortion of the icosahedron, with the B—P bond distances being significantly longer than either the B—C and B—B bonds [average bond distances are 2.02(1), 1.73(2), and 1.79(3) Å, respectively]. Analysis of the isomer distribution of the rearranged chlorophosphacarboranes indicated that the distortions accompanying the substitution of phosphorus for a CH causes a significant change in their rearrangement mechanisms. Thus, the

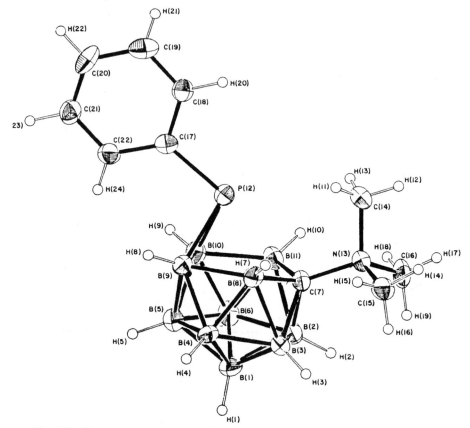

FIG. 35.   Crystal structure of *nido*-C[N(CH$_3$)$_3$]P(C$_6$H$_5$)B$_{10}$H$_{10}$. [Reprinted with permission from W. F. Wright *et al.* (*74*), *J. Organomet. Chem.* **148**, 7 (1978). Copyright 1978 by Elsevier Sequoia S.A.]

similarities of the thermal rearrangements in the unhalogenated phospha-carboranes and dicarbaboranes may only be formal ones.

Substitution of a group 15 atom for a CH unit also affects other properties. Polarographic reduction studies of the isomers of closo arsa- and phosphamonocarbaboranes show that the electron affinity of the cages are substantially greater than in the analogous dicarbaboranes (*78*). Studies of the rates of hydrogen–deuterium exchange between NH$_3$ and 1,7-CDEB$_{10}$H$_{10}$ (E = P, As) suggest that the acidity of the CH group increases in the order 1,7-C$_2$H$_2$B$_{10}$H$_{10}$ < 1,7-CHAsB$_{10}$H$_{10}$ < CHPB$_{10}$H$_{10}$ (*79*).

Treatment of the 1,2 or 1,7 isomers of the phospha-, arsa-, and stibamonocarbaboranes with a mild base such as piperidine causes an abstraction

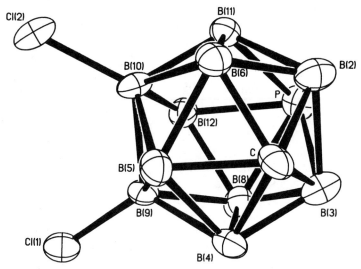

FIG. 36. Crystal structure of *closo*-9,10-Cl_2-1,7-$CHPB_{10}H_8$. [Reprinted with permission from H. S. Wong and W. N. Lipscomb (*77*), *Inorg. Chem.* **14**, 1350 (1975). Copyright 1975 by the American Chemical Society.]

of a BH unit from the icosahedral cage to produce the corresponding anions (*34b,71*). The use of stronger bases, such as $NaOC_2H_5$ or ethanolic KOH, gives rise to either abstraction of a BH unit or removal of the heteroatom (*34b,71*). Further deprotonation of the 1,2- or 1,7-$[CHPB_9H_{10}]^-$ could be achieved by reaction with triethylamine of sodium hydride to give the corresponding 1,2- or 1,7-$[CHPB_9H_9]^{2-}$ dianions (*80*). These are analogous to $[C_2B_9H_{11}]^{2-}$ and should serve as potential η^5 ligands for metals to produce commo and closo complexes. They are also of interest in that the exopolyhedral lone pair of the group 15 atoms could also provide a site for σ bonding to suitable acceptor groups.

Both σ- and π-bonded transition metal complexes of the group 15 monocarbaboranes have been reported (*2,4,67*) and a few examples are discussed. The anionic complex $[2,1,7-(OC)_3MnCHPB_9H_9]^-$ can be prepared by refluxing a THF solution of $Mn(CO)_5Br$ and $[1,7-CHPB_9H_9]^{2-}$ (*80,81*). The neutral complex $1-CH_3-2,1,7-(OC)_3MnCHPB_9H_9$ can be produced by starting with $[1,7-CHP(CH_3)B_9H_9]^-$ (*80,81*). The carbonyls on the Mn can undergo typical ligand-exchange reactions (*82*). The ferrocene analog can be prepared by the reaction of 1,2- or 1,7-$[CHPB_9H_9]^{2-}$ with anhydrous $FeCl_2$ in THF and has been isolated as the tetramethylammonium salt, $[(CH_3)_4N^+]_2[(CHPB_9H_9)_2Fe]^{2-}$ (*80,81*). The neutral *P*-Me derivatives can be prepared either by reaction of the dianionic sandwich compound with MeI or by starting with the monoanion $[CHP(CH_3)B_9H_9]^-$ (81). Demethylation can be accomplished by reaction of $[1,7\text{-CHP-}$ $(CH_3)B_9H_9]_2Fe$ with NaH (81). $[(CHPB_9H_9)_2Fe]^{2-}$ seems to be more stable than its dicarbon analog $[(C_2B_9H_{11})_2Fe]^{2-}$ in that it is not air oxidized in aqueous solution. Treatment of $[(1,7\text{-CHPB}_9H_9)_2Fe]^{2-}$ with ceric ion causes oxidation to a green paramagnetic monoanionic species in which the iron is formally in a +3 oxidation state (81). Both $[(1,7\text{-CHPB}_9H_9)_2Fe]^{2-}$ and its arsenic analog were found to undergo photolytic reactions with $M(CO)_5$ (M = Cr, Mo, W) to form the corresponding E—$M(CO)_5$ σ-bonded mixed-metal derivatives (E = P, As) (83). Thus, it seems that even though the group 15 monocarbaborane dianions can function as both σ- and π-bonding ligands for metals, π bonding is the preferred mode of interaction. The participation of the exopolyhedral lone pairs in metal bonding occurs when the monoanion forms complexes, such as $[(OC)_5\text{-}MoCHAsB_9H_{10}]^-$, $[(OC)_9Mn_2CHPB_9H_{10}]^-$, and $(CO)_2Mo(\eta\text{-C}_7H_7)$-$(CHPB_9H_{10})$ (4,84), or when the π sites are already metallated as in the reaction of $[(1,7\text{-CHEB}_9H_9)_2Fe]^{2-}$ (E = P, As) with $M(CO)_6$ (M = Cr, Mo, W) (83).

It is unfortunate that there are so few X-ray crystal structures determined for the group 15 heteromonocarbaboranes. A partial structure of $[1,7\text{-CHP(CH}_3)B_9H_9]_2Fe$ has been reported. The complex can exist in two isomeric forms as a *dd,ll* racemate and a *dl* meso. An X-ray analysis of a cocrystallite containing both isomers was carried out to an *R* factor of about 13.5% (80). An idealized view is shown in Fig. 37. Although the carbon atom in one cage could be located in the structure, the carbon in the other cage occupies two lattice sites on an approximately statistical basis. From what is known, the configuration about the iron is quite regular; the iron is located above the centers of the two coplanar η^5-bonding pentagonal faces, which were parallel within the limits of experimental error. The distortion introduced by the phosphorus atoms seems to be confined to the cage itself in that the open pentagonal face and the other five-membered ring in the same cage are out of parallel by about 4.5° (80). It has been reported that an X-ray structure was obtained for 3,1,2-$(\eta^5\text{-}C_5H_5)CoCHAsB_9H_9$ (85); however, bond lengths, bond angles, ORTEP drawings, or any other details of the structure were not given.

The chemistry of dicarbaboranes containing group 15 elements has been limited to aza- and arsacarboranes. The reaction of $[7,8\text{-(CH)}_2B_9H_{10}]^-$ with aqueous nitrous acid at 0°C produced $NH(CH)_2B_8H_8$ and $NH_2\text{-}(CH)_2B_8H_9$ in respectable yields (86,87). The azacarborane $NH_2\text{-}(CH)_2B_8H_9$ is thought to be a derivative of 5,6-$(CH)_2B_8H_{10}$, with the NH group replacing one of the bridged hydrogens in the latter compound. Spectral data on $NH(CH)_2B_8H_8$ indicate a nido structure, which was

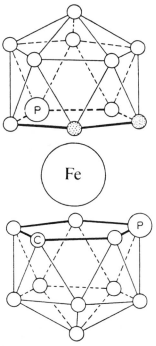

FIG. 37.   Idealized view of the molecular structure of *commo*-[1,7-CHP(CH$_3$)B$_9$H$_9$]$_2$Fe.
[Reprinted with permission from L. J. Todd *et al.* (*80*), *J. Am. Chem. Soc.* **90**, 4489 (1968).
Copyright 1968 by the American Chemical Society.]

confirmed by a crystal structure of its C$_6$H$_5$CH$_2$ derivative, *nido*-10-
(CH$_2$C$_6$H$_5$)-10-(N)-7,8-(CH)$_2$B$_8$H$_8$ (Fig. 38) (*88*). The replacement of the
C$_6$H$_5$CH$_2$N group for BH does not seem to introduce much intracage
distortion. The plane formed by the borons B-2 through B-6 (plane 1) is
parallel to the planar open face of the cage (plane 2) within experimental
error. The molecule does not possess a crystallographic mirror plane
because of a torsional twist about the CH$_2$—N bond; the C-1–C-9–N-10–
B-11 torsion angle is 51.9°, and the plane of the benzyl ring is not perpendicular to plane 2 but forms an angle of 101.9° with it.
    The arsacarboranes have been prepared by the reaction of Tl$_2$[7,8-
(CH)$_2$B$_9$H$_9$] with RAsX$_2$ (R = CH$_3$, X = Br; R = C$_6$H$_5$, *n*-C$_4$H$_9$, X = Cl)
in diethyl ether at 0°C to give the corresponding 3-(R)-3-(As)-1,2-
(CH)$_2$B$_9$H$_9$ in up to 30% yield (*89*). The ^{1}H- and ^{11}B-NMR, infrared, and
mass spectra of these compounds indicate closo icosahedral structures. The
reaction of 3-(C$_6$H$_5$)-3-(As)-1,2-(CH)$_2$B$_9$H$_9$ with BBr$_3$ in dilute CCl$_4$ solution gave high yields (95%) of 3-(Br)-3-(As)-1,2-(CH)$_2$B$_9$H$_9$, which was

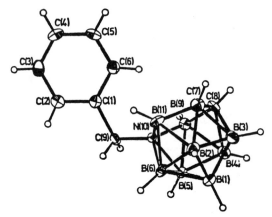

FIG. 38.   Crystal structure of $nido$-10-$(CH_2C_6H_5)$-10-(N)-7,8-$(CH)_2B_8H_8$. [Reprinted with permission from J. Plesek $et\ al.$ (88), $J.\ Chem.\ Soc.,\ Chem.\ Commun.$, 935 (1975). Copyright 1975 by the Royal Society of Chemistry.]

considerably more stable than the alkyl-substituted compound. All of the arsacarboranes react rapidly with ethanolic KOH to remove the arsenic group, producing $K^+[7,8$-$(CH)_2B_9H_{10}]^-$ (89). The production of the arsacarboranes is quite sensitive to the nature of the $[7,8$-$(CH)_2B_9H_9]^{2-}$ counterions. The dilithium salt gave predominantly a yellow gum, which appeared to be a low molecular weight polymer, with very little of the desired arsacarborane. A variety of solvents were used but failed to yield the correct product. However, other investigators have reported the synthesis of $(C_6H_5)As(CH)_2B_9H_9$ in low yields by the reaction of $Na_2[(CH)_2B_9H_9]$ and $C_6H_5AsCl_2$ in toluene (90). The two compounds show different ^{1}H-NMR spectral behaviors, so it is an open question whether the two research groups are dealing with the same compound.

When $(CH_3)_2AsBr$ was reacted with either $Li[7,8$-$(CH)_2B_9H_{10}]$ or $Tl[7,8$-$(CH)_2B_9H_{10}]$, the product obtained was $[(CH_3)_2As]_2(CH)_2B_9H_9$. The same compound could be obtained in 70% yield from the direct reaction of $Tl_2[7,8$-$(CH)_2B_9H_9]$ with $(CH_3)_2AsBr$ in a 1:2 molar ratio (89). Spectral and chemical data are consistent with a 12-vertex nido geometry in which one $(CH_3)_2As$ group is terminally bonded to a boron and the other occupies a bridging position. The terminally bound $(CH_3)_2As$ can be replaced by an ethoxy group by reaction with ethanolic KOH (89). The compound $nido$-$As_2(CH)_2B_7H_7$ was synthesized in moderate yield from the reaction of AsI_3 with $(CH)_2B_7H_9$ in the presence of $(C_2H_5)_3N$ (91). On the basis of ^{1}H- and ^{11}B-NMR spectra, the authors proposed a tentative structure of an 11-atom nido cage with two carbons in the open face separated by two adjacent arsenic atoms and a BH unit. Photolysis of the compound with $(\eta^5\text{-}C_5H_5)Co(CO)_2$ in THF produced $(\eta^5C_5H_5)$-$CoAs_2(CH)_2B_7H_7$ in low yield *(91)*.

IX
HETEROCARBORANES OF GROUP 16 ELEMENTS

Of all the main group heterocarboranes, those of group 16 are the least studied. Todd reviewed these compounds in 1982 *(4)*, and no subsequent work has appeared. Therefore, the reader is referred to Ref. *4* for a discussion of the group 16 heterocarboranes and the pertinent literature citations.

X
SUMMARY

There has been a great deal of recent interest in the preparative chemistry and structural investigations of the main group heterocarboranes. As more structures and high-yield preparative routes become available, bonding trends and mechanistic patterns are developing. However, at this stage of understanding, the results raise more questions than they answer; and a great deal of interesting work needs to be done. With the exception of the Lewis acid character of the metallacarboranes, the reaction chemistry of these systems has hardly been addressed. These future studies hold the promise of being every bit as fascinating, and perplexing, as the previous ones.

ACKNOWLEDGMENTS

We thank the National Science Foundation, the Robert A. Welch Foundation, and the donors of the Petroleum Research Fund, administered by the American Chemical Society, for support of our ongoing research in the area of main group heterocarborane chemistry.

REFERENCES

1. R. N. Grimes, "Carboranes." Academic Press, New York, 1970.
2. R. N. Grimes, ed., "Metal Interactions with Boron Clusters." Plenum, New York, 1982.
3. E. L. Muetterties, ed., "Boron Hydride Chemistry." Academic Press, New York, 1985.
4. L. J. Todd, *in* "Comprehensive Organometallic Chemistry" (G. Wilkinson, F. G. A. Stone, and E. W. Abel, eds.), Vol. 1, Chap. 5.6, p. 543. Pergamon, Oxford, 1982.
5. R. N. Grimes, *Rev. Silicon, Germanium, Tin Lead Compd.* **2**, 223 (1977).

6a. R. N. Grimes, *in* "Comprehensive Organometallic Chemistry" (G. Wilkinson, F. G. A. Stone, and E. W. Abel, eds.), Vol. 1, Chap. 5.5, p. 459. Pergamon, Oxford, 1982.

6b. R. N. Grimes, *in* "Molecular Structure and Energetics" (J. F. Liebman, A. Greenberg, and R. E. Williams, eds.), Vol. 5, Chap. 11, p. 241. VCH, New York, 1988.

7. N. S. Hosmane and J. A. Maguire, *in* "Molecular Structure and Energetics" (J. F. Liebman, A. Greenberg, and R. E. Williams, eds.), Vol. 5, Chap. 14, pp. 297–328. VCH, New York, 1988.

8. M. E. O'Neill and K. Wade, *in* "Comprehensive Organometallic Chemistry" (G. Wilkinson, F. G. A. Stone, and E. W. Abel, eds.), Vol. 1, Chap. 1, p. 1. Pergamon, Oxford, 1982.

9. For an extension to other geometries and/or more extensive discussion, see Ref. *8* and references cited therein.

10. R. K. Bohn and M. D. Bohn, *Inorg. Chem.* **10**, 350 (1971).

11. M. F. Hawthorne, D. C. Young, and P. A. Wegner, *J. Am. Chem. Soc.* **87**, 1818 (1965); M. F. Hawthorne, D. C. Young, T. D. Andrews, D. V. Howe, R. L. Pilling, A. D. Pitts, M. Reintjes, L. F. Warren, Jr., and P. A. Wegner, *J. Am. Chem. Soc.* **90**, 879 (1968).

12a. J. E. Macintyre, exec. ed., "Dictionary of Organometallic Compounds," Suppl. Chapman & Hall, New York, 1986.

12b. J. E. Macintyre, exec. ed., "Dictionary of Organometallic Compounds," 3rd Suppl. Chapman & Hall, New York, 1987.

13. G. Popp and M. F. Hawthorne, *Inorg. Chem.* **10**, 391 (1971).

14. G. Popp and M. F. Hawthorne, *J. Am. Chem. Soc.* **90**, 6553 (1968).

15. J. L. Spencer, M. Green, and F. G. A. Stone, *J. Chem. Soc., Chem. Commun.*, 1178 (1972).

16. H. M. Colquhoun, T. J. Greenhough, M. G. H. Wallbridge, *J. Chem. Soc., Chem. Commun.*, 737 (1977).

17. B. M. Mikhailov and T. V. Potapova, *Izv. Akad. Nauk SSSR, Ser. Khim.* **5**, 1153 (1968).

18. D. A. T. Young, R. J. Wiersema, and M. F. Hawthorne, *J. Am. Chem. Soc.* **93**, 5687 (1971).

19. D. A. T. Young, G. R. Willey, M. F. Hawthorne, M. R. Churchill, and A. H. Reis, Jr., *J. Am. Chem. Soc.* **92**, 6663 (1970).

20. M. R. Churchill and A. H. Reis, Jr., *J. Chem. Soc., Dalton Trans.*, 1317 (1972).

21. P. Jutzi and P. Galow, *J. Organomet. Chem.* **319**, 139 (1987).

22. W. S. Rees, Jr., D. M. Schubert, C. B. Knobler, and M. F. Hawthorne, *J. Am. Chem. Soc.* **108**, 5367 (1986).

23. M. A. Bandman, C. B. Knobler, and M. F. Hawthorne, *Inorg. Chem.* **27**, 2399 (1988).

24. D. M. Schubert, C. B. Knobler, W. S. Rees, Jr., and M. F. Hawthorne, *Organometallics* **6**, 201 (1987).

25. D. M. Schubert, C. B. Knobler, W. S. Rees, Jr., and M. F. Hawthorne, *Organometallics* **6**, 203 (1987).

26. D. M. Schubert, C. B. Knobler, and M. F. Hawthorne, *Organometallics* **6**, 1353 (1987).

27. D. M. Schubert, W. S. Rees, Jr., C. B. Knobler, and M. F. Hawthorne, *Pure Appl. Chem.* **59**, 869 (1987).

28. C. P. Magee, L. G. Sneddon, D. C. Beer, and R. N. Grimes, *J. Organomet. Chem.* **86**, 159 (1975).

29. J. S. Beck and L. G. Sneddon, *J. Am. Chem. Soc.* **110**, 3467 (1988).

30. R. N. Grimes, W. J. Rademaker, M. L. Denniston, R. F. Bryan, and P. T. Greene, *J. Am. Chem. Soc.* **94**, 1865 (1972).

31. E. Canadell, O. Eisentein, and J. Rubio, *Organometallics* **3,** 759 (1984).
32a. R. L. Voorhees and R. W. Rudolph, *J. Am. Chem. Soc.* **91,** 2173 (1969).
32b. R. W. Rudolph, R. L. Voorhees, and R. E. Cochoy, *J. Am. Chem. Soc.* **92,** 3351 (1970).
32c. V. Chowdhry, W. R. Pretzer, D. N. Rai, and R. W. Rudolph, *J. Am. Chem. Soc.* **95,** 4560 (1973).
33. R. W. Rudolph and V. Chowdhry, *Inorg. Chem.* **13,** 248 (1974).
34a. G. S. Wikholm and L. J. Todd, *J. Organomet. Chem.* **71,** 219 (1974).
34b. L. J. Todd, A. R. Burke, H. T. Silverstein, J. L. Little, G. S. Wikholm, *J. Am. Chem. Soc.* **91,** 3376 (1969).
35. D. C. Beer and L. J. Todd, *J. Organomet. Chem.* **50,** 93 (1973).
36. K.-S. Wong and R. N. Grimes, *Inorg. Chem.* **16,** 2053 (1977).
37. W. M. Maxwell, K.-S. Wong, and R. N. Grimes, *Inorg. Chem.* **16,** 3094 (1977).
38. N. S. Hosmane, N. N. Sirmokadam, and R. H. Herber, *Organometallics* **3,** 1665 (1984).
39. N. S. Hosmane, P. de Meester, N. N. Maldar, S. B. Potts, S. S. C. Chu, and R. H. Herber, *Organometallics* **5,** 772 (1986).
40. A. H. Cowley, P. Galow, N. S. Hosmane, P. Jutzi, and N. C. Norman, *J. Chem. Soc., Chem. Commun.,* 1564 (1984).
41. R. D. Barreto, M. Tolle, U. Siriwardane, N. S. Hosmane, J. J. Alexander, and S. G. Shore, unpublished results.
42. M. F. Hawthorne and G. B. Dunks, *Science* **178,** 462 (1972); M. F. Hawthorne, *Pure Appl. Chem.* **29,** 547 (1972); J. N. Francis and M. F. Hawthorne, *Inorg. Chem.* **10,** 863 (1971); G. B. Dunks and M. F. Hawthorne, *in* "Boron Hydride Chemistry" (E. L. Muetterties, ed.). Academic Press, New York, 1975, and references cited therein.
43. J. A. Maguire, G. P. Ford, and N. S. Hosmane, *Inorg. Chem.* **27,** 3354 (1988).
44. U. Siriwardane, N. S. Hosmane, and S. S. C. Chu, *Acta Crystallogr., Sect. C* **C43,** 1067 (1987).
45. N. S. Hosmane, M. S. Islam, U. Siriwardane, J. A. Maguire, and C. F. Campana, *Organometallics* **6,** 2447 (1987).
46. P. Jutzi, P. Galow, S. Abu-Orabi, A. M. Arif, A. H. Cowley, and N. C. Norman, *Organometallics* **6,** 1024 (1987).
47. N. S. Hosmane, J. S. Fagner, H. Zhu, U. Siriwardane, J. A. Maguire, G. Zhang, and B. S. Pinkston, *Organometallics* **8,** 1769 (1989).
48. U. Siriwardane and N. S. Hosmane, *Acta Crystallogr., Sect. C* **C44,** 1572 (1988).
49. R. D. Barreto, T. P. Fehlner, and N. S. Hosmane, *Inorg. Chem.* **27,** 453 (1988).
50. N. S. Hosmane and N. N. Sirmokadam, *Organometallics* **3,** 1119 (1984).
51. N. S. Hosmane, P. de Meester, U. Siriwardane, M. S. Islam, and S. S. C. Chu, *J. Am. Chem. Soc.* **108,** 6050 (1986).
52. N. S. Hosmane, M. S. Islam, B. S. Pinkston, U. Siriwardane, J. J. Banewicz, and J. A. Maguire, *Organometallics* **7,** 2340 (1988).
53. U. Siriwardane, M.S. Islam, J.A. Maguire, and N.S. Hosmane, *Organometallics* **7,** 1893 (1988).
54. N. S. Hosmane, U. Siriwardane, M. S. Islam, J. A. Maguire, and S. S. C. Chu, *Inorg. Chem.* **26,** 3428 (1987).
55. M. S. Islam, U. Siriwardane, N. S. Hosmane, J. A. Maguire, P. de Meester, and S. S. C. Chu, *Organometallics* **6,** 1936 (1987).
56. W. S. Rees, Jr., D. M. Schubert, C. B. Knobler, and M. F. Hawthorne, *J. Am. Chem. Soc.* **108,** 5369 (1986).
57. N. S. Hosmane, P. de Meester, U. Siriwardane, M. S. Islam, and S. S. C. Chu, *J. Chem. Soc., Chem. Commun.,* 1421 (1986).

58. U. Siriwardane, M. S. Islam, T. A. West, N. S. Hosmane, J. A. Maguire, and A. H. Cowley, *J. Am. Chem. Soc.* **109**, 4600 (1987).
59. A. Tabereaux and R. N. Grimes, *Inorg. Chem.* **12**, 792 (1973).
60. M. L. Thompson and R. N. Grimes, *Inorg. Chem.* **11**, 1925 (1972).
61. K. Wade, *Adv. Inorg. Chem. Radiochem.* **18**, 1 (1976); K. Wade, "Electron Deficient Compounds." Nelson, London, 1971.
62. See Ref. *6a*, Sect. 2.5, p. 473.
63. N. S. Hosmane, U. Siriwardane, H. Zhu, G. Zhang, and J. A. Maguire, *Organometallics* **8**, 566 (1989).
64. H. Wadepohl, H. Pritzkow, and W. Siebert, *Organometallics* **2**, 1899 (1983).
65. A. H. Cowley, R. A. Jones, C. A. Stewart, J. L. Atwood, and W. E. Hunter, *J. Chem. Soc., Chem. Commun.*, 921 (1981); M. J. Heeg, C. Janiak, and J. J. Zuckerman, *J. Am. Chem. Soc.* **106**, 4259 (1984); A. H. Cowley, P. Jutzi, F. X. Kohl, J. G. Lasch, N. C. Norman, and E. Schlüter, *Angew. Chem.* **96**, 603 (1984); *Angew. Chem., Int. Ed. Engl.* **23**, 616 (1984); H. Schumann, C. Janiak, E. Hahn, C. Kolax, J. Loebel, M. D. Rausch, J. J. Zuckerman, and M. J. Heeg, *Chem. Ber.* **119**, 2656 (1986).
66. G. Schmid, D. Zaika, and R. Boese, *Angew. Chem.* **97**, 581 (1985); *Angew. Chem., Int. Ed. Engl.* **24**, 602 (1985).
67. R. N. Grimes, *Organomet. React. Synth.* **6**, 192 (1982).
68. J. L. Little, J. T. Moran, and L. J. Todd, *J. Am. Chem. Soc.* **89**, 5495 (1967).
69. L. J. Todd, J. L. Little, and H. T. Silverstein, *Inorg. Chem.* **8**, 1698 (1969).
70. L. J. Todd, A. R. Burke, A. R. Garber, H. T. Silverstein, and B. N. Storhoff, *Inorg. Chem.* **9**, 2175 (1970).
71. L. I. Zakharkin and V. I. Kyskin, *Zh. Obshch. Khim.* **40**, 2241 (1970).
72. L. A. Fedorov, V. I. Kyskin, and L. I. Zakharkin, *Zh. Obshch. Khim.* **42**, 536 (1972).
73. L. I. Zakharkin and V. I. Kyskin, *Izv. Akad. Nauk SSSR, Ser. Khim.*, 2053 (1971).
74. W. F. Wright, J. C. Huffman, and L. J. Todd, *J. Organomet. Chem.* **148**, 7 (1978).
75. E. I. Tolpin and W. N. Lipscomb, *Inorg. Chem.* **12**, 2257 (1973).
76. L. I. Zakharkin and V. I. Kyskin, *Zh. Obshch. Khim.* **40**, 2234 (1970).
77. H. S. Wong and W. N. Lipscomb, *Inorg. Chem.* **14**, 1350 (1975).
78. L. I. Zakharkin and V. I. Kyskin, *Zh. Obshch, Khim.* **41**, 2052 (1971).
79. E. A. Yakovelva, G. G. Isaeva, V. I. Kyskin, L. I. Zakharkin, and A. I. Shatenshtein, *Izv. Akad. Nauk. SSSR, Ser. Khim.*, 2797 (1971).
80. L. J. Todd, I. C. Paul, J. L. Little, P. S. Welcker, and C. R. Peterson, *J. Am. Chem. Soc.* **90**, 4489 (1968).
81. J. L. Little, P. S. Welcker, N. J. Loy, and L. J. Todd, *Inorg. Chem.* **9**, 63 (1970).
82. L. I. Zakharkin and A. I. L'vov, *Zh. Obshch. Khim.* **41**, 1880 (1971).
83. D. C. Beer and L. J. Todd, *J. Organomet. Chem.* **36**, 77 (1972).
84. L. J. Todd, *Pure Appl. Chem.* **30**, 587 (1972).
85. L. J. Todd, A. R. Burke, A. R. Garber, H. T. Silverstein, and B. N. Storhoff, *Inorg. Chem.* **9**, 2175 (1970).
86. J. Plesek, B. Stibr, and S. Hermanek, *Chem. Ind. (London)*, 662 (1974).
87. V. A. Brattsev, S. P. Knyazev, G. N. Danilova, and V. I. Stanko, *Zh. Obshch. Khim.* **45**, 1393 (1975).
88. J. Plesek, S. Hermanek, J. Huffman, P. Ragatz, and R. Schaeffer, *J. Chem. Soc., Chem. Commun.*, 935 (1975).
89. H. D. Smith and M. F. Hawthorne, *Inorg. Chem.* **13**, 2312 (1974).
90. A. R. Siedle and L. J. Todd, *J. Chem. Soc., Chem. Commun.*, 914 (1973).
91. A. M. Barriola, T. P. Hanusa, and L. J. Todd, *Inorg. Chem.* **19**, 2801 (1980).

η² Coordination of Si—H σ Bonds to Transition Metals

ULRICH SCHUBERT

Institut für Anorganische Chemie
Universität Würzburg
Am Hubland
D-8700 Würzburg, Federal Republic of Germany

I

INTRODUCTION

The recent discovery of many metal complexes having η^2-H_2 ligands or "agostic" C—H bonds has focused attention on a type of bonding that had hitherto seemed to be a domain of early main group elements and metal clusters: three-center, two-electron bonds. As more and more examples are found, it has become obvious that such bonds are probably more widespread among transition metal compounds than had been commonly assumed and that it is about time to recognize the features the various species have in common.

To clarify our notions, let us first consider the general case, in which a metal complex moiety (L_nM) and any two fragments X and Y share two electrons. In the remainder of this article, this bonding situation is depicted as in **A** to indicate the general bonding phenomenon. Note that the dashed lines are only a symbol and do not imply any information about the distribution of electrons among L_nM, X, and Y (just as a line does not tell

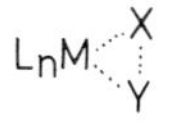

$$L_nM \overset{X}{\underset{Y}{\cdots}} \qquad L_nM \overset{X}{\underset{Y}{\leftarrow}}$$

A 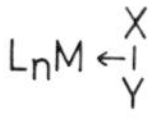   **B**

one anything about electron distribution in a covalent bond). An even distribution of electrons can be expected only if the three groups involved in the three-center interaction are equal, as in H_3^+. This point may be illustrated by the pair of complexes **1** (*1*) and **2** (*2*). The BH_4 complex (**2**) is

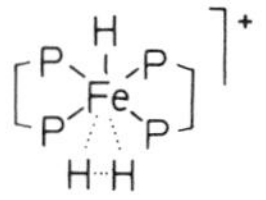

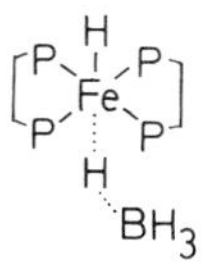

$$P = Ph_2PCH_2CH_2PPh_2$$

$$P' = Me_2PCH_2CH_2PMe_2$$

1   **2**

formally derived from the H_2 complex (**1**) by removal of one hydrogen of the η^2-H_2 ligand as H^+ and replacement by BH_3. Both complexes contain three-center, two-electron bonds. However, in **1** a symmetrical coordination of H_2 (closed three-center bond) is preferred (*3a*), while in **2** the Fe—H—B angle opens to 162°.

A coordination chemist might prefer a formula like **B** and consider the X—Y σ bond η^2 coordinated to the metal fragment. However, someone engaged in the chemistry of either element X or Y might look at it in a different way. Formula **B** stresses the chemical origin of most of these compounds, which are usually obtained by addition of X—Y to a coordinatively unsaturated metal fragment [Eq. (1)].

$$L_nM \;+\; \begin{matrix} X \\ | \\ Y \end{matrix} \;\longrightarrow\; L_nM\!\!\begin{matrix} X \\ \vdots \\ Y \end{matrix} \tag{1}$$

As discussed in more detail below, considering these complexes as "frozen intermediates" in the oxidative addition of X—Y to a fragment L_nM can be useful. Complete oxidative addition would totally cleave the X—Y bond and form both an M—X and a M—Y bond. Why does this not happen in the complexes in question? Why is the oxidative addition process seemingly arrested at some point? This is not completely understood in every case, and theoretical explanations are only slowly emerging (*3a–h*). As shown later in this article, the stage at which addition of X—Y is arrested can be tuned by the substituents at X or Y and by the ligands at the metal.

Formula **B** also indicates some analogy with π complexes, which is indeed supported by a comparison of the crucial orbital interactions in both types of complexes. Figure 1b schematically shows the well-known bonding–backbonding interaction in π complexes. Bonding in η^2-XY complexes is similarly explained by a σ-type interaction between the σ(X—Y) orbital and an empty metal orbital of suitable symmetry, and a π-type interaction (backbonding) between a d-like metal orbital and the σ^*(X—Y) orbital (Fig. 1a). (Note again the coordination chemist's way of dividing the compound into two fragments.) Both types of interaction weaken the X—Y bond. The relative energies of the orbitals involved determine which interaction is dominant and to what degree the X—Y bond is weakened (*3a–h*). In particular, very extensive backbonding splits the X—Y bond and results in formation of the oxidative addition product $L_nM(X)Y$.

Whatever kind of representation one chooses, one has to keep in mind that both **A** and **B** symbolize the same fact, but in a slightly different way. The electrons in a three-center bond are located in molecular orbitals which are composed from contributions of all the involved atoms. Although prorating of electrons to individual two-center interactions is not

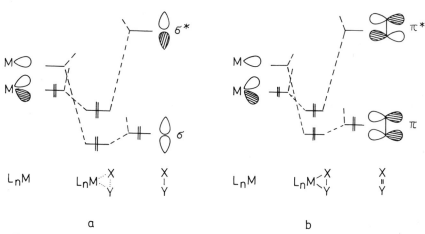

a   b

FIG. 1.   Comparison of the crucial bonding interactions in (a) η^2-X—Y complexes and (b) η^2-X=Y complexes.

very meaningful in molecular orbital terms, the phrase "X—Y interaction" (or M—X interaction, M—Y interaction, respectively) is occasionally used in this article to compare certain properties of the η^2-XY complexes with either uncoordinated X—Y (two-center, two-electron bond) or complexes in which XY is completely oxidatively added to the metal (no XY interaction).

In the vast majority of transition metal complexes having three-center, two-electron bonds, hydrogen is involved (X = H). The spectroscopic opportunities arising from this fact are somewhat offset by the problems in locating hydrogen atoms from X-ray measurements (4) (vide infra). The third center (Y) may be another L_nM fragment ["hydride-bridged" dinuclear complexes (4)], another hydrogen atom [η^2-H$_2$ complexes (5,6a)], or a main group moiety (ER$_n$). Not surprisingly, many complexes are known in which a boron atom is the third center in the M—H—E three-center bond (4,6b,c). Examples of aluminum-containing complexes of this type are less numerous (4).

Although the reasons for participation of main group 4 elements in complexes of type **A** are less obvious than for group 3 elements, examples for all elements of group 4 except Pb are now known (vide infra). However, while complexes containing metal–hydrogen–carbon three-center bonds ("agostic" C—H bonds) have recently received much attention (6a,7a,b), analogies with corresponding silyl complexes have been overlooked or even denied (6a). Ironically, the most systematically investigated type of complex containing a metal–hydrogen–main group element three-center bond is the (η-C$_5$R$'_5$)L$_2$Mn(H)SiR$_3$ system, and much information obtained from this system can be generalized for other complexes containing three-center bonds. The purpose of this article is to compile and review the available information on "η^2-H—SiR$_3$" complexes (and the analogous Ge and Sn compounds) and to make occasional comparisons with complexes having "agostic" C—H bonds. First, the prototypical system CpL$_2$Mn(H)SiR$_3$ (8) is discussed in detail, including complexes which are related to it, and then attention is paid to other types of complexes which have or may have M—H—Si three-center bonds.

II

THE (η-C$_5$R$_5'$)L$_2$Mn(H)SiR$_3$ SYSTEM

A. *Preparation*

The first complexes of type Cp(OC)$_2$Mn(H)SiR$_3$ (R = Ph, Cl) were prepared by Jetz and Graham by photochemical reaction of Cp(OC)$_3$Mn with the corresponding silanes HSiR$_3$ (9) [Eq. (2)]. This route appears to be rather general and allows the synthesis of complexes with a variety of substituents at silicon (10). Instead of using silanes, which are sometimes cumbersome to prepare, reaction of readily available functional hydro silanes with subsequent modification of the coordinated silyl group can sometimes be advantageous (10). Only one Si—H group of a secondary silane adds to a Cp(OC)$_2$Mn fragment, yielding complexes of the type Cp(OC)$_2$Mn(H)SiHR$_2$ (10,11). Even with an excess of Cp(CO)$_3$Mn and more forcing conditions, no SiR$_2$-bridged dinuclear species are formed. The reason for the lack of reactivity of the second Si—H group may be a steric one, because [Cp*(OC)$_2$(H)Mn]$_2$SiH$_2$ (Cp* = η^5-C$_5$Me$_5$) has been obtained (among other products) by the photochemical reaction of Cp*(OC)$_3$Mn with SiH$_4$ (12).

Monosubstituted complexes of the type Cp(OC)LMn(H)SiR$_3$ [L = PR$_3'$, P(OR')$_3$, CNR'] cannot be prepared from the dicarbonyl compounds by CO–ligand exchange, because the conditions necessary for this kind of reaction result in HSiR$_3$ elimination, as discussed in more detail below. However, they are accessible by the photochemical route [Eq. (2)] from Cp(OC)$_2$LMn, if the ligand L is not too bulky (13).

$$(\eta\text{-C}_5\text{R}_5'')(\text{OC})_2\text{LMn} + \text{HSiR}_3 \xrightarrow{h\nu} \text{CO} + (\eta^5\text{-C}_5\text{R}_5'')(\text{OC})\text{LMn(H)SiR}_3$$

$$\text{C}_5\text{R}_5'' = \text{C}_5\text{H}_5,\ \text{C}_5\text{Me}_5,\ \text{C}_5\text{H}_4\text{Me}$$

$$\text{L} = \text{CO, PR}_3',\ \text{P(OR')}_3,\ \text{CNR'} \qquad (2)$$

For the various species MeCp(OC)LMn(H)SiR$_3$ [L = PR$_3'$, P(OR')$_3$] only isomer 3 is obtained: that with the hydrogen atom located between the bulky L and SiR$_3$ ligands. The other possible arrangement of ligands can be enforced by using a phosphinoalkylsilyl ligand. Thus, irradiation of $MeCp(OC)_2Mn$—$PPh_2CH_2CH_2SiR_2H$ (R = Me, Ph) results in intramolecular Si—H addition and formation of **4**. Since the properties of **3** and **4** are not affected by the different geometry, the disposition of ligands in **3** seems to be caused by steric rather than electronic factors (*14*).

3   **4**

The reaction [Eq. (2)] proceeds by photochemically induced dissociative loss of CO and subsequent addition of the silane to the 16-electron species. For the dicarbonyl species $Cp(OC)_2Mn$ and $Cp^*(OC)_2Mn$ and some related molecules, the reaction has been investigated in detail at low temperatures (85–157 K) (*15*). From kinetic measurements an overall second-order rate law [first order in both silane and $Cp(OC)_2Mn$] was found. The activation parameters for the reaction of $Cp(OC)_2Mn$ with a number of trialkylsilanes or $HSiPh_3$ ($\Delta H^{\ddagger}$ 25–30 KJ/mol, $\Delta S^{\ddagger}$ −28 to −40 J/K mol) suggest a loosely bound silane molecule in the activated complex. Similar conclusions had already been reached by a kinetic investigation of the reverse reaction, namely, reductive elimination of $HSiPh_3$ from $Cp(OC)_2Mn(H)SiPh_3$ in the presence of a phosphine [Eq. (3)] (*16*). In this reaction [Eq. (3)] the initial process is the rate-determining dissociation of the silane (in *n*-heptane: $\Delta H^{\ddagger}$ 122 kJ/mol, $\Delta S^{\ddagger}$ 68 J/K mol). However, the absence of a silane exchange between $MeCp(OC)_2Mn(D)SiDPhNp$ (Np = 1-naphtyl) and $H_2SiPhNp$ in benzene at room temperature shows that the 16-electron complex and the silane remain associated and that only strong donors, like PR_3, can displace the silane (*3b,17*). Competition experiments indicate that PPh_3 is 3.6 times as reactive as $HSiPh_3$ toward the intermediate (*16*).

$$Cp(OC)_2Mn(H)SiPh_3 + PPh_3 \ \rightarrow \ Cp(OC)_2MnPPh_3 + HSiPh_3 \qquad (3)$$

B. *Evidence for Three-Center Bonding*

From the very beginning it was realized that some properties of the complexes $Cp(OC)_2Mn(H)SiR_3$ are different from those of other hydrido silyl complexes and that the silane moiety was bonded in an unusual way (*18,19*). We refrain from listing in chronological order the experiments and arguments that eventually led to an understanding of the bonding situation in these complexes. We instead list the chemical, spectroscopic, and structural evidence (in that order) to set the stage for the following discussions.

Complete oxidative addition of $HSiR_3$ to $Cp(OC)_2Mn$ would result in a complex of the type $Cp(OC)_2MnLL'$ having a four-legged piano stool geometry, in which H and SiR_3 could be cis (lateral) (**C**) or trans (diagonal) (**D**). The relative intensities of the two CO bands in the infrared spectra of

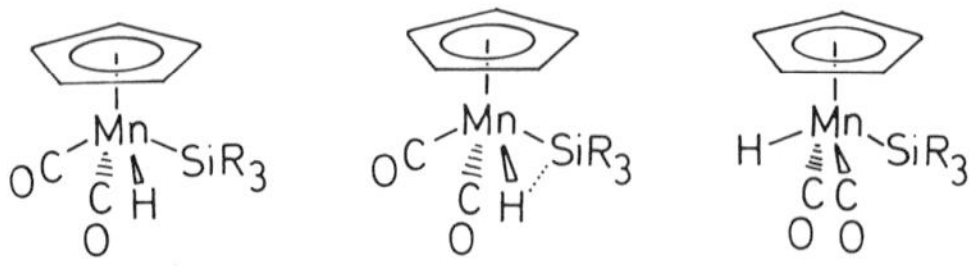

C   **C'**   **D**

$Cp(OC)_2Mn(H)SiR_3$ (any SiR_3 ligand) indicate the exclusive formation of an isomer having a *cis*-dicarbonyl geometry, which is of course a prerequisite for a three-center interaction (**C'**). Depending on the substituents at silicon, the complexes $Cp(OC)_2Mn(H)SiR_3$ are more or less acidic and can be deprotonated to give the corresponding anionic complexes of type **5** [Eq. (4)] (*17,20*).

$$Cp(OC)_2Mn(H)SiR_3 \underset{+H^+}{\overset{-H^+}{\rightleftharpoons}} [Cp(OC)_2MnSiR_3]^- \qquad (4)$$

5

Like other metal carbonylates, the anionic manganese complexes of type **5** react with a variety of organic (*21a,b*), inorganic, or organometallic halides (*20,22a,b*). The corresponding substitution products, as well such complexes as $Cp(OC)_2Re(H)EX_3$ ($EX_3 = GeCl_3$, $GeBr_3$, $SnCl_3$) (*22c*), always have the trans geometry, indicating that this is the sterically more favorable arrangement of ligands. Contrary to this, protonation of $[Cp(OC)_2MnSiR_3]^-$ gives only the cis isomer, never the trans isomer (**D**) (*17*). From these observations it has to be concluded that an electronic factor (possibly formation of the three-center bond, **C'**) overrides the steric preference for **D** in the protonation reaction.

Another important argument is provided by the stereochemistry of substitution reactions at silicon. In the optically active complex **6** nucleophilic substitution by H^-, OH^-, or OMe^- occurs with inversion of configuration at silicon, while retention was observed with the related

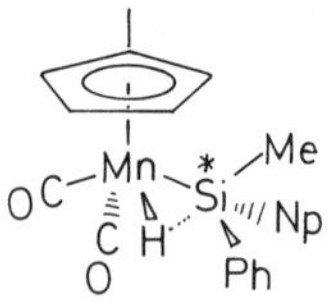

6

complex $Cp(CO)_2FeSiR_3^*$. This change in stereochemistry was explained by the pseudopentacoordination of silicon in **6**, which prevents frontside attack of the nucleophile (*19,23a,b*).

As mentioned above, elimination of $HSiR_3$ from $Cp(OC)_2Mn(H)SiR_3$ occurs readily. For instance, treatment of the dicarbonyl complexes with phosphines does not give phosphine-substituted complexes, because $HSiR_3$ is released from the metal in preference to CO. In contrast, complexes of the type $(OC)_4Fe(H)SiR_3$, having a cis arrangement of H and SiR_3 but no three-center bond, undergo the usual CO displacement with phosphorus ligands (*19,24a*). [Depending on the phosphine other reactions may also occur, but $HSiR_3$ elimination is never observed for tetracarbonyl iron derivatives (*24b*).] The conclusion from this comparison is that in the manganese complexes elimination of $HSiR_3$ is favored because there is already some Si—H interaction in the ground state of the molecules.

The lower acidity of $Cp(OC)_2Mn(H)SiR_3$ complexes, compared with $(OC)_4Fe(H)SiR_3$ having the same SiR_3 ligand, has also been attributed to the different bonding situation (*17,19*). However, the acidity is influenced much greater by the other ligands at the metal. Phosphine substitution lowers the acidity of the iron complexes without affecting the nature of the bonding. In fact, $Cp(OC)_2Mn(H)SiR_3$ and $(OC)_3(PR_3')Fe(H)SiR_3$ have about the same acidities (*22b,25*). Although some influence of the three-center bond on the acidity of the complexes is to be expected, this effect is difficult to quantify and remains obscure. The acidity of the agostic hydride in $(\eta^3\text{-}C_6H_9)Mn(CO)_3$ was shown to be similar to that expected if it were a terminal hydride (*26*).

The infrared absorptions associated with the hydride ligand in $Cp(OC)LMn(H)SiR_3$ are difficult to analyze because the Mn—H stretching mode is usually hidden by the $\nu(CO)$ absorptions and pronouncedly broadened. The latter observation was the first spectroscopic evidence for a Mn—H—Si three-center bond in these complexes. Because the absorptions for hydrogen bridging two metal centers are in some cases considerably broadened, compared with terminal metal–hydrogen stretching modes, Kaesz and colleagues concluded that, in the manganese complexes, "the hydrogen atom must occupy some type of bridging position," although the $\nu(Mn—H)$ band is rather high lying [1890 cm^{-1} in $Cp(OC)_2Mn(H)SiPh_3$] (*27*).

NMR measurements provide the most powerful spectroscopic evidence for three-center bonds in metal complexes of the type $L_nM(H)Y$ (**A**). The chemical shifts of both Y and the hydride ligand are of little diagnostic value with respect to the bonding situation, because their absolute values are in most cases similar to complexes containing terminal hydrides and are to a high degree influenced by the nature of the L_nM moiety and the substituents attached to Y. In only a few cases are pairs of very similar complexes with and without three-center bonds available to allow a meaningful comparison.

The magnitude of the $J(\text{YMH})$ coupling constant, however, is in most cases a reliable indicator for the particular bonding situation. Its upper limit is $^1J(\text{YH})$, if H is directly bonded to Y (two-center, two-electron bond); the lower limit is $^2J(\text{YMH})$ in complexes with no Y—H interaction. $J(\text{YMnH})$ in complexes of type **A** lies between the two extremes, and therefore we do not assign a superscript to J. The magnitude of $J(\text{YMnH})$ depends on the degree to which addition of Y—H to L_nM has occurred (*vide infra*). For instance, while $^1J(\text{HD})$ in HD gas is 43.2 Hz and $^2J(\text{HMD})$ in classic hydride deuteride complexes is less than 2 Hz, $J(\text{HMD})$ values between 22 and 34 Hz are found in η^2-HD complexes (*5,6a*). Similarly, J(CMH) in complexes with nonfluxional M—H—C three-center interactions is smaller than $^1J(\text{CH})$ by about 125 Hz (sp^3-hybridized carbon) but is greater than the coupling constants found in classic hydrido alkyl complexes (<10 Hz) (*6a,7a,b*).

When Y is SiR_3 the upper limit is given by the $^1J(\text{SiH})$ value for coupling in silanes of about 200 Hz. For the lower limit only a few examples can be found in the literature: $^2J(\text{SiMH})$ in $Cp(OC)Fe(H)(SiCl_3)_2$ is 20 Hz (*28*) and in $Cp^*Rh(H)_2(SiEt_3)_2$ is 7.9 Hz (*29*). An even lower value (3.5 Hz) is observed for $Cp_2W(H)Si(SiMe_3)_3$ (*30*). In this three complexes the H and SiR_3 groups are cis, and, at least in the latter two examples, a Si—H interaction can be excluded. Although the magnitude of $^2J(\text{SiMH})$ obviously depends on the nature of the L_nM moiety and on the substituents at silicon, it is safe to assume that no strong Si—H interaction is to be expected, if $J(\text{SiMH})$ drops below 10–20 Hz. Coupling constants of 65 and 69 Hz found for $MeCp(OC)_2Mn(H)SiR_3$ ($SiR_3 = SiPh_3$, SiHPhNp) are distinctly higher and are consistent with the assumption of a three-center, two-electron interaction in these complexes (*17*). The correlation between the magnitude of $J(\text{SiMnH})$ and changes in the bonding situation induced by variations of ligands at the metal and substituents at silicon (*11*) is discussed below.

The first structural evidence for an Si—H interaction in complexes of the type $Cp(OC)_2Mn(H)SiR_3$ was obtained for the $SiPh_3$ derivative (*18,31*). Subsequent structure determinations of other derivatives showed that there are a number of structural features which are typical for the three-center bonding situation (*11*). In this section, these features are discussed in detail for $MeCp(OC)_2Mn(H)SiFPh_2$ (Fig. 2) because the hydrogen atoms in this compound were located by neutron diffraction analysis (*11,32*). The question of how certain structural features correlate with the degree to which $HSiR_3$ is added to the $Cp(OC)LMn$ fragment is deferred to a later section.

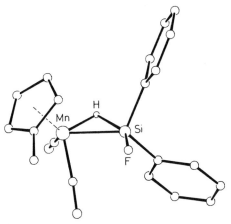

FIG. 2. Structure of MeCp(OC)$_2$Mn(H)SiFPh$_2$ as determined by neutron diffraction. All hydrogen atoms except the hydride ligand are omitted for clarity (*11,32*).

The most prominent structural feature of Fig. 2 is the rather short Si—H distance [180.2(5) pm], which is only about 30 pm (20%) longer than covalent Si—H bond lengths in tetrahedral silanes (148 ± 2 pm). Smith and Bennett have argued that the close approach of the hydrogen to the silicon atom in Cp(OC)$_2$Mn(H)SiR$_3$ derivatives could be only steric in origin and might be caused by repulsion between the hydride ligand and the *cis*-CO ligand (*33*). To evaluate this argument, an estimation of the shortest possible nonbonding Si—H distance has to be made (being aware that there is no sharp border between bonding and nonbonding).

Structural and theoretical studies on 1,3-disiladioxetans have shown that there is no bonding interaction between the two silicon atoms, although they approach each other as close as 230–240 pm (*34*). An even shorter contact appears unlikely; therefore, this distance may be taken as the shortest nonbonding Si—Si distance. In dihydride complexes a H—H separation of 185 pm was considered the limiting distance for H—H interaction (*3e*). Half the sum of both values can therefore be regarded the shortest possible nonbonding Si—H contact (about 210 pm). Because this limit depends somewhat on the substituents at silicon, it is reasonable to assume no significant Si—H interaction at distances greater than 200 pm.

This limit can be probed for some hydrido silyl complexes which are unlikely to display three-center bonding. In the octahedral complex **7** the Si—H distance of 274 pm is far beyond the limit, although the geometry of **7** is slightly distorted to adjust for the smaller size of the hydride ligand (*35*). Even if the equatorial plane becomes more crowded, the Si—H distances in **8** (in which SiMe$_3$, I, both H, and one PMe$_3$ are coplanar) stay above the limit (222 and 231 pm) (*36*). In the very interesting complexes of type **9**, the hydride ligands are symmetrically located between the silyl groups, and the Si—H distances (determined by neutron diffraction studies) are between 221.2 and 238.4 pm (*29,37*). As a marginal note it is worth drawing attention to the fact that in **9** both silane units are completely oxidatively added, although the resulting high formal oxidation state of the metal could have been avoided by formation of three-center bonds.

7 (*35*)   8 (*36*)   9a (M = Rh) (*29*)
9b (M = Ir) (*37*)

Returning to Cp(OC)$_2$Mn(H)SiR$_3$ complexes, it appears unlikely that the Si—H distance is reduced to 180 pm only by steric repulsion between H and CO while at the same time the OC—Mn—CO angle remains close to 90° (*vide infra*).

Evaluation of the Mn—H distance in Fig. 2 [156.9(4) pm] is more difficult. The only other precise (i.e., by neutron diffraction) determined Mn—H distance is that of (OC)$_5$MnH [160.1(16) pm] (*38*). It is hard to estimate, however, how changes in the ligand sphere of the metal, particularly exchange of three CO ligands for a Cp ligand, affect the Mn—H distance. Transition metal–hydrogen distances generally seem to be relatively insensitive structural parameters. Even if a hydride bridges two metals, the Mn—H distance is only about 10–15 pm longer than in complexes with terminal hydride ligands (*4*).

Similar difficulties are encountered if the Mn—Si distance of Fig. 2 [235.2(4) pm] is to be assessed. The only structurally characterized complexes containing Me—Si bonds are derivatives of (OC)$_5$MnSiR$_3$ [Mn—Si 240–256 pm (*39a–e*)], but, again, the influence of the different ligand environments on this bond length is unclear. For this reason is Fig. 2 better compared with another complex in which (1) the steric and electronic properties of the metal fragment are as similar as possible to the MeCp(OC)$_2$Mn fragment, and (2) the same silyl group is bonded as in Fig. 2, but (3) the silyl ligand is bonded to the metal by a "normal" two-center bond. Under these premises the obvious choice is Cp-(OC)$_2$FeSiFPh$_2$. By X-ray structure analysis, an Fe—Si distance of 227.8(1) pm was found (*40*), which is 7.4 pm longer than the Mn—Si distance in Fig. 2. Bearing in mind the similarity of both metal fragments, this difference is much too large to be explained by only an increase in the bonding radius of the metal. Instead, the obvious lengthening of the Mn—Si bond in Fig. 2 must be attributed to the Mn—H—Si three-center bond.

Apart from the short Si—H distances and the lengthening of the Mn—Si bonds, analysis of the geometry of the Cp(OC)$_2$Mn fragment and the SiR$_3$ group gives additional support for a delocalized Mn—H—Si bond. Without an Si—H interaction, Fig. 2 would be of the *cis*-Cp(CO)$_2$MLL′ type (**E**) ('four-legged piano stool'). In complexes of this type, C(CO)—M—C(CO) bond angles of about 74–78° are found, which are rather independent of M, L, and L′. In Fig. 2 the C(CO)—Mn—C(CO) angle is 89.7(1)°.

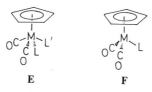

This value is typical for complexes of the type Cp(OC)$_2$ML (**F**) ('three-legged piano stool'). A vector originating from Mn and perpendicular to both CO ligands points in between H and Si. Therefore, Fig. 2 (and all derivatives thereof) is of the type **F**, in which the Si—H bond acts as a single ligand.

In all complexes of type Cp(OC)$_2$Mn(H)SiR$_3$ (and also PR$_3$-substituted derivatives) the orientation of the silane relative to the Cp(OC)$_2$Mn fragment is the same. If the Mn—H—Si triangle is taken as a reference plane, both carbonyl ligands are below this plane at about the same distance, one substituent at silicon (R^1) is within this plane, and the other two substituents at silicon (R^2) are above and below. The substituent R^1 is the most electronegative one (in Fig. 2, the fluorine atom); how much this orientation of the silane is prone to steric effects has not yet been probed.

The geometrical distortions induced on the SiR$_3$ group by the η^2 coordination of the silane can be described in different ways: Starting with a tetrahedral silane HSiR^{1}R$^2_2$, the angle H—Si—R^1 opens on coordination to the metal [in Fig. 2, H—Si—F 148.8(2)°], while H—Si—R^2 and R^1—Si—R^2 become smaller [in Fig. 2, H—Si—C(Ph) 91.9(2) and 98.0(2)°, F—Si—C(Ph) 102.0(1) and 103.4(2)°]. The silyl ligand SiR^{1}R$^2_2$ appears to be tilted relative to the Mn—Si axis in such a way that a hypothetical sp^3 orbital at silicon does not point at the metal atom (as would be the case in a two-center bond), but in between Mn and H. This tilting of the SiR$_3$ ligand can be observed if the three substituents at silicon are identical. For instance, in MeCp(OC)$_2$Mn(H)SiCl$_3$ the Mn—Si—Cl angles are 117.46(8), 117.22(8), and 113.95(8)°, the latter angle corresponding to the chlorine atom located in the Mn—H—Si plane (R^1) (*41*). In Cp(OC)$_2$Mn(H)SiPh$_3$ (*18,31*) the tilting is about twice as large.

On the basis of the structural data, one would expect the two CO ligands in the dicarbonyl complexes to be inequivalent. However, in ^{13}C-NMR spectra of Cp(OC)$_2$Mn(H)SiR$_3$ complexes there is only one carbonyl resonance, even at low temperatures. In the case of $MeCp(OC)_2Mn(H)$-SiMePhNp, the expected diastereomers (owing to the chirality of both Mn and Si) were not observed (17), and the prochiral methyl groups in $MeCp(OC)(PMe_2Ph)Mn(H)SiHPh_2$ appear equivalent in the ^1H-NMR spectrum (11).

Fluxional behavior in solution is typical for hydrido silyl complexes, even in the absence of a three-center bond (as in the octahedral iron complex 7 and derivatives thereof) (35,42). In $SiHR_2$ derivatives there is no exchange on the NMR time scale between the hydride ligand involved in the three-center bond and the hydrogen atom terminally bonded to silicon; in other words, two separate and sharp signals are observed for these atoms in the ^1H-NMR spectra (13,17). On the other hand, proton exchange was observed in the following experiment: reaction of both $[MeCp(OC)_2Mn$—SiHPhNp]$^-$ with DCl and $[MeCp(OC)_2Mn$—SiDPhNp]$^-$ with HCl gave an 1:1 mixture of $MeCp(OC)_2Mn(H)SiDPhNp$ and $MeCp(OC)_2$-Mn(D)SiPhNp (17). The mechanism of the dynamic process therefore remains unclear, and further experiments are needed. By the aforementioned failure to exchange the silyl groups between $MeCp(OC)_2Mn(D)$-SiDPhNp and $H_2SiPhNp$ (17), only mechanisms involving complete dissociation of the silane can be excluded. A possible mechanism could involve species in which a momentarily dissociated silane remains within the solvent cage and recombines with the same metal atom (17), species containing an "open" Mn—H—Si three-center bond without Mn—Si interaction (3b), or intramolecular hydrogen migration (43).

C. Theoretical Interpretation

As pointed out in Section I, there are several ways to depict the bonding situation in Fig. 2 in valence bond terms, all of them having drawbacks for the sake of simplification. A more precise description can only be given by molecular orbital interaction diagrams.

Saillard et al. carried out molecular orbital calculations of the extended Hückel type for the model compound $Cp(OC)_2Mn(H)SiH_3$ (3b), the geometry of which was based on the experimentally determined structure of $MeCp(OC)_2Mn(H)SiCl_3$ (41). The frontier molecular orbital interaction diagram is given in Fig. 3. It shows the frontier orbitals of the well-known 16-electron $Cp(OC)_2Mn$ fragment on the left-hand side, with a hybrid σ-type vacant orbital lying above a group of three occupied, d-type (t_{2g}) orbitals. The HOMO and the LUMO of the distorted H $\cdots$ SiH$_3$ fragment are the Si—H bonding σ and antibonding σ^* orbitals.

Between these two fragments two major interactions are observed (3b): an interaction between σ(SiH) and the vacant hybrid orbital of

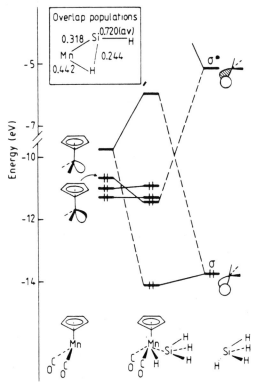

FIG. 3. Frontier molecular orbital interaction diagram for $Cp(OC)_2Mn(H)SiH_3$. [Reproduced from Rabaâ *et al.* (*3b*) with permission of Elsevier Science Publishers, Amsterdam.]

$Cp(OC)_2Mn$, corresponding to delocalization of the $\sigma(SiH)$ bonding pair to the Mn atom and leading to a loss of Si—H bonding character, and an interaction between $\sigma^*(SiH)$ and the highest t_{2g}-type occupied orbital of the $Cp(OC)_2Mn$ fragment, corresponding to delocalization of a metal π lone pair into the $\sigma^*(SiH)$ orbital ("backbonding" in terms of coordination chemistry) and leading to a gain of Si—H antibonding character. According to extended Hückel calculations, the Si—H overlap population is reduced to 0.24 by both interactions, compared with 0.74 for uncoordinated SiH_4 (*3b*). This means that on coordination to the metal the Si—H bond is weakened but not fully broken.

Photoelectron (PE) spectra provide direct and quantitative information on the electron distribution within a compound. Recent investigation of the PE spectra of a number of $MeCp(OC)_2Mn(H)SiR_3$ derivatives confirmed the theoretical results (the PE spectrum of the $SiCl_3$ derivative is discussed below in another context) (*44*): if SiR_3 is $SiXPh_2$ (X = H, F), the metal ionization band corresponding to the t_{2g} orbitals is not split much and only slightly stabilized compared with $MeCp(OC)_3Mn$. From this result and the fact that the ionization potential of the phenyl π system is not much lower than in uncoordinated $HSiXPh_2$, the conclusion was drawn that by coordination of the Si—H bond the metal is not oxidized, i.e., that the Si—H bond is not "oxidatively" added (44).

D. Tuning the Three-Center Interaction

From the molecular orbital diagram it is evident that the three-center interaction can be tuned by changing the energies of both fragments relative to each other and by varying the HOMO–LUMO gap of the silane. Strengthening of any of the two orbital interactions discussed leads to a weaker Si—H interaction and to a higher degree of oxidation of the metal. In practice, variation of the energy of the orbitals in question is achieved by changing substituents at silicon and ligands at the metal (the complementary possibility of replacing Mn or Si by neighboring elements is discussed below).

Apart from PE spectra three physical parameters are particularly sensitive measures of changes in the Mn—H—Si interaction: the activation enthalpy for the reductive elimination of the silane ($\Delta H^{\ddagger}$), the Si—Mn—H coupling constant [J(SiMnH)] in the NMR spectrum, and the Mn—Si bond length [d(Mn—Si)] (11,45). We first discuss how substituents at *silicon* influence the degree to which the silane is added to the metal and then look at the influence of the electronic and steric properties of the *metal* moiety. The general idea is to change only one group at a time while keeping the remainder of the molecule constant and to study the *relative* changes in $\Delta H^{\ddagger}$, J(SiMnH), and d(Mn—Si). From a series of such observations, information on the progress of the oxidative addition reaction can be obtained.

If $HSiR_3$ is eliminated from $Cp(OC)_2Mn(H)SiR_3$, dissociation of the silane is the rate-determining process (16), and therefore thermal stability of these complexes is governed by the magnitude of the activation parameters for this step. Electronegative substituents at silicon increase the stability considerably: while H_2SiEt_2 elimination from $MeCp(OC)_2Mn(H)$-$SiHEt_2$ is so favorable that the complex decomposes in solution at room temperature within a few minutes, $MeCp (OC)_2Mn(H)SiHPh_2$ or $MeCp$-$(OC)_2Mn(H)SiPh_3$ are stable under these conditions. At 100°C $HSiCl_3$ is eliminated 10^5 times more slowly from $Cp(OC)_2Mn(H)SiCl_3$ than $HSiPh_3$ is from $Cp(OC)_2Mn(H)SiPh_3$ (16). Although activation parameters were determined only for $MeCp(OC)_2Mn(H)SiPh_3$ (16) and $MeCp(OC)_2Mn$-$(H)SiHPh_2$ (45) in this series, the magnitude of $\Delta G^{\ddagger}$ (and $\Delta H^{\ddagger}$) clearly decreases in the order $SiCl_3 \gg SiPh_3 > SiHEt_2$. It is tempting to assume that elimination of the silane takes place more easily the stronger the Si—H interaction. The kinetic parameters can indeed be qualitatively correlated with the bonding situation in the ground state, as the following discussion of $d(Mn–Si)$ and $J(SiMnH)$ shows.

The upper part of Table I (11,18,23b,31,32,41,46) lists a series of six complexes differing only by the kind of SiR_3 ligand (8). All have the same overall geometry and show the aforementioned structual features typical for a three-center bond. Thermal stability of the complexes decreases from $SiR_3 = SiCl_3$ to $SiR_3 = SiPh_3$ and SiMePhNp; that is, in the latter two complexes the silane is more readily eliminated than in the complexes listed above them. As stability decreases, the Mn—Si distance increases by 17 pm! Direct correlation of this considerable difference in bond lengths with the bonding situation is tempting but premature at this point.

If one considers the structures in Table I to be snapshots showing the silanes in different stages of the oxidative addition reaction, a reasonable interpretation of the data is that the Mn—Si distance becomes shorter as the Si—H bond of the entering silane becomes weaker as the addition reaction progresses. However, we are not watching the same silane in all structures, and we have to make a correction to be on common ground. It is known from organosilanes that the "covalent radius" of silicon is strongly influenced by its substituents. In the series $R_{4-n}SiCl_n$ (R = organic moiety) the Si—C bond increases about 3 pm for each replacement of a Cl

TABLE I

SELECTED DISTANCES (pm) OF η-$C_5R_5'(OC)LMn(H)SiR_3$

η-C_5R_5'	L	SiR_3	Mn—Si	Mn—H	Si—H	Reference(s)
MeCp	CO	$SiCl_3$	225.4(1)	147(3)	179(4)	41
Cp	CO	$SiCl_2Ph$	231.0(2)	149(6)	179(6)	46
MeCp	CO	$SiFPh_2^a$ (7)	235.2(4)	156.9(4)	180.2(5)	11,32
MeCp	CO	$SiHPh_2$	236.4(2)	—	—	11
Cp	CO	$SiPh_3$	242.4(2)	155(4)	176(4)	18,31
MeCp	CO	$SiMePhNp^b$	246.1(7)	—	—	23b
C_5Me_5	CO	$SiHPh_2$	239.5(1)	152(3)	177(3)	11
C_5Me_5	CO	μ-SiH_2	243.4(3)	—	—	12
MeCp	PMe_3	$SiHPh_2$	232.7(1)	149(4)	178(4)	11
MeCp	$Ph_2PCH_2CH_2SiMe_2$ (4)	245.7(2)	153(4)	175(4)	14

^{a} Neutron diffraction.
^{b} Np, 1-Naphthyl.

atom by an R group (47). The same trend is found in silyl metal complexes in which the SiR_3 ligand is bonded to the metal by a normal two-center bond: the Fe—Si distance in $Cp(OC)_2FeSiCl_3$ [221.6(1) pm] is 6.2 pm shorter than in $Cp(OC)_2FeSiFPh_2$ [227.8(1) pm] (40). Based on these numbers one would expect a shortening of the Mn—Si bond length in $Cp(OC)_2Mn(H)SiR_3$ of about 9 pm, if $SiPh_3$ is replaced by $SiCl_3$, only because of the different covalent radii of silicon in $SiPh_3$ and $SiCl_3$ groups. The actual shortening is about twice as high, and the additional decrease in the Mn—Si distance must be attributed to a steady change in the bonding situation within this series.

Being suitably cautious because of the high standard deviations of the Mn—H and Si—H distances, a few trends can be recognized if the halogen atoms of $Cp(OC)_2Mn(H)SiCl_3$ are successively replaced by organic groups (Table I): (1) the Mn—Si distance (after correction of the Si radius) becomes distinctly longer, (2) the Mn—H distance becomes a little longer, and (3) the Si—H distance remains about the same. If the increasing bond radius of silicon is taken into account, the Si—H bond length actually decreases on a relative scale. According to the structure correlation method (48), a trajectory for the reaction of silanes with the $Cp(OC)_2Mn$ fragment can be proposed from these observations [Eq. (5)] (11,41). The

$$L_nM + \overset{H}{\underset{SiR_3}{}} \rightleftharpoons L_nM \overset{H}{\underset{SiR_3}{}} \rightleftharpoons L_nM \overset{H}{\underset{SiR_3}{}} \rightleftharpoons L_nM \overset{H}{\underset{SiR_3}{}} \quad (5)$$

$$\qquad\qquad\qquad\qquad\qquad G \qquad\qquad\qquad H$$

silane probably approaches the metal with the hydrogen ahead (G). As the addition of the silane proceeds (G → H), the Si—H bond pivots, strongly increasing the Mn—Si interaction, only slightly strengthening the Mn—H interaction, and further weakening the Si—H bond. In the manganese complexes oxidative addition is arrested at a stage corresponding to H.

It is gratifying that later molecular orbital calculations (3a) and structure correlations (49) for the addition of C—H σ bonds to transition metals gave similar results. The results obtained by structure correlation appear more straightforward for the Si—H addition, because they are derived from a single type of complex and the correction of the silicon radius is more precise than the very crude correction of the metal radius, which was applied in the C—H case.

As discussed for the Mn—Si bond lengths, two factors contribute to differences in $J(SiMnH)$ in a pair of $Cp(OC)_2Mn(H)SiR_3$ complexes having different SiR_3 groups, say, $SiCl_3$ and $SiHPh_2$. According to structural and kinetic data, the Si—H interaction within the Mn—H—Si three-center bond is larger in the $SiHPh_2$ derivative. Therefore, the coupling constant should be larger than in the $SiCl_3$ derivative (vide supra). On the other hand, exchange of the electronegative chlorine atoms by phenyl groups decreases the s contribution in the silicon orbital directed toward the metal. This effect should decrease the absolute value of the coupling constant (50). Both factors are superimposed and have a countercurrent effect on J(SiMnH). Therefore, only small changes in J(SiMnH) can be expected if only the SiR_3 group is changed. Nevertheless, J(SiMnH) in $MeCp(OC)_2Mn(H)SiCl_3$ is significantly smaller than in $MeCp(OC)_2$-$Mn(H)SiHPh_2$ (Table II). A more detailed investigation of the influence of substituents at silicon on the magnitude and the sign of the SiMH coupling constants in hydrido silyl complexes *without* a M—H—Si three-center bond is necessary for a more quantitative evaluation of the available NMR data.

In summary, comparison of the physical data of complexes of type $Cp(OC)_2Mn(H)SiR_3$ clearly shows that addition of $HSiR_3$ to the Cp-$(OC)_2Mn$ fragment is "frozen" at a later stage if the R groups at silicon are electronegative. Vice versa, electron-donating R substituents favor a stronger Si—H interaction in these complexes. To what extent the different size of the SiR_3 group also affects the three-center interaction remains to be determined.

There is no doubt in the series of complexes of type $Cp(OC)_2Mn(H)SiR_3$ having different SiR_3 groups that addition of $HSiCl_3$ is most advanced. An interesting question is whether the Si—H bond in $MeCp(OC)_2Mn(H)$-$SiCl_3$ is fully broken or whether there is still a small bonding interaction. Although the structure of $MeCp(OC)_2Mn(H)SiCl_3$ shows all the previously discussed geometrical features typical for three-center bonding, the Mn—Si

TABLE II

SELECTED ^{29}Si-NMR DATA OF η-$C_5R_5'(OC)LMn(H)SiR_3{}^a$

C_5R_5'	L	SiR_3	J(SiMnH) (Hz)	^1J(SiH) (Hz)
C_5Me_5	CO	$SiHPh_2$	65.4	200.3
MeCp	CO	$SiHPh_2$	63.5	205.2
MeCp	$CNBu^n$	$SiHPh_2$	57.5	194.1
MeCp	$P(OPh)_3$	$SiHPh_2$	45	199
MeCp	$P(p\text{-}ClC_6H_4)_3$	$SiHPh_2$	46.5	196
MeCp	PPh_3	$SiHPh_2$	43	191
MeCp	$P(p\text{-}MeC_6H_4)_3$	$SiHPh_2$	44.5	191
MeCp	PMe_3	$SiHPh_2$	38	188
MeCp	CO	$SiCl_3$	54.8	—
MeCp	PMe_3	$SiCl_3$	20	—

^{a} From Ref. *11*. Spectra were obtained at 30°C.

distance [225.4(1) pm] is only about 4 pm longer than the Fe—Si distance in $Cp(OC)_2FeSiCl_3$ [221.6(1) pm] (40) (remember that the difference between both distances is 7.4 pm for the $SiFPh_2$ derivatives). This is about the difference in bonding radii of Mn and Fe one would expect if both metals are engaged in the same kind of bonding and have about the same ligand environment. While the magnitude of $J(SiMnH)$ in $MeCp(OC)_2Mn(H)SiCl_3$ (Table II) suggests at least some Si—H interaction (11), PE spectra and Fenske–Hall molecular orbital calculations signify that this complex has proceeded very far toward formation of a Mn(III) center and individual Mn—Si and Mn—H bonds (44) [contrary to $MeCp(OC)_2Mn(H)SiXPh_2$, as mentioned above]. The metal ionization band corresponding to the t_{2g} orbitals is now extensively split, because one of the orbitals is stabilized by a strong interaction with $\sigma^*(SiH)$. The He(I)/He(II) intensity comparison and the shift of the entire metal band to higher ionization energy relative to $Cp(OC)_3Mn$ are indicative of the metal approaching the +3 oxidation state (d^4). The chlorine lone pair ionization bands are shifted 1 eV to lower energy compared with uncoordinated $HSiCl_3$, as would be expected for an oxidative addition (44). The close proximity of the silicon and hydrogen atoms in this particular complex is explained by directed hybrid orbitals involving equally weighed $d(yz)$ and $d(z^2)$ orbitals (HOMO and LUMO) of the metal complex (44).

Let us now turn to the question how, and to what extent, the three-center interaction can be tuned by variation of the ligands at the metal. By the same rationale as before the silyl group is now kept constant, while the steric and electronic properties of the metal fragment are varied by substituting one CO ligand. For NMR spectroscopic reasons (discussed below), the $SiHPh_2$ group was chosen. Already on a qualitative level it is obvious that a PR_3 ligand increases the thermal stability of the complexes, i.e., renders the elimination of the silane more difficult. While $MeCp(CO)_2$-$Mn(H)SiHEt_2$ readily eliminates H_2SiEt_2 at 25°C, phosphine-substituted derivatives of the type $MeCp(OC)_2Mn(H)SiHEt_2$ are stable at this temperature (45).

Kinetic investigation of a series of complexes of type $MeCp$-$(OC)(PR_3)Mn(H)SiHPh_2$ with different PR_3 ligands shows that H_2SiPh_2 is eliminated according to the same first-order rate law as for the corresponding dicarbonyl derivatives (vide supra) (45). There is both a steric and an electronic influence of the PR_3 ligand on $\Delta H^{\ddagger}$: activation enthalpy increases with increasing electron density at the Mn atom but decreases with increasing bulkiness of the PR_3 ligand. The highest $\Delta H^{\ddagger}$ value (139 kJ/mol) is observed for $MeCp(OC)(PMe_3)Mn(H)SiHPh_2$ (compared with 106 kJ/mol for the dicarbonyl derivative), owing to the high basicity but low steric demand of the PMe_3 ligand. Because the $\Delta H^{\ddagger}$ values show the same trend as the NMR coupling constants, the relative changes in the magnitude of $\Delta H^{\ddagger}$ can roughly be correlated with the degree of Si—H interaction in the ground state of these complexes (45).

In the series of complexes of type MeCp(OC)LMn(H)SiR$_3$ having different ligands L, the SiHPh$_2$ group was used in order to have an internal reference on how ^1J(SiH) changes on variation of L (11). The problem is analogous to that discussed before: if the electronic and steric properties of the MeCp(OC)LMn fragment are varied, the coupling constants would also be affected to some degree even in the absence of a three-center interaction. Increasing or decreasing the Si—H interaction *additionally* influences the coupling constants. Changes in ^1J(SiH) (Table II) should mainly reflect changes in the electronegativity of the MeCp(OC)LMn fragment on variation of L, while J(SiMnH) undoubtedly is subject to both influences. Table II clearly shows that within a series of complexes MeCp(OC)LMn(H)SiHPh$_2$ both ^1J(SiH) and J(SiMnH) decrease as the basicity of L increases. However, while ^1J(SiH) changes by only 8% from L = CO to L = PMe$_3$, J(SiMnH) decreases to a much greater extent (11). Although this is only a qualitative argument, the strong decrease of J(SiMnH) is probably due to a weakening of the Si—H interaction. It may be noted in passing that ^2J(PMH) coupling constants in phosphine-substituted metal hydrides are rather insensitive to changes of the ligands at the metal for a given orientation of H and P.

Both kinetic data and coupling constants show that oxidative addition of a given silane to Cp(OC)LMn can be promoted by small and basic ligands at the manganese atom. This corresponds to the general experience that oxidative addition reactions are favored by small metal moieties having high electron density at the metal.

We can question again whether in MeCp(OC)(PMe$_3$)Mn(H)SiHPh$_2$ the Si—H bond is fully broken or not. The latter seems to be the case, because the J(SiMnH) value becomes even smaller when SiHPh$_2$ is replaced by SiCl$_3$ [20 Hz in MeCp(OC)(PMe$_3$)Mn(H)SiCl$_3$] (11). On the other hand, this is also another indication that there is still some Si—H interaction in MeCp(OC)$_2$Mn(H)SiCl$_3$. Unfortunately, the structure of MeCp(OC)-(PMe$_3$)Mn(H)SiCl$_3$ is not yet available to test the degree of Si—H interaction by structural methods.

In the series of PR$_3$-substituted derivatives only the crystal structures of MeCp(OC)(PMe$_3$)Mn(H)SiHPh$_2$ and the cyclic compound **4** (R = Me) are known. Comparison of MeCp(OC)(PMe$_3$)Mn(H)SiHPh$_2$ with the corresponding dicarbonyl derivative shows a striking similarity of the geometry of both complexes even in details. The main difference is a 3.7 pm shorter Mn—Si distance in the PMe$_3$ derivative (11). Because no suitable complexes are available for a meaningful comparison, it is difficult to estimate the influence of the PMe_3 ligand on the bonding radius of the manganese atom as well as the degree to which the Mn—Si bond is shortened due to a more advanced addition of the silane to the metal center.

Interestingly, the Mn—Si distance is distinctly lengthened [245.7(2) pm] in the cyclic compound **4** (R = Me) (14). This lengthening must be attributed to the presence of three alkyl substituents at silicon. As mentioned before, the complexes $MeCp(CO)_2Mn(H)Si(alkyl)_3$ are not stable, because elimination of $HSiR_3$ is favored by the electron-donating alkyl groups. The phosphine ligand in **4** has an opposite effect and stabilizes the complex toward reductive elimination of the silane. The chelate effect probably has an additional stabilizing effect. The resulting Mn—Si distance reflects the influence of both the PR_3 and the $Si(alkyl)_3$ ligands.

Another way of modifying the properties of the metal fragment is to substitute the Cp ligand. The differences between $MeCp(OC)_2$-$Mn(H)SiHPh_2$ and $Cp^*(OC)_2Mn(H)SiHPh_2$ turned out to be rather small: substitution of MeCp by Cp^* results in a small but significant decrease of $\Delta H^{\ddagger}$ for the reductive elimination of H_2SiPh_2, in the lengthening of the Mn—Si distance by about 3 pm, and in a 1.9 Hz larger $J(SiMnH)$ value (11). These results are consistent with the assumption that in the Cp^* derivative the Si—H interaction is slightly larger than in the MeCp derivative; that is the addition of H_2SiPh_2 is somewhat obstructed by the Cp^* ligand, probably for steric reasons.

The data given in this section unequivocally show that the three-center, two-electron Mn—H—Si interaction can be finely tuned by both the substituents at silicon and the ligands at the metal. High electron density at the metal, small ligands, and electronegative substituents at silicon weaken the Si—H interaction and favor oxidative addition of the silane. By suitable choice of substituents and ligands one can aim to obtain a series of complexes which correspond to different points on the reaction coordinates of oxidative addition, from a position where the hydrogen is still strongly bonded to silicon to a fully oxidatively added silane. This possibility makes the $CpL_2Mn(H)SiR_3$ system unique with respect to studying this type of reaction in detail by physical methods.

E. *Related Complexes*

Another way of modifying the bonding properties of the metal complex fragment is to change the metal itself while retaining the general type of complex. The first examples which were investigated in this respect are the analogous rhenium complexes $Cp(OC)_2Re(H)SiR_3$, prepared by Graham and colleagues (22c). Both chemical behavior and the crystal structure of

Cp(OC)$_2$Re(H)SiPh$_3$ (**10**) (*51*) indicate that, contrary to the corresponding manganese complexes, there is no metal–hydrogen–silicon three-center bond. [However, molecular orbital calculations indicate that there still might be some Si—H interaction (*3b*).] Complex **10** is obtained by UV irradiation of Cp(OC)$_3$Re and HSiPh$_3$ as the cis isomer [Eq. (6)]. Protonation of [Cp(OC)$_2$ReSiPh$_3$]$^-$, however, gives the trans isomer, which is

stable in the solid state but in solution slowly converts to the cis isomer [Eq. (6)] (*22c*). The complex Cp(OC)$_2$Re(H)Si(CH$_2$Ph)$_3$ exists as a cis–trans mixture in solution (*22c*).

The deprotonation–reprotonation experiment shows that, in contrast with the manganese complexes, the driving force for the preferred formation of the sterically less favorable cis isomer, namely, the three-center M—H—Si interaction, is diminished or absent in the rhenium complexes. Consequently, the structure of **10** is closer to the four-legged piano stool type (**E**), with H and SiPh$_3$ being two individual ligands. The CO—Re—CO angle closes to 83.5° (compared with about 90° in all the manganese complexes), and the Si—H distance lengthens to 219 pm, which is distinctly above the limit discussed earlier. [Although the hydride ligand was not located, its position was reasonably determined by minimization of intramolecular contacts (*51*).] The Re—Si distance in **10** [249(1) pm] is only about 6.5 pm longer than the Mn—Si distance in Cp(OC)$_2$Mn(H)SiPh$_3$. If the silyl group were bonded in the same manner in both complexes, a much larger difference is expected, similar to the pairs of complexes (OC)$_5$MSiH$_3$ [Mn—Si 240.7(5) pm (*39d*), Re—Si 256.2(12) pm (*52a*)] or (OC)$_5$MSi(SiMe$_3$)$_3$ [Mn—Si 256.4(6) pm (*39c*), Re—Si 266.5(9) pm (*52b*)].

On the basis of the previous discussion of bond distances in complexes containing M—H—Si three-center bonds, the relative shortening of the Re—Si distance in **10** compared with Mn—Si in Cp(OC)$_2$Mn(H)SiPh$_3$ must be attributed to the absence or at least a strong weakening of a three-center interaction. Therefore, the silane appears to be fully oxidatively added in **10**. This is in agreement with the general experience that oxidative addition reactions proceed more readily with the heavier transition metals.

Considering related complexes, one has to consider members of the $(\pi\text{-arene})(OC)_2M(H)SiR_3$ family (M = first row transition metal) as potential candidates that might contain M—H—Si three-center bonds. Both $cis\text{-}(\eta^4\text{-}C_4H_4)(OC)_2Fe(H)SiR_3$ *(15)* and $cis\text{-}(\eta^6\text{-}C_6R_6')(OC)_2Cr(H)SiR_3$ *(9,15,53a,b)* have been prepared by the photochemical route [as in Eq. (2)], but only the chromium compounds were examined with respect to bonding *(53a)*. The derivative $\eta^6\text{-}C_6Me_6(OC)_2Cr(H)SiHPh_2$ (Fig. 4) turned out to be very similar to $\eta\text{-}C_5Me_5(OC)_2Mn(H)SiHPh_2$, and a Cr—H—Si three-center bond was clearly established for Fig. 4 *(53a)*. The solid state structures of both compounds closely resemble one another. In particular, Fig. 4 also shows the typical geometrical features associated with a three-center bond. In the Cr—H—Si triangle the relevant distances are Cr—Si 245.6(1), Cr—H 161(4), and Si—H 161(4) pm, compared with Mn—Si 239.5(1), Mn—H 152(3), and Si—H 177(3) pm in Cp*-$(OC)_2Mn(H)SiHPh_2$ *(53a)*. Since the covalent radius of Cr(0) should be only 1–2 pm longer than that of isoelectronic Mn(I) in a similar electronic and steric ligand environment, the observed difference in the metal–silicon bond lengths of about 6 pm suggests that bonding of H_2SiPh_2 in the chromium complex corresponds to an earlier stage of the oxidative addition reaction than in the manganese complex. The assumption of a weaker metal–silicon interaction in Fig. 4 and, hence, a stronger Si—H interaction is also supported by the shorter Si—H and longer Cr—H distances and a larger coupling constant $J(\text{SiCrH})$ (70.8 Hz). The stronger Si—H interac-

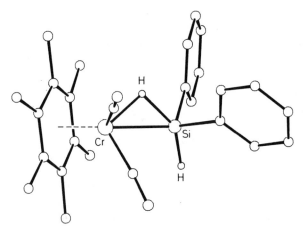

FIG. 4.   Molecular structure of Cp*$(OC)_2Cr(H)SiHPh_2$ *(53a)*.

tion in Fig. 4 is probably due to the larger size of the π-arene ligand rather than the different metal, in agreement with the steric influence of the ligands found for the series of phosphine-substituted manganese complexes (53a). In a series of complexes of type η^6-C$_6$R$_6$(OC)$_2$Cr(H)SiHPh$_2$, J(SiCrH) was shown to increase if electron-donating substituents R are replaced by electron-withdrawing ones. At present, the largest J(SiMH) coupling constant is observed in [η^6-C$_6$H$_4$(COOMe)$_2$](OC)$_2$Cr(H)SiHPh$_2$.

While it is evident from the structure of η^6-C$_6$Me$_6$(OC)$_2$Cr(H)SiHPh$_2$ (Fig. 4) that other metals (at least first row transition metals) may replace manganese in the M—H—Si three-center bond, the question arises whether replacement of silicon by higher homologs results in retention of the three-center bond. Germyl complexes of the type MeCp(OC)$_2$-Mn(H)GeR$_3$ have been prepared either by the photochemical route [analogous to Eq. (2)] or by thermal reaction of GeR$_3{}^2$–with MeCp(OC)$_3$-Mn followed by protonation of the thus obtained anionic complexes [MeCp(OC)$_2$MnGeR$_3$]$^-$ (23b). The compound MeCp(OC)$_2$Mn(H)SnPh$_3$ has also been prepared according to Eq. (2) (54). Independent of the method of preparation, only the cis isomer is obtained, and the preference for this isomer was also confirmed by the previously discussed deprotona-tion–reprotonation experiments (23b,54). By the same reasoning as above, the absence of a trans isomer in both the germyl and stannyl complexes may be related to a bonding interaction between hydrogen and germanium or tin.

Further evidence for Mn—H—Sn three-center bonding in the tin com-pound was obtained by NMR spectroscopy and an X-ray structure deter-mination (54). In alkyltin hydrides the ^1J(SnH) values are typically 1500–1800 Hz, while those of ^2J(SnCH) are 50–70 Hz. In cis-(OC)$_4$Os(H)SnCl$_3$, in which there should be no interaction between the hydride and the SnCl$_3$ ligand, ^2J(^{119}SnOsH) = 136 Hz and ^2J(^{117}SnOsH) = 129.5 Hz (55). The distinctly higher values of J(^{119}SnMnH) = 270 Hz and J(^{117}SnMnH) = 252 Hz in MeCp(OC)$_2$Mn(H)SnPh$_3$ strongly favor some Sn—H interaction.

The solid state structure of MeCp(OC)$_2$Mn(H)SnPh$_3$ closely resembles the structures of analogous silyl complexes and shows again all the pre-viously discussed structural features typical for a three-center interaction. The Sn—H distance [216(4) pm] is only about 45 pm (21%) longer than in methyl-substituted stannanes and therefore indicative of a bonding interac-tion between both atoms. It should be noted that the Si—H distances in the compounds Cp(OC)LMn(H)SiR$_3$ are also 15–20% longer than in tetrahedral silanes. To evaluate the Mn—Sn distance, the complex is again better compared with Cp(OC)$_2$FeSnPh$_3$(Fe—Sn 253.3 and 254.0 pm, two independent molecules) (56a) than the species (OC$_5$)MnSnR$_3$. As discussed for the silyl derivatives, the bonding radius of Mn in the Cp(OC)$_2$Mn fragment is at most 4 pm larger than that of Fe in the $Cp(OC)_2Fe$ fragment. Because the metal–tin distances in $MeCp(OC)_2Mn(H)SnPh_3$ and $Cp(OC)_2FeSnPh_3$ differ by as much as 10 pm, the $SnPh_3$ moiety must be differently bonded in both complexes. The relative lengthening of the Mn—Sn distance must of course be attributed to the three-center bond.

The very similar geometries of $MeCp(OC)_2Mn(H)SnPh_3$ and the corresponding silyl complexes, particularly $Cp(OC)_2Mn(H)SiPh_3$ (18,31), allow an interesting comparison with respect to the question of to what extent the HER_3 molecule has been added to the metal atom. While the bonding radius of manganese should be the same in both complexes, the difference in the bond radii of Sn and Si is 27 pm, as judged by the difference between the mean Sn—C and the mean Si—C distances. If both complexes were in the same stage of oxidative addition, the difference between the distances Mn—Si in $Cp(OC)_2Mn(H)SiPh_3$ and Mn—Sn in $MeCp(OC)_2Mn(H)SnPh_3$ would also be about 27 pm. Actually, this difference is only about 21 pm [Mn—Sn 263.6(1) pm], i.e., the relative approach of the tin atom to the transition metal is closer. This means that addition of $HSnPh_3$ to the $MeCp(OC)_2Mn$ fragment is in a later stage than that of $HSiPh_3$ in $Cp(OC)_2Mn(H)SiPh_3$. Moreover, a qualitative comparison of J (SnMnH) with J(SiMnH) is consistent with this interpretation.

It is also worth noting that the Mn—Sn distance in $MeCp(OC)_2Mn(H)SnPh_3$ is much greater than the Re—Si distance in $Cp(CO)_2Re(H)SiPh_3$. The increased size of the atoms involved (relative to Mn and Si), therefore, cannot be responsible for the absence of a three-center bond in the rhenium silyl compound. The preference of the $Cp(OC)_2Mn$ moiety for this kind of bonding is manifested by the dinuclear complexes of type **11** [with Mn–Pt 280.4(3), Mn—H 169(15), Pt—H 200(15) pm in **11b**] (22a).

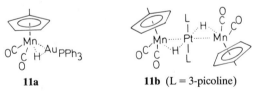

11a   **11b** (L = 3-picoline)

III

DINUCLEAR METAL COMPLEXES WITH BRIDGING
M—H—Si INTERACTIONS

At the time this article was written, the only mononuclear complexes for which a metal–hydrogen–silicon three-center bond was unambiguously proved belong to the (π-arene) $(OC)_2M(H)SiR_3$ family. Some less certain cases, which may contain a three-center bond, are discussed later. However, there are at least three undoubted examples (12–14) for the same kind of bonding in dinuclear metal compounds. [Attention has also been drawn to an osmium cluster with a bridging Os—H—Sn interaction (56b).]

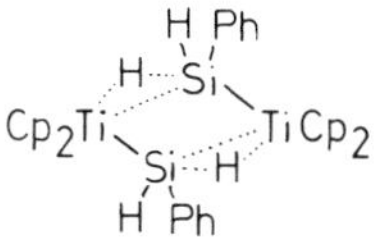

12 (57) 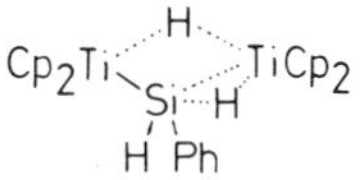   13 (57) 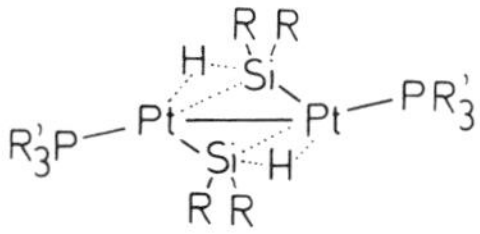   14 (59)
(SiR$_2$ = SiMe$_2$, SiPh$_2$)

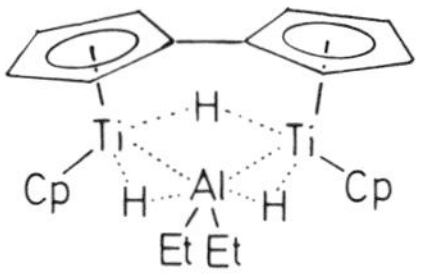

15 (58)

The distinction between between mono- and dinuclear metal complexes is only a formal one in terms of bonding (consider one substituent at silicon the second metal atom instead of an organic group, halide, or hydride substituent). The structural and spectroscopic features related to the three-center bond should therefore be more or less the same as previously discussed.

The complexes 12 and 13 were discovered during an investigation of the dehydrogenative coupling of primary organosilanes to yield linear polysilanes under the catalytic influence of dialkyltitanocenes. Complex 12 was detected by ^{1}H NMR in an actively polymerizing PhSiH$_3$ reaction containing a catalytic amount of Cp$_2$TiMe$_2$ and was subsequently prepared in high yield by reaction of Cp$_2$TiMe$_2$ with an approximately 3-fold excess of PhSiH$_3$. Compound 12 is unstable in solution and decomposes to 13 and polysilane under ambient conditions. Addition of PhSiH$_3$ to a solution of 13 regenerates 12. Complex 13 can also be synthesized by reaction of Cp$_2$TiMe$_2$ and PhSiH$_3$ in an 1:1 molar ratio in diethyl ether (57).

The crystal structures of 13 (Fig. 5) and 12 provide unambiguous proof for the presence of Ti—H—Si three-center bonds (57). To emphasize the similarities with mononuclear complexes, we consider one Si—H bond of the silane "(Cp$_2$Ti)PhSiH$_2$" to be added to the titanium center of another Cp$_2$Ti—SiH$_2$Ph in 12 and to a Cp$_2$TiH molecule in 13 (type I). In both complexes oxidative addition is arrested at a point corresponding to H in Eq. (5).

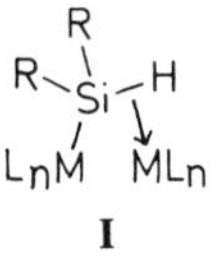

I

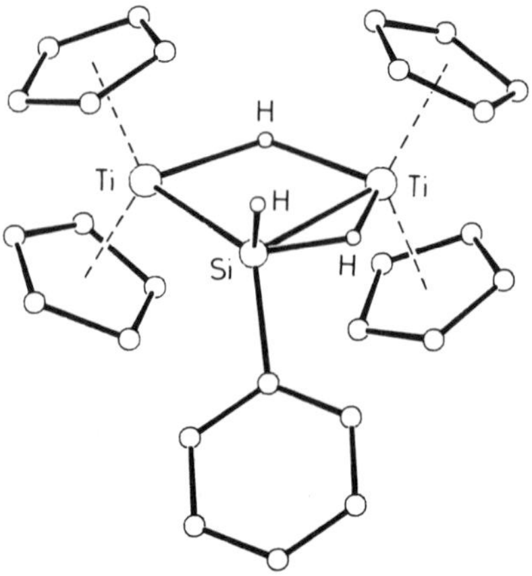

FIG. 5. Molecular structure of the dimeric titanium compound **13** (*57*).

While the Si—H bond lengths for the terminal hydride substituents in **12** and **13** are as expected [147(3) pm], the Si—H distances within the Ti—H—Si three-center bonds are 158(3) pm in **12** and 156(7) pm in **13**. Note that these distances are considerably shorter than in the η-C_5R_5'(OC)-LMn(H)SiR$_3$ series (Table I) and only 7% longer than covalent Si—H bonds (compared to 15–20% in the manganese series). Consequently, the Ti—Si distances within the Ti—H—Si triangles are rather long. Because the silicon atoms in **12** and **13** are bonded to one of the titanium atoms by a conventional two-center bond [Ti–Si 260.4(2) and 258.3(2) pm in **12** and 261(2) pm in **13**] and to the other by a three-center Ti—H—Si bond [Ti—Si 289.1(2) and 285.1(2) pm in **12** and 278(2) pm in **13**], the difference in bond lengths directly reflects the different bonding situation. The relative lengthening of the Ti—Si distance within the three-center bond is about 10% in **12** and 6.5% in **13**, compared with at most 4% in the series of manganese complexes. The Si—H and Ti—Si distances of **12** and **13** therefore clearly show that in these complexes addition of the silane to the metal [Eq. (5)] is arrested at a much earlier stage than in any of the previously discussed mononuclear metal complexes. This interpretation is confirmed by the bond angles at silicon. Looking only at the "parent silane" (Cp$_2$Ti)PhSiH$_2$moiety, the tetrahedral geometry at silicon is preserved to a much higher degree on coordination of the Si—H bond to the metal than in the manganese complexes. The structural and electronic similarity between **13** and the aluminum compound **15** (*58*) should be noticed.

In the ^{1}H-NMR spectra of **12** and **13**, separate signals in the Si—H region and in the metal hydride region are observed (*57*). The molecules are therefore not fluxional on the NMR time scale with respect to hydrogen scrambing. A ^{29}Si-NMR spectrum of **13** revealed coupling of the silicon nucleus to three different hydrogen atoms, with coupling constants of 148, 58, and 14 Hz (*57*), demonstrating uniquely the lower limit [^2J(SiTiH) = 14 Hz], the upper limit [^1J(SiH) = 148 Hz], and the intermediate case [J(SiTiH) of the three-center bond = 58 Hz] in a single compound. In agreement with the structural data, the three-center Si—H coupling constant is closer to ^1J(SiH) than in the Cp(OC)LMn(H)SiR$_3$ complexes (Table II), corresponding to an earlier stage of the addition reaction. Of course, the absolute values of J cannot be directly compared between both types of complexes because of the different metal moieties involved.

The platinum complexes of type **14** (*59*) are formally related to the titanium complex **12**. One can again treat these complexes as examples of incipient oxidative addition of a metallosilane to another metal atom (which incidentally is the other metal atom of a dinuclear complex) (**I**). Complexes of type **14** with various phosphine ligands and dimethylsilyl or diphenylsilyl moieties are best prepared from diethylsilane or diphenylsilane and Pt(C$_2$H$_4$)$_2$PR$_3'$ according to Eq. (7) (*59*). Compound **14** [R = Me, R' = PCy$_3$(Cy = C$_6$H$_{11}$), subsequently referred to as **14a** in Fig. 6] is also obtained, but in lower yield, by thermolysis of [Pt(μ-H)(SiMe$_2$Ph)(PCy$_3$)]$_2$ or by reaction of Pt(C$_2$H$_4$)$_2$PCy$_3$ with HSi$_2$Me$_5$ (*59*).

$$2 \text{ Pt}(C_2H_4)_2PR_3' + 2 \text{ H}_2SiR_2 \rightarrow \textbf{14} + 4 \text{ C}_2H_4 + H_2 \qquad (7)$$

Although the IR and ^{1}H-NMR spectra of **14** show some features unusual for platinum hydrides, the occurrence of a Pt—H—Si three-center bond is mainly proved by a structure determination of **14a** (Fig. 6) (*59*). The hydrogen atom was located from a difference Fourier map (but was not

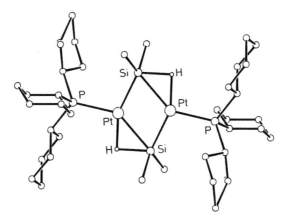

Fig. 6. Molecular structure of the dimeric platinum compound **14a** (*59*).

refined) at a distance of 178 pm from Pt and 172 pm from silicon within the greater of the two P—Pt—Si angles [144.6(3) versus 104.5(3)°] and bridging the longer of the two Pt—Si bonds [242.0(2) versus 232.4(2) pm]. While the shorter Pt—Si distance of **14a** (Fig. 6) is comparable to Pt(II)—Si(alkyl)₃ bond lenghts (*60a–c*), lengthening of the second Pt—Si distance by about 4% is attributed to the Pt—H—Si three-center bond. The relative increase in the Pt—Si distance owing to three-center bonding is distinctly smaller than in the titanium complexes **12** and **13** but comparable to the less stable complexes within the Cp(OC)₂Mn(H)SiR₃ series, indicating a similar degree of Si—H interaction.

Graham and Hoyano have prepared a number of dinuclear metal complexes (**16–19**, Fig. 7) by reaction of W(CO)₆ or Re₂(CO)₁₀ with diethyl- or diphenylsilane, and these products probably also contain metal–hydrogen–silicon three-center bonds (*18,61*). Since spectroscopic evidence is lacking [only for **17** was three-center interaction also suggested by an analysis of the vibrational spectrum (*27*)] and the crucial hydrides were not located in the X-ray structure determinations, all arguments for and against three-center bonds are solely based on the geometry of the heavy atom skeleton. In every case, there are two mutually trans CO ligands perpendicular to the plane of the M₂Si triangle or M₂Si₂ rectangle, respectively. Because the M—M—CO bond angles (M = W, Re) of these carbonyl ligands are close to 90°, the hydride ligands are certainly located within the M₂Si (M₂Si₂) plane (Fig. 7).

Among the complexes **16–19** a three-center interaction appears most likely for the tungsten compound W₂(CO)₈(H)₂(SiEt₂)₂ (**16**) (*61*). There is a

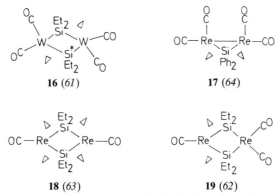

FIG. 7. Heavy atom skeletons of compounds **16–19**. At each metal atom two additional CO ligands above and below the plane of the paper are omitted for clarity. Δ indicates possible positions of the hydride ligands.

conspicuous similarity with the platinum compound **14**: in the centrosymmetric W_2Si_2 core the W—Si distances are pairwise inequivalent [258.6(5) and 270.3(4) pm], and the CO ligands in the W_2Si_2 plane are not symmetrical with respect to the W—W axis. The hydride ligand is obviously located between the longer of the two W—Si vectors (W—Si* in Fig. 7) and the adjacent CO ligand [OC—W—Si* 109.1(5)°], causing the other ligands in this plane to close up. The resulting OC—W—CO angle of 78.5° is of the same magnitude as those in *cis*-Cp(OC)$_2$MLL′ complexes. The seemingly small Si—W—CO angle of 66.6° corresponds to a C(CO)···Si contact of 256 pm, which is about the same as found in MeCp(OC)$_2$Mn(H)SiFPh$_2$ (Fig. 2).

Cowie and Bennett argued that an unreasonably short C(CO)···H contact can be avoided only if there is a short Si—H distance, i.e., a W—H—Si three-center bond (*62*). This argument can be probed in the following way: if the hydride is placed in the coordination plane containing both Si* and the adjacent CO ligand, at a distance of 170 pm from the tungsten atom and 200 pm from silicon, which would be the limit for a nonbonding Si—H contact (*vide supra*), a C(CO)···H distance of 194 pm is calculated (corresponding to a OC—W—H angle of 61.4°). In the manganese complex (Fig. 2), the distance between the hydride ligand and the *cis*-carbonyl ligand is 209 pm. Any increase in the nonbonding C(CO)—H distance of **16** shortens the Si—H distance. Therefore, from steric considerations the assumption of a W—H—Si three-center bond seems reasonable.

The rhenium compounds **17–19** (*62–64*) are less crowded because they contain one ligand less in the coordination plane containing the hydride ligand. Therefore the OC—Re—Si angle enclosing the hydride ligand opens to 125–127.5°, providing enough space to accommodate the hydride ligand without the contacts with the neighboring CO and SiR$_3$ ligands being too short. The similarity of the Re—Si bond lengths in Re$_2$(CO)$_8$-(μ-SiPh$_2$)$_2$ (*62,63*), Re$_2$(CO)$_8$(H)$_2$(SiPh$_2$) (**17**), Re$_2$(CO)$_7$(H)$_2$(SiEt$_2$)$_2$ (**19**), and Re$_2$(CO)$_6$(H)$_4$(SiEt$_2$)$_2$ (**18**) was taken as evidence against a Re—H—Si three-center bond in the latter three complexes (*62*). However, this argument does not take into account (1) the different ligand environments at the rhenium atoms; (2) the different substituents at silicon; (3) the fact that lengthening of the M—Si distance due to three-center bonding may be small, if the addition of the silane is in an early stage; or (4) that the various influences on the M—Si bond length may compensate each other. At the present time, the available data are therefore not sufficient to decide whether the rhenium complexes **17–19** contain Re—H—Si three-center bonds or terminal hydride ligands.

IV

OTHER COMPLEXES POSSIBLY CONTAINING M—H—Si
THREE-CENTER BONDS

Apart from the dinuclear metal complexes **17–19** there are also some mononuclear metal compounds for which the occurrence of M—H—Si three-center bonds appears possible, judging from currently available data. The first complex in this category, **20**, was obtained by Jetz and Graham among other products by reaction of trichlorosilane with $Cp_2(OC)_4Fe_2$ (*65*).

20

Subsequently three complexes of this type [$SiR_3 = SiCl_3$ (*66*), SiF_2Me (*67*), and $SiMe_2Ph$ (*68*)] were structurally characterized, but only in one of them ($SiR_3 = SiF_2Me$) was the hydrogen located.

The overall geometries of the three complexes are very similar, namely, basically of the four-legged piano stool type, the hydride ligand being trans to CO and cis to both SiR_3 ligands. The silyl ligands (Si—Fe—Si between 112.5 and 115.3°) are bent toward the hydride site, and the OC—Fe—Si angles (81.4–85.5°) are therefore larger than typical angles between cis ligands in complexes of the type CpL_4M. In the SiF_2Me derivative a crystallographic mirror plane bisects the molecule, making both silyl groups crystallographically equivalent. The hydride ligand was located within this mirror plane at a distance of 149(6) pm from iron and 206(7) pm from each silicon atom (OC—Fe—H 109°) (*67*).

According to the previous discussion, these Si—H distances appear too long for a bonding interaction (at most it could be a very weak interaction). However, some caution with respect to the hydrogen position seems merited. The famous example of $[(OC)_5Cr—H—Cr(CO)_5]^-$ must be recalled, for which a linear Cr—H—Cr linkage was inferred from the crystallographic inversion symmetry (*69a*), whereas a later neutron diffraction study (*69b*) showed the bridging hydrogen atom to be disordered between two crystallographically related sites 60 pm apart from each other, resulting in a Cr—H—Cr angle of 158.9(6)°. The available crystallographic data for $Cp(OC)Fe(H)(SiR_3)_2$ (**20**) are not sufficient to exclude the possibility that the hydrogen position found in the SiF_2Me derivative is an average of two sites displaced from the mirror plane, corresponding to a double-minimum potential well. In this case the complexes **20** would contain both a classically bonded SiR_3 group and a Fe—H—Si three-center bond, and they would be isoelectronic and isostructural with $Cp(OC)(PR_3)Mn(H)SiR_3$ (3). In this connection the similarity of the bond angles around the manganese atom in $MeCp(OC)(PMe_3)Mn(H)SiHPh_2$ (11) [P—Mn—Si 113.6(6)°, P—Mn—CO 87.4(2)°, Si—Mn—CO 80.0(2)°] and the corresponding angles in the $Cp(OC)Fe(H)(SiR_3)_2$ derivatives (*vide supra*) is noteworthy. Obviously, a neutron diffraction study and detailed NMR spectroscopic investigations are necessary to resolve the question of how the hydride ligand is bonded in **20**.

Another iron compound which possibly contains a Fe—H—Si three-center bond is $(dppe)(OC)Fe(H)_3SiR_3$ (70). Despite the fact that the hydride ligands could not be located in an X-ray structure determination of the $Si(OEt)_3$ derivative, the heavy atom skeleton suggests some unusual kind of bonding. The iron atom is approximately trigonal bipyramidally surrounded by the two phosphorus atoms, the $Si(OEt)_3$ group, and the CO ligand. Both the apical CO ligand and the equatorial silyl ligand are symmetric relative to the dppe ligand. This geometry suggests that at least some of the hydride ligands must be trans to the carbonyl group. If there were three terminal hydride ligands and a classically bonded silyl group, a formal oxidation state of +4 of the iron atom would result, which is rather unlikely. Either a η^2-H_2 ligand (leaving a terminal hydride and a terminal silyl ligand) or a η^2-$HSiR_3$ ligand (leaving two terminal hydrides) is more reasonable resulting in the formal oxidation state +2 and providing an explanation for the unusual geometry. The high fluxionality of these compounds even at low temperatures prevented detailed interpretation of the NMR data. However, the T_1 value of 438 mseconds at room temperature of the $Si(OEt)_3$ derivative (71) appears too high for a η^2-H_2 complex (72), even if classic and nonclassic sites exchange rapidly.

The M—H—Si three-center interactions may not only be an important property of ground-state structures of some hydrido silyl complexes, but may also play a role in dynamic processes. Many hydrido silyl complexes are highly fluxional, even if the hydride and the silyl ligand do not interact in the ground state. The compound $(dppe)(OC)_2Fe(H)SiR_3$ is more dynamic than $(dppe)(OC)_2FeH_2$ or $(dppe)(OC)_2Fe(SiR_3)_2(ER_3)$ (E = Si, Sn), indicating that at least in these cases the combination of a hydride and silyl ligand favors fluxionality (35). Complexes of the type *cis*-$(PR'_3)_2Pt(H)SiR_3$ exhibit two dynamic processes: and intramolecular interchange of the PR'_3 positions (during which the P—Pt—H spin correlation in the NMR

$$R_3'P{\diagdown}{\underset{R_3'P{\diagup}}{Pt}}{\diagdown}{\overset{H}{\underset{SiR_3}{}}}$$

J

spectra is retained) and a reversible addition–elimination equilibrium. One of the possible mechanisms for the intramolecular PR_3' exchange includes rotation of a η^2-HSiR$_3$ ligand (J) (73a,b).

The importance of M—H—Si three-center interactions in reaction intermediates is only slowly emerging. Kinetic and mechanistic studies of silane alcoholysis catalyzed by $[IrH_2S_2(PPh_3)_2]SbF_6$ (S = solvent) indicate that the silane is activated by η^2 coordination (73c). Among the two isomers mer-(CO)$_3$(PPh$_3$)Fe(H)SiPh$_3$, the isomer with a trans arrangement of H and PPh$_3$ undergoes HSiPh$_3$ elimination with PPh$_3$ 183 times faster than the isomer with SiPh$_3$ and PPh$_3$ trans to each other. The effect of structure on the rate constants of HSiPh$_3$ elimination was explained by a steric interaction between PPh$_3$ and SiPh$_3$ in the former isomer, forcing close contact between H and SiPh$_3$, which could be a Fe—H—Si three-center interaction (25a).

V

η^2-SiH AND η^2-CH COMPLEXES: A COMPARISON

For a more general comparison between complexes containing a three-center bond involving either a silicon or a carbon atom, let us return to the notion that in complexes of this type the E—H bond (E = C, Si) is not completely oxidatively added to the metal moiety (L$_n$M) and that the addition process [Eq. (8)] is instead arrested at some point along the

$$L_nM + \overset{H}{\underset{ER_3}{|}} \rightleftharpoons L_nM\overset{H}{\underset{ER_3}{\cdots|}} \rightleftharpoons L_nM\overset{H}{\underset{ER_3}{\diagup}} \tag{8}$$

reaction coordinates. The trajectory for both E = Si and E = C is very similar, as discussed earlier. Comparing the relevant physical properties of complexes containing M—H—E three-center bonds with those of either uncoordinated HER$_3$ molecules or classic hydrido alkyl or hydrido silyl complexes, we can assign complexes of this type to a particular stage of the oxidative addition (reductive elimination) reaction.

The main difference between mononuclear complexes containing either a M—H—C or a M—H—Si three-center bond is that most η^2-CH complexes correspond to an earlier stage of the addition reaction than do the η^2-SiH complexes: J(CMH) coupling constants are usually closer to the values for ^1J(CH), while J(SiMH) values are closer to ^2J(SiMH), and the relative lengthening of the C—H distance on η^2 coordination is usually smaller than that of coordinated Si—H bonds. For example, in the representative iron complex 21 [the structure of which was determined by neutron diffraction analysis (74)], the coordinated C—H bond

[116.4(3) pm] is lengthened by about 7% relative to the other two hydrogen atoms at the same carbon atom [C—H 109.2(3) pm]. This compares with an increase of Si—H bond length of about 20% in the series of complexes Cp(OC)LMn(H)SiR$_3$ (*vide supra*).

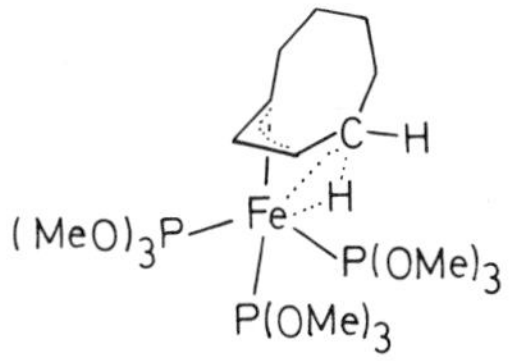

21 (74) 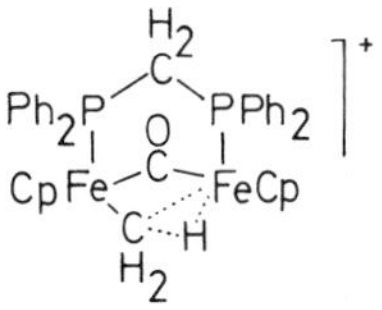   22 (75)

The picture we have at present may be incomplete or biased because it is based only on available compounds, which are limited in number. As more and more complexes of this kind are found, it may be that there is a smooth reaction profile for both C—H and Si—H addition, corresponding to a gradual change of the physical parameters.

A few dinuclear complexes are known, in which the α-CH bond of a metal-bonded alkyl group is η^2 coordinated to another metal moiety, similar to the previously discussed dinuclear silyl complexes. In the iron complex **22** (*75*) both metal moieties are the same, and therefore the Fe—C distances [202.5(3) and 211.3(3) pm] can be directly compared without correction. The relative lengthening of the M—C distance of the Fe—H—C three-center bond (4.5%) is comparable to that in the silane-bridged titanium complex **13** (6.5%), corresponding to a similar stage of the oxidative addition.

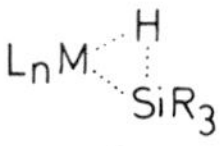

K 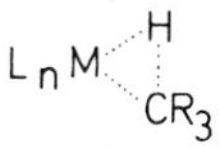   L

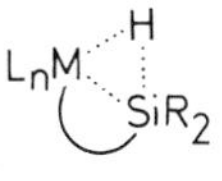

K' 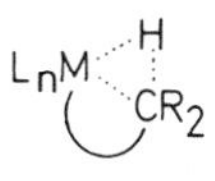   L'

In all complexes of the CpL$_2$Mn(H)SiR$_3$ family, except **4**, the silane is bonded to the metal only by the three-center bond (type **K**). Contrary to this, the "agostic" C—H bond in all known complexes with M—H—C three-center bonds is part of a ligand, which is also classically bonded to the same metal at another coordination site (type **L'**), either by a covalent bond or a π-bond (as in **21**) or by a metal–metal bond (as in **22**). Although this difference probably has no influence on the principal characteristics of the three-center bond, it may be very important with respect to the stability of the complexes. If the Si—H interaction in complexes of type **K** becomes too strong, the silane is easily eliminated, and the complex "decomposes" because the electron-deficient L_nM moiety is usually not stable [i.e., the equilibrium Eq. (8) is shifted to the left owing to the instability of L_nM]. In complexes of type **K′** or **L′**, the Si—H or C—H linkages remain close to the metal whether it is part of a two-center or a three-center bond. It cannot dissociate away from proximity to the metal. This means that complexes **L′** are probably stabilized by the chelate effect and they are accessible despite a rather weak interaction between the metal and the C—H bond (according to an early stage of the oxidative addition).

The reason why all the known η^2-HSiR$_3$ complexes correspond to a late stage of the oxidative addition process and complexes with "agostic" C—H bonds to a rather early stage, may simply be a coincidence. By making use of the chelate effect it should be possible to obtain stable complexes of type **K′** with stronger Si—H interactions than in the hitherto known complexes **K**. Vice versa, by proper choice of the L_nM moiety and the R groups at the carbon atom, complexes with weaker C—H interactions should be accessible, in which the auxiliary bond is no longer necessary (type **L**).

A final remark concerns complexes, represented by Green's methyltitanium complex **23** (76) or Schrock's alkylidene complexes [e.g., **24** (77)], in

23

24

which the three-center bond and the auxiliary bond originate from the same carbon atom. If oxidative addition of the C—H bond to the metal proceeded to completion, a hydrido methylene complex would result in the former case and a hydrido alkylidyne complex in the latter. Because of the reluctance of a silicon atom to become trigonally planar or even two coordinate, it will certainly be very difficult to get similar complexes with silicon instead of carbon.

ACKNOWLEDGMENTS

This article was written during a sabbatical leave at the University of Southern California in Los Angeles. I thank R. Bau and colleagues for providing a stimulating atmosphere during my stay. Our own work was done with a number of enthusiastic co-workers, whose names are given in the references, and supported by the Deutsche Forschungsgemeinschaft. I am also indebted to D. Lichtenberger for providing a preprint and for discussions.

REFERENCES

1. R. H. Morris, J. F. Swyer, M. Shiralian, and J. D. Zubkowski, *J. Am. Chem. Soc.* **107**, 5581 (1985).
2. R. Bau, H. S. H. Yuan, M. V. Baker, and L. D. Field, *Inorg. Chim. Acta* **114**, L27 (1986).
3. Selected articles dealing with theoretical interpretations of η^2 coordination of X—H bonds: a. J.-Y. Saillard and R. Hoffmann, *J. Am. Chem. Soc.* **106**, 2006 (1984) (C—H, H—H); b. H. Rabaâ, J.-Y. Saillard, and U. Schubert, *J. Organomet. Chem.* **330**, 397 (1987) (Si—H); c. D. L. Lichtenberger and G. E. Kellogg, *J. Am. Chem. Soc.* **108**, 2560 (1986) (C—H); d. Y. Jean, P. Eisenstein, F. Volatron, B. Maouche, and F. Sefta, *J. Am. Chem. Soc.* **108**, 6587 (1986) (H—H); e. P. J. Hay, *J. Am. Chem. Soc.* **109**, 705 (1987) (H—H); f. J. K. Burdett, J. R. Phillips, M. R. Pourian, M. Poliakoff, J. J. Turner, and R. Upmacis, *Inorg. Chem.* **26**, 3054 (1987) (H—H); g. O. Eisenstein and Y. Jean, *J. Am. Chem. Soc.* **107**, 1177 (1985) (C—H); h. N. J. Fitzpatrick and M. A. McGinn, *J. Chem. Soc., Dalton Trans.*, 1637 (1985) (C—H).
4. R. G. Teller and R. Bau, *Struct. Bonding (Berlin)* **44**, 1 (1981).
5. G. J. Kubas, *Acc. Chem. Res.* **21**, 120 (1988).
6a. R. H. Crabtree and D. G. Hamilton, *Adv. Organomet. Chem.* **28**, 299 (1988).
6b. R. N. Grimes, "Boron Clusters with Transition Metal–Hydrogen Bonds," p. 269. Plenum, New York, 1982.
6c. C. E. Housecroft and T. P. Fehlner, *Adv. Organomet. Chem.* **21**, 57 (1982).
7a. M. Brookhart and M. L. H. Green, *J. Organomet. Chem.* **250**, 395 (1983).
7b. R. H. Crabtree, *Chem. Rev.* **85**, 245 (1985).
8. In more general discussions no distinction is made between η^5-C$_5$H$_5$ (Cp) and η^5-MeC$_5$H$_4$ (MeCp) ligands. The presence of the methyl group does not affect the properties of the corresponding complexes. MeCp(OC)$_3$Mn is often used instead of Cp(OC)$_3$Mn because it is less expensive.
9. W. Jetz and W. A. G. Graham, *Inorg. Chem.* **10**, 4 (1971).
10. E. Colomer, R. J. P. Corriu, and A. Vioux, *Inorg. Chem.* **18**, 695 (1979).
11. U. Schubert, G. Scholz, J. Müller, K. Ackermann, B. Wörle, and R. F. D. Stansfield, *J. Organomet. Chem.* **306**, 303 (1986).
12. W. A. Herrmann, E. Voss, E. Guggolz, and M. L. Ziegler, *J. Organomet. Chem.* **284**, 47 (1985).
13. U. Schubert, G. Kraft, and C. Kalbas, *Transition Met. Chem. (Weinheim, Ger.)* **9**, 161 (1984).
14. U. Schubert, K. Bahr, and J. Müller, *J. Organomet. Chem.* **327**, 357 (1987).
15. R. H. Hill and M. S. Wrighton, *Organometallics* **6**, 632 (1987).
16. A. J. Hart-Davis and W. A. G. Graham, *J. Am. Chem. Soc.* **94**, 4388 (1971).
17. E. Colomer, R. J. P. Corriu, C. Marzin, and A. Vioux, *Inorg. Chem.* **21**, 368 (1982).
18. W. A. G. Graham, *J. Organomet. Chem.* **300**, 81 (1986).
19. E. Colomer and R. J. P. Corriu, *Top. Curr. Chem.* **96**, 79 (1981).
20. W. Jetz and W. A. G. Graham, *Inorg. Chem.* **10**, 1647 (1971).
21a. E. Colomer, R. J. P. Corriu, and A. Vioux, *J. Organomet. Chem.* **267**, 267 (1984).
21b. F. Carré, E. Colomer, R. J. P. Corriu, and A. Vioux, *Organometallics* **3**, 970 (1984).
22a. E. Kunz, M. Knorr, J. Willnecker, and U. Schubert, *New J. Chem.* **12**, 467 (1988).
22b. E. Kunz and U. Schubert, *Chem. Ber.* **122**, 231 (1989).

22c. D. F. Dong, J. K. Hoyano, and W. A. G. Graham, *Can. J. Chem.* **59**, 1455 (1981).
23a. E. Colomer, R. J. P. Corriu, and A. Vioux, *J. Chem. Res., Synop.*, 168 (1977); E. Colomer, R. J. P. Corriu, and A. Vioux, *J. Chem. Res., Miniprint*, 1939 (1977).
23b. F. Carré, E. Colomer, R. J. P. Corriu, and A. Vioux, *Organometallics* **3**, 1272 (1984).
24a. G. Bellachioma, G. Cardaci, E. Colomer, R. J. P. Corriu, and A. Vioux, *Inorg. Chem.* **28**, 519 (1989).
24b. U. Schubert and M. Knorr, *Inorg. Chem.* **28**, 1765 (1989).
25. U. Schubert, E. Kunz, M. Knorr, and J. Müller, *Chem. Ber.* **120**, 1079 (1987).
26. S. S. Kristjánsdóttir, A. E. Moody, R. T. Weberg, and J. R. Norton, *Organometallics* **7**, 1983 (1988).
27. M. A. Andrews, S. W. Kirtley, and H. D. Kaesz, *Adv. Chem. Ser.* **167**, 229 (1978).
28. W. Jetz and W. A. G. Graham, *Inorg. Chem.* **10**, 1159 (1971).
29. M. J. Fernandez, P. M. Bailey, P. O. Bentz, J. S. Ricci, T. F. Koetzle, and P. M. Maitlis, *J. Am. Chem. Soc.* **106**, 5458 (1984).
30. U. Schubert and A. Schenkel, *Chem. Ber.* **121**, 939 (1988).
31. W. A. G. Graham and M. J. Bennett, *Chem. Eng. News* **48**(24), 75 (1970); W. L. Hutcheon, Ph.D. thesis. University of Alberta, Edmonton, Alberta, Canada, 1971.
32. U. Schubert, K. Ackermann, and B. Wörle, *J. Am. Chem. Soc.* **104**, 7378 (1982).
33. R. A. Smith and M. J. Bennett, *Acta Crystallogr., Sect. B* **33**, 1118 (1977).
34. M. J. Michalczyk, M. J. Fink, K. J. Haller, R. West, and J. Michl, *Organometallics* **5**, 531 (1986).
35. M. Knorr, J. Müller, and U. Schubert, *Chem. Ber.* **120**, 879 (1987).
36. A. R. Barron, G. Wilkinson, M. Montevalli, and M. B. Hursthouse, *J. Chem. Soc., Dalton Trans.*, 837 (1987).
37. J. S. Ricci, T. F. Koetzle, M.-J. Fernandez, P. M. Maitlis, and J. C. Green, *J. Organomet. Chem.* **299**, 383 (1986).
38. S. J. LaPlaca, *Inorg. Chem.* **8**, 1928 (1969).
39a. M. C. Couldwell and J. Simpson, *J. Chem. Soc., Dalton Trans.*, 714 (1976).
39b. M. C. Couldwell, J. Simpson, and W. T. Robinson, *J. Organomet. Chem.* **107**, 323 (1976).
39c. B. K. Nicholson, J. Simpson, and W. T. Robinson, *J. Organomet. Chem.* **47**, 403 (1973).
39d. D. W. H. Rankin and A. Robertson, *J. Organomet. Chem.* **85**, 225 (1975).
39e. G. L. Simon and L. F. Dahl, *J. Am. Chem. Soc.* **95**, 783 (1973).
40. U. Schubert, G. Kraft, and E. Walther, *Z. Anorg. Allg. Chem.* **519**, 96 (1984).
41. U. Schubert, K. Ackermann, G. Kraft, and B. Wörle, *Z. Naturforsch. B: Anorg. Chem. Org. Chem.* **38**, 1488 (1983).
42. M. Knorr and U. Schubert, *J. Organomet. Chem.* **365**, 151 (1989).
43. P. Meakin, E. Muetterties, and J. P. Jesson, *J. Am. Chem. Soc.* **95**, 75 (1973); P. Meakin, E. Muetterties, and J. P. Jesson, *J. Am. Chem. Soc.* **94**, 5271 (1972).
44. D. L. Lichtenberger and A. Rai-Chaudhuri, personal communication.
45. G. Kraft, C. Kalbas, and U. Schubert, *J. Organomet. Chem.* **289**, 247 (1985).
46. K. A. Simpson, Ph.D. thesis. University of Alberta, Edmonton, Alberta, Canada, 1973.
47. W. Hönle and H. G. von Schnering, *Z. Anorg. Allg. Chem.* **464**, 139 (1980).
48. H. B. Bürgi, *Angew. Chem.* **87**, 461 (1975); H. B. Bürgi, *Angew. Chem., Int. Ed. Engl.* **14**, 460 (1975).
49. R. H. Crabtree, E. M. Holt, M. Lavin, and S. M. Morehouse, *Inorg. Chem.* **24**, 1986 (1985).
50. R. Krentz and R. K. Pomeroy, *Inorg. Chem.* **24**, 2976 (1985).
51. R. A. Smith and M. J. Bennett, *Acta Crystallogr. Sect. B* **33**, 1113 (1977).

52a. D. W. H. Rankin and A. Robertson, *J. Organomet. Chem.* **105**, 331 (1976).
52b. M. C. Couldwell, J. Simpson, and W. T. Robinson, *J. Organomet. Chem.* **107**, 323 (1976).
53a. U. Schubert, J. Müller, and H. G. Alt, *Organometallics* **6**, 469 (1987).
53b. E. Matarasso-Tchiroukhine and G. Jaouen, *Can. J. Chem.* **66**, 2157 (1988).
54. U. Schubert, E. Kunz, B. Harkers, J. Willnecker, and J. Meyer, *J. Am. Chem. Soc.* **111**, 2572 (1989).
55. J. R. Moss and W. A. G. Graham, *J. Organomet. Chem.* **18**, P24 (1969).
56a. R. F. Bryan, *J. Chem. Soc. A*, 192 (1967).
56b. C. J. Cardin, D. J. Cardin, H. E. Parge, and J. M. Power, *J. Chem. Soc., Chem. Commun.*, 609 (1984).
57. C. T. Aitken, J. F. Harrod, and E. Samuel, *J. Am. Chem. Soc.* **108**, 4059 (1986).
58. L. J. Guggenberger and F. N. Tebbe, *J. Am. Chem. Soc.* **95**, 7870 (1973).
59. M. Auburn, M. Ciriano, J. A. K. Howard, M. Murray, N. J. Pugh, J. L. Spencer, F. G. A. Stone, and P. Woodward, *J. Chem. Soc., Dalton Trans.*, 659 (1980).
60a. M. Ciriano, M. Green, J. A. K. Howard, J. Proud, J. L. Spencer, F. G. A. Stone, and C. A. Tsipis, *J. Chem. Soc., Dalton Trans.*, 801 (1978).
60b. R. D. Holmes-Smith, S. R. Stobart, T. S. Cameron, and K. Jochem, *J. Chem. Soc., Chem. Commun.*, 937 (1981).
60c. H. Yamashita, T. Hayashi, T. Kobayashi, M. Tanaka, and M. Goto, *J. Am. Chem. Soc.* **110**, 4417 (1988).
61. M. J. Bennett and K. A. Simpson, *J. Am. Chem. Soc.* **93**, 7156 (1971).
62. M. Cowie and M. J. Bennett, *Inorg. Chem.* **16**, 2325 (1977).
63. M. Cowie and M. J. Bennett, *Inorg. Chem.* **16**, 2321 (1977).
64. M. Elder, *Inorg. Chem.* **9**, 762 (1970).
65. W. Jetz and W. A. G. Graham, *Inorg. Chem.* **10**, 1159 (1971).
66. L. Manojlović-Muir, K. W. Muir, and J. A. Ibers, *Inorg. Chem.* **9**, 447 (1970).
67. R. A. Smith and M. J. Bennett, *Acta Crystallogr. Sect. B* **331**, 1118 (1977).
68. Reference *46*, as cited in Ref. *67*.
69a. L. B. Handy, J. K. Ruff, and L. F. Dahl, *J. Am. Chem. Soc.* **92**, 7312 (1970).
69b. J. Roziere, J. M. Williams, R. P. Stewart, J. L. Petersen, and L. F. Dahl, *J. Am. Chem. Soc.* **99**, 4497 (1977).
70. M. Knorr, S. Gilbert, and U. Schubert, *J. Organomet. Chem.* **347**, C17 (1988).
71. H. Rüegger, personal communication.
72. D. G. Hamilton and R. H. Crabtree, *J. Am. Chem. Soc.* **110**, 4126 (1988).
73a. H. Azizian, K. R. Dixon, C. Eaborn, A. Pidcock, N. M. Shuaib, and J. Vinaixa, *J. Chem. Soc., Chem. Commun.*, 1020 (1982).
73b. H. C. Clark and M. J. Hampden-Smith, *Coord. Chem. Rev.* **79**, 229 (1987).
73c. X.-L. Luo and R. H. Crabtree, *J. Am. Chem. Soc.* **111**, 2527 (1989).
74. R. K. Brown, J. M. Williams, A. J. Schultz, G. D. Stucky, S. D. Ittel, and R. L. Harlow, *J. Am. Chem. Soc.* **102**, 981 (1980).
75. G. M. Dawkins, M. Green, A. G. Orpen, and F. G. A. Stone, *J. Chem. Soc., Chem. Commun.*, 41 (1982).
76. Z. Dawoodi, M. L. H. Green, V. S. B. Mtetwa, and K. Prout, *J. Chem. Soc., Chem. Commun.*, 1410 (1982).
77. R. R. Schrock, *Acc. Chem. Res.* **12**, 98 (1979).

Isoelectronic Organometallic Molecules

ALLEN A. ARADI and THOMAS P. FEHLNER

Department of Chemistry
University of Notre Dame
College of Science
Notre Dame, Indiana 46556

I

INTRODUCTION

One of the oldest, but still useful, means of interrelating compounds is the comparative analysis of series of isoelectronic species (*1–4*). In a strict sense, the term isoelectronic refers to species containing the same number of electrons. A generally accepted premise is that such species will exhibit similar modes of bonding as, for example, reflected by geometric structure. The validity of this premise is supported by the usefulness of the isoelectronic concept in ordering and relating compounds of different elements (*5*). The emphasis, particularly when the isoelectronic concept is used in a pedagogical sense, is on the similarities exhibited by a given series. However, the compounds found in such series often behave very differently in terms of chemical properties, e.g., reactivity. Thus, in principle, the real variation in electronic structure throughout an isoelectronic series also provides a means of understanding how systematic perturbation of a nuclear framework, while keeping the total number of electrons in the system constant, can be used to vary chemical properties in predictable ways.

Our purpose here is to compare isoelectronic compounds containing direct main group element (including carbon)–transition metal bonding interactions. In terms of boron versus carbon it is a theme we have pursued for several years (*6*). A wide variety of main group element–transition metal compounds have also been characterized in many other laboratories. These novel compounds constitute a set of isoelectronic species such that a more generalized, if not complete, comparison is now possible. Similar considerations helped establish the isolobal principles used so effectively by Hoffmann (*7*), Mingos (*8*), Wade (*9*), and others. Indeed, one could well argue that little can be added to their seminal work and that yet another review can hardly be justified. However, our outlook is different in one essential feature. Rather than point out similarities in the way related fragments combine, we explore in what ways the members of an isoelectronic series differ! In true isoelectronic comparisons differences can be rather easily understood in terms of the perturbation of nuclear charge

189

within the molecular framework. Hence, varying the metal permits systematic variation in main group fragment properties, while varying the main group atom allows variation in the transition metal fragment properties. In less rigorous uses of the term isoelectronic, e.g., isolobal clusters, the differences have a more complex origin and are much less useful in this sense.

There have been a number of reviews in recent years covering parts of the topic of this article. Of special interest is an insightful review by Schmid (10) on main group–metal bonding appearing over a decade ago. Also of particular note are the excellent summaries by Whitmire (11), Herrmann (12), Adams and Horvath (13), Bottomley and Sutin (14), Vahrenkamp (15), Huttner and Knoll (16), Housecroft (17), Kennedy (18), and Johnson and Lewis (19), as well as more general sources ($20,21$). The reader should keep in mind that we have made no effort to produce a comprehensive compilation of compounds containing main group element–transition metal interactions. Such information can be found in the topical or general reviews, often under the main group element of interest. We restrict ourselves to a somewhat eclectic selection of isoelectronic compounds for which sufficiently detailed information exists.

II

ISOELECTRONIC DEFINITIONS

There are a number of ways in which the term isoelectronic is used to intercompare compounds. Strictly speaking, the term applies only to species containing the same total number of electrons, i.e., N_2 and CO are isoelectronic but N_2 and P_2 are not. Often, however, core electrons are neglected in such comparisons, and, in terms of valence electrons, N_2 and P_2 are isoelectronic, i.e., valence isoelectronic. The idea is stretched further when it is applied to a particular type of interaction occurring in a series of molecules. For example, $(OC)_3Fe(C_2H_4)_2$ and $CpCo(C_2H_4)_2$ ($Cp = \eta^5$-C_5H_5) might be said to be isoelectronic with respect to the metal–olefin interaction even though the isolobal (7) CpCo and $(OC)_3Fe$ fragments contain 34 and 38 valence electrons, respectively ($22,23$). Likewise, $[B_6H_6]^{2-}$ and $H_2Ru_6(CO)_{18}$ can be considered to be isoelectronic in terms of the number of electrons assigned to skeletal bonding (Section V,A) ($24,25$).

The uses of the term isoelectronic depend on the validity of the separation of the electrons into different types. Separation of valence and core electrons is good for most chemical purposes, although both must be considered when attempting to utilize photoelectron spectroscopic information to examine bonding in molecules (*26*). In addition, changes in the size of the atom core significantly perturb electronic structure and have dramatic effects on structure and reactivity. On the other hand, the separation implied in the isolobal principle (third example) and the electron counting rules (fourth example) involves valence electrons and may not be so clear-cut. Indeed, as seen below, the assumption that such separations hold *in detail* has given rise to some interesting discussions in the literature.

In this article we use the term isoelectronic to refer to species that contain the same total number of valence electrons. It will be necessary, however, to relax this definition for molecules that contain different substitutents, e.g., Me rather than Et. Otherwise, the number of comparisons possible becomes very small. Likewise, some clusters are compared that are only isoelectronic in the sense of skeletal electron count in order to show the limits of the validity of such a separation.

There are a number of ways of working the isoelectronic game, but an effective one is to simply realize that isoelectronic compounds differ only in the location of one or more protons. In thought experiments, the transfer can take place between nuclei, e.g., N_2 to CO, out of one nucleus to a position elsewhere in the molecule, e.g., CO to HBO, or out of one nucleus to infinity, e.g., CO to $[BO]^-$. Obviously the reverse can also be contemplated, and the net charge can vary without changing the arguments, e.g., $[H_3O]^+$ to $[H_2F]^+$. In this view, the connection between the electronic structures of isoelectronic species as expressed in the language of molecular orbitals (MOs) becomes readily appreciated. We have already presented some elementary aspects of such an approach and elaborate on this theme before each of the categories of E–M systems considered in this article (*27*). A general theoretical approach is presented in the book by Albright *et al.* (*1*).

Molecular orbitals are characterized by energies and amplitudes expressing the distribution of electron density over the nuclear framework (*1–3*). In the linear combination of atomic orbital (LCAO) approximation, the latter are expressed in terms of AO coefficients which in turn can be processed using the Mulliken approach into atomic and overlap populations. These in turn are related to relative charge distribution and atom–atom bonding interactions. Although in principle all occupied MOs are required to describe an observable molecular property, in fact certain aspects of structure and reactivity correlate rather well with the nature of selected filled and unfilled MOs. In particular, the properties of the highest occupied MO (HOMO) and lowest unoccupied MO (LUMO) permit the rationalization of trends in structural and reaction properties (*28*). A qualitative predictor of stability or, alternatively, a predictor of electron count for a stable species is the magnitude of the energy gap between the HOMO and LUMO. Stablility is associated with large gaps. Acid–base properties also correlate with LUMO–HOMO properties. Hence, the position of highest amplitude in the HOMO often coincides with position of attack of a Lewis acid, and the position of highest amplitude in the LUMO often predicts the site of attack of a Lewis base.

Two molecules with the same number of electrons will have the same number of filled MOs but the total number of MOs will depend on the number of nuclei constituting the molecular framework. A change in the nuclear charge, Z, of an atom will result in a change in the atomic orbitals (size and energy), and a perturbation in those MOs in which the AO coefficients of the particular atom are nonzero. If the atom affected participates extensively in the HOMO and/or LUMO, the HOMO–LUMO gap (stability) and acid–base properties will be changed. The effective nuclear charge can be changed by variation in group (increase or decrease in Z by one unit with a compensating change at another nuclear center in the molecule or a change in the overall charge on the species) or by variation in period (change in effective Z caused by differing numbers of core electrons). The effect of these changes are illustrated for each category of E–M species. This is done first with model compounds using the Fenske–Hall nonparameterized quantum chemical approach to define changes within an isoelectronic series (29). The advantage of the quantum chemical approach here is that although one is tied to practical chemistry one is not restricted by it. That is, one is free to roam among presently uncharacterized but perfectly reasonable molecules that flesh out an interesting series. As this technique is now available on the MAC II (as well as other computing systems) and as it handles most elements in the periodic table, it is a valuable resource for the working chemist. The calculations are followed by a comparison of the properties of series of selected, known isoelectronic molecules and ions.

III

SIMPLE LIGANDS

The interactions between main group species and transition metal fragments range from coordinate covalent bonds, where the main group species acts primarily as an electron donor and the transition metal fragment as an electron acceptor, to cluster systems where the concept of the main group fragment as a simple ligand is difficult to apply (30). We begin the discussion with a description of the changes in the electronic structures of free ligands as protons are moved about. Then similar perturbations for ligands bound to a transition metal are examined. In both cases eigenvalue spectra (set of MO energies) and MO compositions derived from Fenske–Hall calculations are used to demonstrate trends.

A. Free Ligands

In Fig. 1, the eigenvalue spectra of N_2, CO, and BF are shown. These 10-valence electron diatomic molecules are related by the transfer of a proton between the nuclei. Note that if continued one more step, the unbound BeNe "molecule" would result. Figure 1 is a good illustration of the fact that atoms in molecules remember their origins. The contribution of the more electropositive element to the HOMO increases across the series while its contribution to the lower lying MOs decreases. A type of "phase separation," as it were, takes place in that as one proceeds to BeNe the LUMO pair (and one high-lying empty MO not shown) become the empty Be $2p$ AOs while the HOMO becomes the filled Be $2s$ AO. The other filled MOs become the filled Ne $2p$ and $2s$ AOs. As the nuclear charges of the atoms making up the diatomic become more and more different, the filled MOs gain more character of the element with higher Z. However, the HOMO becomes more centered on the element of lower Z and rises in energy. The LUMO also becomes more centered on the more

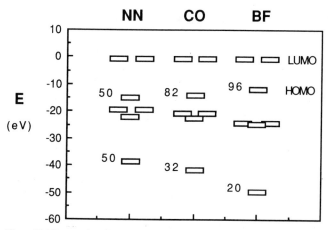

Fig. 1.   Plot of MO energies (Fenske–Hall) for selected 10-electron diatomics related by proton transfer between nuclei (the highest lying unfilled MO is not shown). The numbers adjacent to selected levels give the percentage character of the electropositive atom in the corresponding MO.

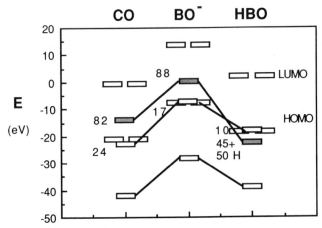

FIG. 2.   Plot of MO energies (Fenske–Hall) for selected 10-electron systems related by proton transfer away from the nuclei. The numbers adjacent to selected levels give the percentage electropositive element character in the corresponding MO unless identified otherwise. Correlation lines are drawn for the σ manifold. The shaded level for CO corresponds to the highest filled σ level which becomes the BH bonding MO of HBO.

electropositive element, and, hence, both the σ base and π acid properties of, for example, CO are primarily associated with the carbon atom.

In Fig. 2, a proton is removed from the carbon atom of CO to produce [BO]⁻. Aside from a rise in energy of all the MOs because of the negative charge, there are small changes in relative orbital energies and compositions. When the proton is brought back into a position on the molecular axis of [BO]⁻ opposite the boron atom, a dramatic change takes place. The formerly shielded proton now selectively stabilizes that MO of [BO]⁻ (or CO) with a large amplitude in the region of space that it now occupies, i.e., the HOMO of [BO]⁻ (or CO). As a consequence, the HOMO–LUMO gap of HBO is larger, and the nature of the HOMO of HBO differs from that of [BO]⁻ (or CO). The basic pair in [BO]⁻ or (CO) becomes the BH bond pair. Obviously, deprotonation of HBO leads to a high-lying MO with large amplitude in the region of space formerly occupied by the proton, i.e., a basic site.

Figure 3 constitutes a variation on the theme presented in Fig. 1. In going from HBO to HCN there is a transfer of proton between the heavy nuclei. The results parallel those in Fig. 1 (except in reverse) in all regards except one. The EH orbital decreases in energy in going from HBO to HCN because of the increased nuclear charge of the more electropositive element. Hence, HCN is expected to be a better base than HBO but a poorer H donor.

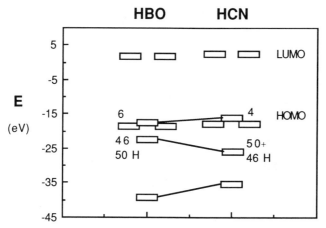

FIG. 3. Plot of MO energies (Fenske–Hall) for HBO and HCN, which are related by proton transfer between nuclei (the two highest lying unfilled MOs are not shown). The numbers adjacent to selected levels give the percentage electropositive element character in the corresponding MO unless identified otherwise. Correlation lines are drawn for the σ manifold.

B. Metal Complexes

Although the changes in electronic structure illustrated above are elementary, and perhaps obvious, parallel changes occur in more complex series of isoelectronic molecules. This is illustrated with the isoelectronic metal complexes in Figs. 4 and 5. In Fig. 4, $Ni(CO)_4$ is compared with $Co(NO)(CO)_3$. Here the molecules are related by the transfer of a proton between the metal nucleus and one nucleus of a ligand. Now the number of MOs is truly formidable; however, by classifying them according to their origin, considerable simplification is possible. For example, 20 of the filled MOs of $Ni(CO)_4$ (only 16 shown in Fig. 4) are associated with the carbonyl ligands. Those MOs derived from the HOMO of CO have significant metal character and a substantial spread in energy which is a consequence of the σ-donor–acceptor interaction of CO with the metal. The highest filled MOs of $Ni(CO)_4$ are mainly metal in character with some CO character resulting from the π-donor–acceptor interaction of the metal with the CO ligands.

As one goes from $Ni(CO)_4$ to $Co(NO)(CO)_3$ the MOs with large metal character rise because they are centered on a nucleus of reduced Z, while at the same time one ligand MO in each ligand block decreases in energy. The latter MOs are associated with the nitrosyl ligand, and the energy change is a reflection of the argument presented in Fig. 1. Because the LUMO now has significant NO character, it also decreases in energy. Both

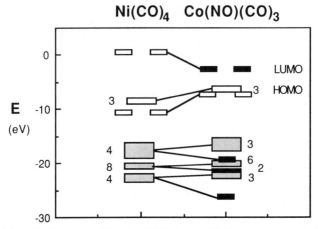

FIG. 4.   Plot of MO energies (Fenske–Hall) for two 50-electron metal–tetraligand complexes related by proton transfer between metal and ligand nuclei (the 4 lowest lying filled MOs and the 14 highest lying unfilled MOs are not shown). Small blocks correspond to a single MO. The larger blocks contain the number of MOs indicated, and the energies of the upper and lower edges are defined by the highest and lowest MO energies for a particular block. The stippled blocks correspond to MOs with large CO ligand contributions, whereas the solid blocks correspond to MOs with large contributions from the NO ligand.

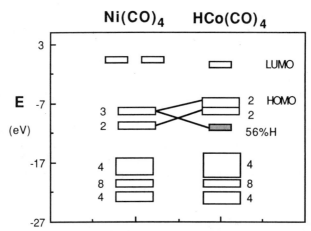

FIG. 5.   Plot of MO energies (Fenske–Hall) for two 50-electron metal tetracarbonyls related by proton transfer away from the metal nucleus (the 4 lowest lying filled MOs and the 16 highest lying unfilled MOs are not shown). Small blocks correspond to a single MO. The larger blocks contain the number of MOs indicated, and the energies of the upper and lower edges are defined by the highest and lowest MO energies for a particular block. The stippled block corresponds to the Co—H bonding MO.

these changes cause a reduction in the HOMO–LUMO gap. Both changes should also be reflected in the chemistry of the two complexes; e.g., one would expect nucleophiles to attack the NO ligand in preference to a CO ligand. A more comprehensive and rigorous discussion of the differences between metal carbonyls and nitrosyls including evidence from photoelectron spectroscopy can be found in works by Lichtenberger and Kellogg (31) and by Green (32).

In Fig. 5, the eigenvalue spectra of $Ni(CO)_4$ and $HCo(CO)_4$ are compared. Here the molecules are related by the transfer of a proton from the metal to an adjacent position. In this case, the CO ligands must rearrange to provide space for the new hydride ligand. Again, the reduction in the nuclear charge of the metal results in an increase in energy of all the metal-centered orbitals except one. This orbital has now become highly mixed with the H $1s$ AO and constitutes the MH bonding MO. Because of the significant geometric difference between $Ni(CO)_4$ and $HCo(CO)_4$ there is no longer a simple one-to-one correspondence between MOs as seen in Fig. 2 for CO and HBO. However, the net consequence is the same. Deprotonation of $HCo(CO)_4$ with no structural rearrangement should lead to a species with a high-lying filled MO in the region of space previously occupied by the proton. In fact $HCo(CO)_4$ is rather acidic, which may reflect in part the ease of CO rearrangement to the energetically more favorable tetrahedral structure of the anion.

In terms of systematic variation of the properties of main group element or transition metal, variation within a given group is as important as the "proton transfers" discussed above. Here the source of change is the difference in effective nuclear charge as the core electron density increases. To illustrate trends we use the evidence provided by photoelectron spectroscopy, since calculational techniques become more suspect as the number of electrons increases. For small molecules, and even for some large systems, the photoelectron spectra can be assigned empirically, and therefore the information derived on the radical cation states of a given molecule is independent of the approximations of quantum chemistry (33). Thus, the trends discussed below are empirical in nature rather than theoretical, i.e., in the context of Greenwood and Earnshaw's preface to *Chemistry of the Elements* (5), these ionization data are some of the facts of chemistry. Koopmans' theorem (34), however, provides a connection between the eigenvalue spectra discussed above and the photoelectron spectroscopic results to be discussed below. That is, in a closed shell molecule each filled MO can be identified with a radical cation state of the molecule, and the trends already discussed above are verified by photoelectron spectroscopic data. In the frozen orbital approximation, one equates the MO energies with the negative of the ionization potentials

(26). Although the quantitative use of Koopmans' theorem is rarely justified, it is useful as a language for discussing trends in ion states in terms of filled MOs. We use it in this sense.

With the photoelectron spectroscopic technique one gains information that relates only to the occupied MOs; hence, what follows is restricted to the lowest ionization potentials corresponding to the highest occupied MOs. A most striking change in going down a group is the decrease in the ionization potentials of the valence electrons. As shown in Fig. 6 for the series N_2 to P_2, this decrease is about 5 eV (35). Concomitant with decreasing ionization potential is an increase in MO size. Thus, the expected increase in basic properties of the molecule is modified by a greater diffuseness of the electron density. A second important difference is the fact that the σ and π levels do not change by the same amount. As a consequence, the highest lying ionizations of P_2 lie closer together than those of N_2, i.e., the separation between the highest σ and π levels decreases appreciably. Hence, one might expect the potential sites of basicity in a heavier congener to be either more competitive or even different than those for the first row species. To emphasize this point, the valence ionizations of H_2CEH (E = N, P) are shown schematically in Fig. 7 (36). Here there is actually an inversion of the "lone pair" and π ionizations in going from N to P.

Similar changes occur when one compares CO and CS (Fig. 8) (37). Further, in comparing PN and CS one sees that the energies change in the same manner as the energies of the two highest filled MOs of N_2 and CO (Fig. 1) but that the scale of the energy change is much smaller. Finally, comparing CS with HBS (Fig. 8) shows that the large change caused by

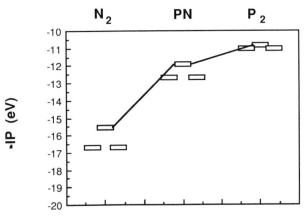

FIG. 6. Correlation of the two lowest ionization potentials (IP) for N_2, PN, and P_2 corresponding to the highest filled σ and π MOs (28,35).

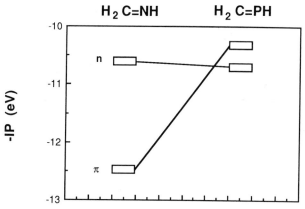

FIG. 7. Correlation of the "lone pair" and π ionizations of H_2CNH and H_2CPH (*36*).

relocating a shielded proton elsewhere in the molecular framework causes a perturbation of essentially the same magnitude as in the first row example (Fig. 2) (*38*).

The electronic structure of these small molecules could serve as the basis of a full article in another context, and our short summary of trends in MOs (or negative ionization potentials) lacks the depth the topic deserves. However, an understanding of these preliminary considerations is requisite for understanding the trends in structure and bonding observed in main group element–transition metal compounds.

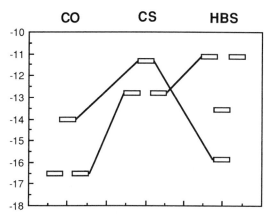

FIG. 8. Correlation of the two lowest ionization potentials for CO and CS (corresponding to the highest filled σ and π MOs) and correlation of the lowest σ ionization of the latter with the BH ionization of HBS (see also Fig. 2) (*38*).

C. Selected Examples

In this section, we focus on isoelectronic series containing the main group hydrides tabulated below. We chose these species, first, because they are the simplest main group fragments that can form bonds with transition metals and, second, because the consequences of these fragments interacting with transition metals have a direct bearing on important chemical processes, such as H_2O splitting, catalytic oxidation of organic molecules, CO reduction followed by C—C bond formation, nitrogen fixation, and desulfurization reactions.

—CH₃ (—CH₂)	—NH₂ (—NH)	—OH (—O)	
—SiH₃ (—SiH₂)	—PH₂ (—PH)	—SH (—S)	—Cl
—GeH₃ (—GeH₂)	—AsH₂ (—AsH)	—SeH (—Se)	

Of importance in each series is comparison of structure, unusual physical properties (if any), and chemical reactivity. A general observation is that the main group–hydride fragments when bonded to transition metals tend to lose hydrogen more easily as one progresses down each main group family. Hence, only a few examples of molecules containing SnH_3, SbH_2, or TeH fragments exist. Molecules containing the main group 13 fragments, e.g., $—BH_4^-$, might have also been included, but we have chosen not to do so because there is no direct main group element–transition metal bond. However, the scientific literature is rich in complexes of this nature and reviews are available (39).

1. 130-Valence Electron Mononuclear Complexes

The first set of examples are drawn from the work of Bercaw and colleagues (40). The four 130-electron compounds $(Cp^*)_2W(Cl)Me$ (1), $(Cp^*)_2W(Me)_2$ (2), $(Cp^*)_2Ta(PH_2)(Me)H$ (3), and $(Cp^*)_2Ta(SiH_3)(Me)H$ (4) are typical for early transition metals ($Cp^* = \eta^5\text{-}C_5Me_5$).

a. Structure and Bonding. Although no crystal structures have been published for compounds 1–4 a distorted tetrahedral geometry similar to that of $Cp_2Hf(Me)_2$ (41) is reasonable for 1 and 2. On the other hand, the latter two most likely have the Cp^* ligands axial in a distorted trigonal bipyramidal arrangement like that published for the structure of $(Cp)_2Ti(CO)(\eta^2\text{-PhCCPh})$ (see Section IV,B). As well established for this type of molecule (42), the Cp^* ligands form a "bent sandwich" with the metal at the base of the resulting **V** structure. The Cp^* rings are usually planar with the methyl groups bent away from the metal.

Compounds **3** and **4** have the PH_2 and the SiH_3 fragments positioned centrally equatorially with the H and Me occupying lateral positions coplanar with the Ta atom (*40*). Models of the bonding of similar metallocenes such as $(Cp)_2Ta(H)_3$ have been developed by Lauher and Hoffmann (*43*) using calculational techniques and by Green *et al.* (*44*) using photoelectron spectroscopic data. The three $(Cp)_2Ta$ fragment valence orbitals (of symmetry $2a_1$, $2b_1$, and $3a_1$) lying in a single plane are utilized to bond the three hydride ligands. Thus, the photoelectron spectrum of $(Cp)_2Ta(H)_3$ has no ionization band below 7.5 eV, confirming the absence of nonbonding pairs on the metal. In contrast, **1** and **2** are expected to show one ionization band below 7.5 eV, and closely related compounds indeed do so.

b. Chemical Reactivity.   Because of its reactivity toward nucleophiles, compound **1** is a useful starting material for a number of other permethylmetallocenes. It can be converted to **2** by treatment with MeLi, and the chlorine atom can be substituted by a hydride atom by reaction with $LiAlH_4$ to give $(Cp^*)_2W(Me)(H)$. Compound **2** can be synthesized from **1** by alkylation with MeLi. Conversely, **1** can be regenerated from **2** by treatment with Me_3SiCl and water (*39*). Reaction with PhLi is unusual in that an intermediate cation, $[(Cp^*)_2W(=CH_2)(H)]^+$, may be involved. The electrophilic center is localized on the methylenic carbon atom, which subsequently becomes alkylated to form $(Cp^*)_2W(H)(CH_2Ph)$. This has been viewed as activation of a methyl hydrogen by interaction with a metal via a C–H–M bridge (*42*). Compound **1** reacts with the carbene source $LiCH_2PMe_2$ not by addition of an alkylidene to the metal as would be expected, but rather by metallation of a Cp* ring. Displacement of the chlorine atom results in bridging to the metal atom. An analogous product is observed on refluxing a toluene solution of $(Cp^*)_2Ti(Me)_2$, but here, however, methane is evolved (*45*).

2. *128-Valence Electron Mononuclear Complexes*

A related isoelectronic group of compounds is $(Cp^*)_2Hf(Me)_2$ (**5**) (*41*), $[(Cp^*)_2Ta(Me)_2]^+$ (**6**), $(Cp^*)_2Ta(=CH_2)(Me)$ (**7**) (*46*), $(Cp^*)_2Ta(\eta^2\text{-}C_2H_4)H$ (**8**) (*47*), $[(Cp^*)_2W(=O)Me]^+$ (**9**) (*39*), $(Cp^*)_2Ti(EH)_2$ (**10**) (E = S, Se, O) (*48*), $(Cp^*)_2Hf(NHMe)(H)$ (**11**) (*49*), $(Cp^*)_2M(OH)_2$ (**12**) (M = Zr, Hf) (*50*), $(Cp^*)_2Hf(OH)(H)$ (**13**) (*50*), and $(Cp^*)_2M(NH_2)(H)$ (**14**) (*50*).

a. Structure and Bonding.   The bonding of the main group fragment to the metal in metallocenes **5–14** involves the utilization of two of the three valence orbitals of the Cp_2M fragment. Evidence comes from structural studies. The structures of **5, 7, 10**, and **11** have been determined by X-ray crystallography and are shown schematically in Fig. 9. Compound **5** has distorted tetrahedral structure with the cyclopentadienyl groups tilted 132° (*41*). The two Hf—Me bond lengths of 2.318 and 2.382 Å are significantly different, and, in addition, the Me groups are disposed at a near right angle (94.8°). By comparison, the Cp ligands in **7** are tilted almost to an identical degree (135.7°) (*46*), but the Ta—Me bond length is shorter (2.246 Å), the Ta—CH_2 bond length is short enough to be considered a tantalum–carbon double bond (2.026 Å). The plane defined by the CH_2 group is almost perpendicular (88°) to that containing the $C(H)_3$—Ta—$C(H)_2$ atoms. This places the $p\pi$ orbital of the methylene in an orientation suitable for π bonding to a metal orbital that lies in this plane. The angle between the Ta—Me and Ta—CH_2 bonds is a little larger than that observed in **5** (95.6°).

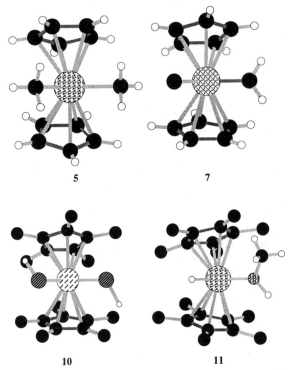

FIG. 9. Structures of $(Cp^*)_2Hf(Me)_2$ (**5**) (*41*), $(Cp^*)_2Ta(=CH_2)(Me)$ (**7**) (*46*), $(Cp^*)_2$-$Ti(SH)_2$ (**10**) (*48*), and $(Cp^*)_2Hf(NHMe)(H)$ (**11**) (*49*) generated from the respective X-ray parameters. Note the orientations of $=CH_2$ (88°), —SH (65°), and —NHMe (60°) from the horizontal plane, to maximize backbonding with the metal atoms.

Compound **11** is closely related to **7**. If the Me group on N and the H on Hf were interchanged, **11** would be related to **7** by the transfer of a proton between the metal and main group atom nuclei. The consequences are evidenced in the structural data. As observed for the previous two compounds, the tilt angle of the Cp ligands is 139.1°. Compound **11** has an Hf—N bond length of 2.027 Å, which is very close to that of the methylidene in **7**. Likewise the N—Me bond is out of the H—Hf—N plane, suggesting an interaction of the N $p\pi$ orbital with a metal in-plane orbital. However, the angle achieved (63°) is much less owing to either steric interactions of the methyl group with the Cp ring and/or a much more polarized and weaker π interaction. Consistent with this interpretation is the larger H—Hf—N angle of 101° (*49*).

Additional information on electronic structure comes from a NMR study of a closely related series of compounds of the type $(Cp^*)_2Hf(X)(H)$, where X is H, Me, OH, NH_2, NHMe, or NMe_2. Bercaw and co-workers observed that in the ^{1}H-NMR spectrum the metal-bonded H atom becomes more shielded as the π donor ability of X increases (*51*). The hydride chemical shifts corresponding to X are δ 15.6, 13.1, 10.2, 9.3, 9.1, and 11.5, respectively. Clearly, the π interaction of the main group ligand with the metal is important in compounds of this type.

$(Cp^*)_2Ti(SH)_2$ (**10**) has Ti—S bond lengths of 2.409 and 2.418 Å. The hydrogens on the sulfur atoms are trans to one another and form angles of 66 and 64.5° with the S—Ti—S plane. This disposition suggests that here, as in **7** and **11**, overlap between a filled orbital on sulfur and an empty orbital on titanium is important (*48*).

b. Chemical Reactions.   Complexes **5–11** are all somewhat air sensitive in solution. The lighter analog of **5**, $(Cp)_2Zr(Me)_2$, has been shown to oxidatively add selenium in the Zr—Me bonds to yield $Cp_2Zr(SeMe)_2$, to photolytically insert C_2D_4 in a Zr—Me bond to give Cp_2Zr-$(CD_2CD_2CH_3)(Me)$, and to add CO to give the acetyl derivative Cp_2Zr-$[\eta^2C(=O)CH_3](Me)$ (*52*).

While $(Cp)_2Zr(Me)_2$ reacts mainly with electrophiles by oxidative addition to the M—Me bond, $(Cp^*)_2Ta(=CH_2)(Me)(7)$ has nucleophilic character situated at the methylidenic carbon atom. Lewis acids such as $AlMe_3$ form adducts at this position, and **7** can be regenerated by reaction with NEt_3. Compound **7** attacks CD_3I, and the intermediate $(Cp^*)_2$-$Ta(Me)(CH_2CD_3)$ eliminates CH_3D to yield $(Cp)_2Ta(\eta^2\text{-}CH_2CD_2)$. The carbene carbon ends up incorporated in an ethene molecule. On the other hand, reaction with CO or C_2H_4 leads to products where the methylidene ligand has been replaced.

Compounds **11** and **14** react with moisture to lead to the hydroxide $(Cp^*)_2Hf(OH)(H)$ (**13**) (*50*), presumably through the dihydroxyl intermediate $(Cp^*)_2Hf(OH)_2$ (*12*). The nitrogen atom in **14** is exchange-able with ^{15}N when stirred in $^{15}NH_3$ (*53*). Compound **10** in solution exists in an equilibrium with its dimer $[(Cp^*)_2Ti](\mu-S)_2$ as a minor component. It also reacts with sulfur transfer agents to form metallacyclohexasulfanes (*54*).

3. *122-Valence Electron Mononuclear Complexes*

The final set of examples in this series are the isoelectronic 122-total valence electron complexes $(Cp^*)_2Hf(Me)H$ (**15**), $(Cp^*)_2V(Me)$ (**16**), $(Cp^*)_2Ta(=CH_2)H$ (**17**), $[(Cp^*)_2W(Me)]^+$ (**18**), $[(Cp^*)_2W(=CH_2)H]^+$ (**19**), $(Cp^*)_2Ta(=O)H$ (**20**), and $[(Cp^*)_2W(=O)H]^+$ (**21**) which have been studied mainly by Schrock (*55*), Green (*56*), Marks and Kolb (*39*), and Bercaw and colleagues (*57*).

a. Structure. To the best of our knowledge, no crystal structures have been published for compounds **15–21**, and the bonding is directly related to that discussed above for the metallocenes **7–14**. Conspicuously absent in the series **15–21** are the heavier congeners such as the chalcogens containing SH.

b. Chemical Reactivity. Of particular interest in terms of chemical reactivity are compounds **18** and **19**, which differ only in the location of a hydrogen atom. In fact **18** and **19** are in equilibrium and demonstrate an "agostic" metal–hydrogen activation (*42*). The methylidenes **17** and **19** are the more stable, but the alkyl isomers $[(Cp^*)_2W(CH_3)]^+$ (**18**) and $(Cp^*)_2Ta(CH_3)$ can be trapped by addition of CO, CH_2PMe_3, PH_3, or SiH_4 to give $(Cp^*)_2M(CO)(Me)$, $(Cp^*)_2M(=CH_2)(Me)$ (**7**), $(Cp^*)_2Ta$-$(PH_2)(Me)H$ (**3**), and $(Cp^*)_2Ta(SiH_3)(Me)H$ (**4**), respectively (*40*). The analogous structural form of **20** and **21** with an OH fragment has not been observed. Instead, the hydrogen atom prefers to bond to the metal rather than the oxygen atom. This trend is the same as that observed for polynu-clear metal cluster systems (see Section V,C). Compounds **17** and **19** convert to **20** and **21** on treatment with water. These latter oxo complexes exchange the oxo ligand with labeled H_2^*O (*39*). The cation $[(Cp^*)_2W-(=O)H]^+$ (**21**) reacts with KOH to give the neutral compound $(Cp^*)_2$-$W=O$ (**22**) (*39*), which is isoelectronic with the hypothetical structure $(Cp^*)_2W(=CH_2)$. Complex **22** is electrophilic, and reacts with MI to give $(Cp^*)_2W(=O)(Me)$ and with $LiAlH_4$ by reduction to the hydroxo $(Cp^*)_2W(OH)(H)$. Facile reduction of transition metal–oxo complexes has an important bearing on catalytic CO cleavage.

4. *Dinuclear Complexes*

Although there are many isoelectronic series in the category of dinuclear complexes, we discuss only one set of examples taken from Herrmann's work (58), in order to demonstrate the diverse range of reactivities possible. This set includes the compounds $[Cp^*Mn(CO)_2(H)]_2(\mu\text{-}SiH_2)$ (23) (59), $[CpFe(CO)_2]_2(\mu\text{-}Se)$ (24) (60), $[Cp^*Re(CO)_2]_2(\mu\text{-}O)$ (25) (61), $[Cp^*Re(CO)_2]_2(\mu\text{-}Te)$ (26) (62), $[Cp^*Mn(CO)_2]_2(\mu\text{-}AsH)$ (27) (63a,b), $[Cp^*Mn(CO)_2]_2(\mu\text{-}Te)$ (28) (64), and $[(MeCp)Mn(CO)_2]_2(\mu\text{-}CH_2)$ (29) (65) (Fig. 10). Compounds 23 and 24 are isoelectronic, and members of the entire set are related in the sense of having isoelectronic bridging main group fragments.

a. Structure and Bonding. X-Ray structures have been published for compounds 23–29 (59–65). Compound 29 (see Fig. 10) might be considered as the parent molecule in the series 23–29. It has a methylene bridge and, hence, has been thoroughly studied. The Mn—Mn distance of 2.779 Å and the Mn—C—Mn angle of 87.3° in 29 shows that it has a single metal–metal bond.

A bonding scheme for metal dimers bridged by isoelectronic fragments begins with 29 and is based on molecular orbital calculations (66) and the photoelectron spectrum (67). Analysis of the photoelectron spectrum below the 11-eV range reveals two bands that can be assigned to metal–main group fragment interactions. The band found at 10.2 eV results from ionization of an MO corresponding to the main group fragment σ donor interaction with the metal center. The band at lower ionization potential (8.3 eV) corresponds to a metal fragment donor interaction with the π acceptor orbital on the main group fragment. The calculations provide more detail. The LUMO of the $[CpMn(CO)_2]_2$ fragment is bonding between the metal atoms and has the appropriate symmetry and energy to overlap with the σ (a_1) orbital of the CH_2 fragment. The HOMO of the $[CpMn(CO)_2]_2$ fragment is antibonding between the metal atoms but has the appropriate symmetry and energy to overlap and interact with the π (b_1) orbital of the CH_2 fragment. Hence, both σ donation and π acceptance of electron density by CH_2 strengthens the Mn—Mn interaction.

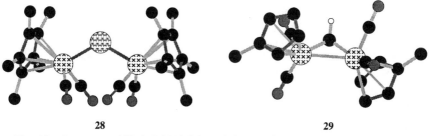

28   29

FIG. 10.   Structures of $[Cp^*Mn(CO)_2]_2(\mu\text{-}Te)$ (28) and $[(MeCp)Mn(CO)_2]_2(\mu\text{-}CH_2)$ (29) from X-ray structures, showing a series of transition metal dimers linked by isoelectronic main group element fragments (63b).

Hence, the $=CH_2$ fragment is viewed as interacting with the two metals in a six-electron, three-center triangular bonding scheme analogous to that in cyclopropane (67). Finally, there is net charge transfer from the $[CpMn(CO)_2]_2$ fragment to the bridging methylene ligand, suggesting increased nucleophilic character for the main group fragment.

Compound 23 has a distorted tetrahedral geometry at the Mn atom. The large metal–metal distance of 4.306 Å indicates the absence of a direct bond. Other important parameters are a Mn—Si distance of 2.434 Å and a Mn—Si—Mn angle of 124.4°. The cyclopentadienyl ligands are trans to one another. In the ^{1}H-NMR spectrum of 23 the SiH_2 fragment protons are found at δ 4.59. Consistent with the model developed for CH_2, the downfield resonance suggests significant charge transfer from the metals to the SiH_2 fragment. However, in 23 there can be no synergistic interaction between the main group metal and metal–metal interactions since, owing to the hydride ligands on the metal atoms, there is no metal–metal bond. In compound 24, which is isoelectronic to 23, the Fe—Se distance is 2.449 Å, and the Fe—Se—Fe angle is 115.27°, which is about 9° smaller than that subtended at the silicon atom in 23. The geometry at Fe is distorted trigonal pyramidal with the cyclopentadienyl ligand at an apex. As in 23, the large Fe—Fe distance precludes a metal–metal bond.

The oxo-bridged dimer $[Cp^*Re(CO)_2]_2(\mu\text{-O})$ (25) (61) has a distorted square pyramidal configuration at rhenium. Consistent with the two fewer electrons relative to 23, the Re—Re bond length of 2.817 Å demonstrates bonding between the two metal atoms. Similarly, the tellurium-bridged $[Cp^*Re(CO)_2]_2(\mu\text{-Te})$ (26) (62) has an Re—Re bond length of 3.140 Å which indicates a significant Re—Re interaction. As expected, the Re—Te distance of 2.679 Å is much longer than the Re—O bond length of 1.973 Å in 25, and the Re—Te—Re angle of 71.76° is smaller than the Re—O—Re angle of 91.1°.

Surprisingly, in going from the valence isoelectronic rhenium to manganese compounds (26–28) the M—Te—M angle opens up considerably to 123.8° and the Mn—Mn distance of 4.21 Å shows that there is little or no direct metal–metal bonding in 28. These structural changes are accompanied by rotation of the $Cp^*Mn(CO)_2$ fragment about the M—E bond axis as shown in Fig. 10 as well as by a decrease in the M—Te distance to a value (2.459 Å) indicative of Mn—Te multiple bonding. Apparently, going from Re to Mn stabilizes the open structure with added M—E bonding over the closed structure with a M—M bond.

The valence isoelectronic compound to 28, $[Cp^*Mn(CO)_2]_2(\mu\text{-AsH})$ (27) (63a,b), has an Mn—As bond length of 2.247 Å, an As—H bond length of 1.52 Å, and an Mn—As—Mn angle of 139.3°. In terms of bonding 28 is much more closely related to 27 than to 26 or the "parent"

29. These examples demonstrate that the bonding in these ostensibly similar compounds can be significantly varied by changing either the metal or main group atom cores.

  b. Chemical Reactions.   Thermolysis of **23** does not lead to hydrogen evolution, bond formation between the metal atoms, and generation of the silicon analog of **29**. Instead, the compound fragments to yield [Cp*Mn]$_2$(μ-CO)$_3$ and Cp*Mn(CO)$_3$ (*59*). A similar reaction occurs with the oxo-bridged compound **25**.

  As expected from the discussion of model compounds, the formal conversion of the SiH$_2$ fragment to Se yields sites of Lewis basicity. Hence, the selenium and tellurium bridges of **24** and **26** can be protonated by HBF$_4 \cdot$OEt$_2$ and the process reversed using weak bases such as NHEt$_2$. Alkylation with CF$_3$SO$_3$CH$_3$ also readily takes place.

  An interesting cluster-building reaction is observed with the chromium analog-[CpCr(CO)$_2$]$_2$(μ-Se) and Fe$_2$(CO)$_9$, leading to the cluster framework Cr$_2$Fe(μ^3-Se) (*68*). A similar "tetrahedrane" cluster with (μ^3-Te) forms from reaction of **28** and Fe$_2$(CO)$_9$ (*69*). This tendency toward novel transition metal cluster building, with main group elements as anchoring units is important in the design of homogeneous cluster catalysts because of the superior strength of the transition metal–main group metal bond relative to metal–metal bonds. In a related reaction, hydrogen is lost when [Cp*Mn(CO)$_2$]$_2$(μ-AsH) (**27**) is heated, and a new complex, [Cp*Mn(CO)$_2$]$_2$(μ,η^2-As$_2$), with a "butterfly" framework forms (*63a,b,70*). Here the As$_2$ unit binds in an acetylenic fashion to the metal framework. The molybdenum and tungsten analogs have been made previously (*71*). The compound is related to the series [CpMo(CO)$_2$]$_2$(μ,η^2-C$_2$H$_2$) (*72*), [(OC)$_3$Co]$_2$(μ,η^2-P$_2$) (*73*), [(OC)$_3$Co]$_2$(μ,η^2-As$_2$) (*74*), and [(OC)$_3$Fe]$_2$(μ,η^2-S$_2$) (*75*), all of which have been characterized by X-ray crystallography.

5. *Complexes with Three-Center, Two-Electron Si—H—M Bonds*

  A general review by Cundy *et al.* (*76*) on silicon–transition metal complexes has appeared, and again, we select only a few molecules to highlight different properties and chemistry within an isoelectronic group. In the four examples (Cp)Mn(CO)$_2$(μ-H)Si(F)(Ph)$_2$ (**30**), (Cp)Fe(CO)$_2$Si-(F)(Ph)$_2$ (**31**), (Cp)Mn(CO)$_2$(μ-H)Si(Cl)$_3$ (**32**), and (Cp)Fe(CO)$_2$Si(Cl)$_3$ (**33**), compounds **30–31** and **32–33** are isoelectronic pairs based on the total number of valence electrons. Compounds **30** and **32** also form a related pair with respect to the three-center, two-electron Si—H—Mn interactions.

a. Structure.   Compounds **30** and **31** (or **32** and **33**) are formally inter-converted by moving a proton from the Fe core to a bridging position on the metal–silicon axis. This leads to an apparent shortening of the M—Si bond from 2.352 Å (Mn—Si) to 2.278 Å (Fe—Si) in the **30–31** pair (Δ 0.074 Å) and from 2.254 to 2.216 Å in the **32–33** pair (Δ 0.038 Å). To interpret these distances one should take into account the difference in the radii of Mn and Fe atoms (*77*). Since Mn is about 0.04 Å larger than Fe, the M—Si interaction is unaffected in **32** and **33** but clearly increased (0.035 Å) in going from **31** to **30**. Likewise, an increase in the electronegativity of the substituents on silicon results in a 0.098 Å decrease in the M—Si bond in going from **30** to **32** and 0.062 Å in going from **31** to **33** (*78*). The larger change in the former pair can be attributed to the position of the H atom relative to M—Si, i.e., in **30** it is closer to the silicon than in **32**, and one expects this to be reflected in a difference in properties like acidity.

b. Chemical Reactions.   In **30** the hydride is closer to the silicon atom and, presumably, more covalently bonded. On the other hand, in **32** it is closer to the metal atom and more like a metal hydride. Metal hydrides in general exhibit considerable Brönsted acidity. Thus, the hydride atom in **32** is easily abstracted by Me$_3$N, but that in **30** is unaffected by this reagent (*79*).

The isoelectronic group of compounds represented by **30** and **31** react differently with nucleophiles. For example, the M—Si bond in **30** is cleaved by phosphines while in **31** only CO substitution occurs (*80*). Incorporation of phosphines into **30** has been accomplished only when attached to the metal before addition of the silicon fragment. Conversely, **31** undergoes Fe—Si bond cleavage when treated with electrophiles, such as Cl$_2$ and PCl$_5$, whereas hydride bridged complexes **30** remain intact (*79*).

Compounds similar to **30**, e.g., CpMn(CO)$_2$(μ-H)Si(Ph)(1-Np)Cl (1-Np = 1-naphthyl), are reduced by hydride reagents, such as NaH and LiAlH$_4$. Reaction proceeds first by hydride–halogen exchange followed by further reduction to yield the anion [CpMn(CO)$_2$Si(H)(Ph)(1-Np)]$^-$ containing an intact Mn—Si bond (*79*). On the other hand, compounds similar to **31**, e.g., CpFe(CO)$_2$(Me)Si(Ph)(1-Np), are cleaved by LiAlH$_4$.

IV

POLYHAPTO LIGANDS

Between the simple metal–main group atom interactions discussed above and clusters discussed below one finds systems in which the main group fragment can be considered a polyhapto ligand. Although the metal–main group interactions are more complex, for molecules with a single metal center the metal–ligand interaction is easily distinguished. Hence, it constitutes a useful classification. For polyhapto systems with more than one metal center, the cluster description found below is often an equal or more effective description. For this reason, the multimetal–main group systems are discussed in the section on clusters.

A. *Model Compounds*

To introduce this section we illustrate some of the features of the bonding of one of the simplest π ligands, ethylene, bound to a transition metal fragment. The structure of $(OC)_4FeC_2H_4$, the exemplar of an olefin–metal complex, is shown in Fig. 11 in comparison with the proposed structure of $[(OC)_4FeB_2H_5]^-$ (*81,82a,b*). The electronic structures of these two molecules have been examined in detail as have the photoelectron spectra of $(OC)_4FeC_2H_4$ and the closely related $CpFe(CO)_2B_2H_5$ (*82a,b*). Hence, the nuclear charge perturbations are well defined by empirical as well as by calculational techniques.

In Fig. 12, we compare Fenske–Hall MO energies for $(OC)_4FeC_2H_4$, $[(OC)_4FeB_2H_4]^{2-}$, and $[(OC)_4FeB_2H_5]^-$. The first two are related by the transfer of two protons away from the dihapto ligand (one from each carbon atom). All the energies increase because of the double negative charge, and, as should be appreciated from the discussion of the simple ligands above, the MO energies of the E_2H_4 ligand rise relative to the metal fragment MOs as E goes from C to B. The primary ligand-to-metal

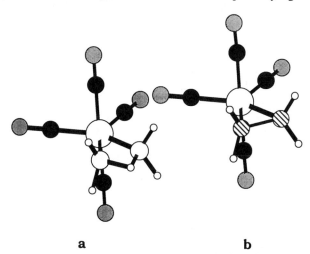

a   **b**

FIG. 11. Schematic drawing of the structures of (a) $[(OC)_4FeB_2H_5]^-$ (*82a,b*) and (b) $(OC)_4FeC_2H_4$ (*81*).

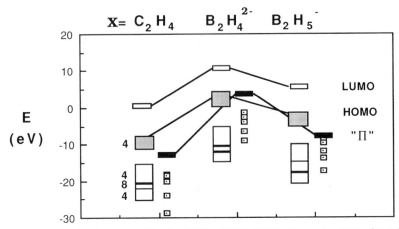

FIG. 12.   Plot of MO energies (Fenske–Hall) of $(OC)_4FeX$ for X equal to C_2H_4, $[B_2H_4]^{2-}$ (hypothetical), and $[B_2H_5]^-$ which are related by proton transfer within the E_2H_n ligand (the 4 lowest lying filled MOs and 21 of the unfilled MOs are not shown). Small rectangles or squares correspond to a single MO. The larger blocks contain the number of MOs indicated, and the energies of the upper and lower edges are defined by the highest and lowest MO energies for a particular block. The stippled blocks correspond to the metal "nonbonding" MOs, whereas the solid rectangles correspond to the MOs derived from the π MOs of the E_2H_4 ligands for $(OC)_4FeC_2H_4$ and $[(OC)_4FeB_2H_4]^{2-}$ and the BHB MO for $[(OC)_4FeB_2H_5]^-$.

bonding MO (derived from the π MO of the free E_2H_4 ligand and emphasized in black in Fig. 12) is now of comparable energy to the metal d band. The metal–main group interaction may still be significant because, even though ligand donor orbital is more diffuse than for carbon, there is a better energy match of the boron and metal AOs. There is a corresponding rise in the secondary metal-to-ligand bonding MO (derived from the π^* MO of the free E_2H_4 ligand) which reduces the net metal–main group bonding. Hence, $[B_2H_4]^{2-}$ acts more like a pure donor than C_2H_4.

In going from $[(OC)_4FeB_2H_4]^{2-}$ to $[(OC)_4FeB_2H_5]^-$ one of the protons is returned to a bridging position in the E_2H_4 ligand. All the MOs are stabilized because of the reduced negative charge but, as expected from the example illustrated in Fig. 2, the MO correlating with the π MO of the E_2H_4 ligand is stabilized preferentially since this proton is situated so that it would be buried in one of the lobes of the π orbital of the free ligand (see Fig. 10). This bridging proton enhances the metal–boron interaction by reorienting the $p\pi$ lobes so that they point more directly at the metal center. The two situations are compared in Fig. 13. When E is C the M—C bonding density in the ligand-to-metal MO is outside the C_2Fe triangle, whereas when E is B the M—B bonding density is inside the triangle. In the latter case this provides compensation for the lack of an interaction

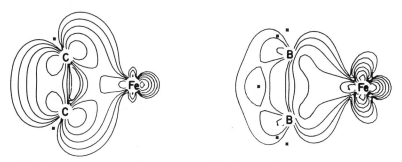

Fig. 13. Plot of relative electron densities for the principal ligand-to-metal donor–acceptor interaction for $(OC)_4FeC_2H_4$ and $[(OC)_4FeB_2H_5]^-$ (*82b*).

corresponding to the metal-to-C_2H_4 π^* back donation, which is so important in the bonding of olefins to metal centers.

To conclude these preliminary considerations we go to the right of carbon and compare N_2H_2 with C_2H_4 as potential ligands. As indicated in Fig. 14, there are three major differences. First, although both species have π MOs, that of N_2H_2 lies at lower energy because of the increase in the effective nuclear charge associated with the p AOs. Second, two MOs of C_2H_4 with large CH character have now become "lone pair" (n_N) orbitals. Taking a proton from a CH bonding region and placing it in a shielded nucleus results in a large destabilization of the corresponding MO. Hence, these "lone pair" orbitals lie at higher energy despite the fact that the nuclear charge of N is greater than that of C. Indeed the symmetric n_N combination is now the HOMO. Finally, the LUMO which is the π^* MO

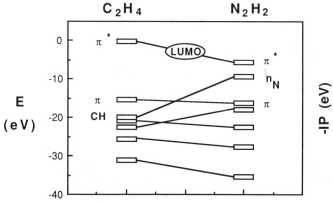

Fig. 14. Correlation of the MOs (Fenske–Hall) of *trans*-N_2H_2 and C_2H_4.

falls in energy. Hence, compared to C_2H_4, one expects N_2H_2 to be a poorer π donor and better π acceptor but to act primarily as a σ donor.

B. Selected Examples

In the following, three comparisons of isoelectronic systems are presented. The first consists of a comparison of the isoelectronic ligands ethyne (HCCH) and phosphaethyne (HCP) for different transition metals and in different states of "aggregation." The second compares cyclopentadienyl and pentadienyl complexes, while the third relates complexes of benzene and hexaphosphabenzene. Although narrow in terms of breadth of compounds, we hope there is sufficient detail in these systems to convey some appreciation of the systematic variation possible with isoelectronic substitution.

1. Monoalkyne and Phosphaalkyne Complexes

Transition metal chemistry with ethyne and substituted ethynes is well developed. Although phosphaethyne is difficult to make (flash pyrolysis of $MePCl_2$ at 900–1100°C) and polymerizes above −124°C unless stored under reduced pressure, the derivative 2,2-dimethylpropylidyne phosphane, Bu^{t}CP, is much more convenient to use, and the number of its transition metal complexes is growing rapidly (83–85). The bonding capabilities of the RCP molecule as a ligand have been discussed based on calculational as well as photoelectron spectroscopic results. It appears that the HOMO of Bu^{t}CP is the π MO (ionization potential of 9.7 eV) and that the σ and π separation is larger than in RCN (see Fig. 7 for a closely related example). Hence, η^2 coordination is favored as will be evident from the examples presented below. However, coordination via the σ "lone pair" on phosphorus can be forced, e.g., trans-(DIPHOS)$_2$ Mo-(Bu^{t}CP)$_2$ (84,85). The phosphine ligands on the octahedral Mo are disposed such that the only available sites for the Bu^{t}CP ligands are axial with sufficient space to accommodate only the thin end of the ligand.

a. Early Transition Metal Complexes. In $(Cp)_2M(PMe_3)(\eta^2$-Bu^{t}CP) (**34**) (M = Ti, Zr) (86) and $(Cp)_2Ti(CO)(\eta^2$-PhCCPh) (**35**) (87), the alkynes are bound similarly in a sideways (η^2) fashion (Fig. 15). X-Ray structures reveal a lengthening of the C—C and C—P bonds on coordination, and the η^2 ligand, the metal, and the other ligand are nearly coplanar. The Cp ligands are situated above and below this plane and bent back at an angle of about 134°.

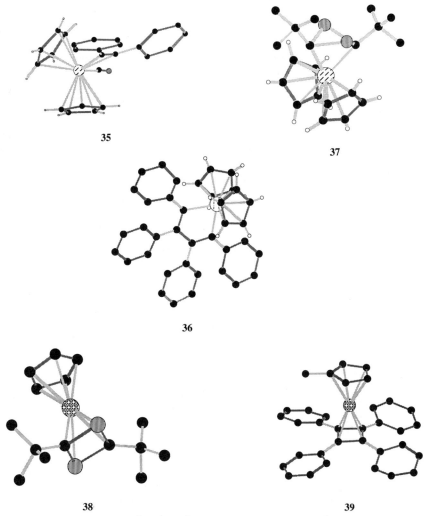

FIG. 15.   Structures of $(\eta^5\text{-}Cp)_2Ti(\eta^2\text{-}PhCCPh)(CO)$ **(35)** *(87)*, $(\eta^5\text{-}Cp)_2Zr(C_4Ph_4)$ **(36)** *(92)*, $(\eta^5\text{-}Cp)_2Zr(Bu^tCP)_2$ **(37)** *(91)*, $(\eta^5\text{-}Cp)Co(\eta^4\text{-}Bu^tCP)_2$ **(38)** *(94,95)*, and $(\eta^5\text{-}MeCp)_2Co(C_4Ph_4)$ **(39)** *(96)* as generated from X-ray bond lengths and angles. Compounds **36** and **37** are isoelectronic with respect to the metal–ligand interaction. Similarly, **38** and **39** are isoelectronic.

While very few details have been published about the reactions of **34** *(88,89)*, compound **35** is an effective hydrogenation catalyst for carbon–carbon unsaturated bonds, but is inactive toward N_2. The coordinated alkyne appears to be an important integral part of the catalyst *(90)*.

Compound **35** does not form above 30°C when an aromatic solvent is used; instead, two alkynes are dimerized to give a titanacyclopentadiene having the structure of 1,1-bis(η^5-cyclopentadienyl)-2,3,4,5-tetraphenylzirconole (**36**) (Fig. 15). It is an important intermediate in alkyne oligomerization reactions on transition metal complexes.

b. Metal Dialkyne Complexes.   The complex **34** adds a second mole of phosphaalkyne in the same manner as **35** adds another alkyne, but the product, a folded 1,3-diphosphabicyclo[1.1.0]butane (**37**), is radically different. Compound **37** (Fig. 15) (*91*) is isoelectronic with **36** (*92*). The characteristic singlet (δ −247 ppm) in the ^{31}P-NMR spectrum eliminates a zirconacyclobutadiene structure. The zirconium atom in **37** is tetrahedrally coordinated. The metal–carbon σ bonds of 2.213 and 2.210 Å are about 0.04 Å shorter than what is normal in related Cp$_2$Zr complexes (i.e., 2.265 and 2.250 Å for the latter zirconole) (*92,93*). Also the long Zr—P distances (2.906 and 2.909 Å) indicate an absence of Zr—P bonds.

To highlight what one would expect in reactions of the diphosphazirconole **37**, it is instructive to examine the η^4-1,3-diphosphacyclobutadiene complex (**38**) (*94,95*), whose X-ray structure is compared in Fig. 15 with that of the isoelectronic η^4-cyclobutadiene complex **39** (*96*). Compound **38** is readily obtained from reaction of (Cp)Co(η^2-C$_2$H$_4$)$_2$ and 2 equiv of Bu^{t}CP. The same reaction with a pure alkyne does not stop at a cyclodimer but leads to cyclotrimerization (*97*). In fact, transition metal–cyclobutadiene complexes normally form only at temperatures above 80°C, presumably from a metallole intermediate, by a double reductive elimination process. It is noteworthy how readily this cyclodimerization to complex **38** takes place with phosphaalkynes.

The four P—C bonds in the diphosphacyclobutadiene ring of **38** are almost equal (1.795 Å), and the two carbon and phosphorus atoms are 2.057 and 2.243 Å away from the metal atom (Fig. 15). In addition the *tert*-butyl groups are bent away from the metal by 4–6°. The ^{31}P-NMR spectrum of **38** shows a singlet at δ 38.1 ppm, indicating that the ligand remains symmetrical in solution. In contrast, the average C—C bond length in **39** is 1.468 Å in the square planar cyclobutadiene ring. This ring is 1.983 Å away from the metal atom, and the phenyl substituents are bent slightly away from the cobalt atom (Fig. 15) (*96*). The bending away of substituents from the metal is a common feature of transition metal–cyclobutadiene complexes (*98–101*).

Like its cyclobutadiene analog **39**, the η^4-1,3-diphosphacyclobutadiene complex **38** has high thermal stability, is stable to air, and generally has low reactivity. Reagents that normally release cyclobutadiene from transition metal–cyclobutadiene complexes by oxidative demetallation [i.e., HCN, (NH$_4$)$_2$Ce(NO$_3$)$_6$, and alkali metals] do not react with compound **38**.

However, an important reaction does take place at the phosphorus atoms as each displaces an ethene molecule from $(Cp)Co(\eta^2\text{-}C_2H_4)_2$ to form a trimetal complex. Thus, complexation of the phosphaalkyne via the π system does not eliminate the Lewis basicity of the phosphorus "lone pairs." Presumably, the same reaction can be envisaged for the 1,3-diphosphabicyclo[1.1.0]butane zirconole (**37**). The phosphorus atom provides an additional bonding dimension otherwise not encountered in complexes formed from pure alkynes.

In addition to **38** and **39**, other similar η^4-1,3-diphosphacyclobutadiene complexes have been made with iron, rhodium, and iridium (*40*). Interesting chemistry is observed with rhodium, where two of the molecules appear to have dimerized with a metathetical extrusion of di-*tert*-butylacetylene to give a sandwich complex with the two rhodium atoms complexing on opposite sides of a twisted 1,2,4,5-tetraphosphacyclohexatriene ring (*40*).

Platinum also coodinates Bu^tCP in a sideways manner to give $(Ph_3)_2Pt(\eta^2\text{-}Bu^tCP)$ (**40**) which has been characterized by X-ray diffraction (*102*). A crystal structure of the analogous alkyne derivative $(Ph_3)_2Pt[\eta^2\text{-}HCC(C_6H_{11}O)]$ (**41**) has been published (*103*). In both structures, the platinum atom exhibits a distorted square planar geometry. The former structure has a C—P bond length of 1.672 Å (which is 0.128 Å longer than that of the free Bu^tCP), a Pt—P distance of 2.32 Å, and a Pt—C length of 1.973 Å; as compared with a C—C bond length of 1.302 Å, a Pt—CH separation of 2.069 Å, and a Pt—C distance of 2.306 Å, for the η^2-coordinated 1-ethynylcyclohexanol in **41**.

The DIPHOS analog of **40** reacts with $Fe_2(CO)_9$ in a cluster-building step, eliminating three CO molecules to give a bimetallic trinuclear compound with the phosphaalkyne π bonding to two $Fe(CO)_3$ fragments and coordinating to the platinum through its lone pair (*104,105*). It also reacts with $[Re_2H_2(CO)_8]$, but this time via a carbonyl transfer to the carbon end of the CP fragment to give a ketene functionality, while the phosphorus end σ bonds to all three metal atoms (*40*). Compound **40** reacts with $[Pd(PPh_3)_4]$ to give a giant cluster with three Pt and two Pd atoms linked together by three phosphaalkynes (*105*). In contrast, the only reaction reported for $(Ph_3P)_2Pt[\eta^2\text{-}HCC(C_6H_{11}O)]$ is with a second alkyne by double oxidative addition of the platinum atom into the terminal C—H bonds. The resulting octahedral complex has two axial hydrides and the two alkynyl groups situated trans-equatorial in a linear geometry.

These reactions show how the reactivity of a molecule can be altered using the isoelectronic substitution of a CH fragment for a P atom. The higher propensity for cluster formation when a CH fragment in a molecule is replaced by a group 15 atom is important in the syntheses of labile cluster molecules.

2. Cyclopentadienyl Sandwich Complexes

Formal reductive cleavage of the two rings of nickelocene (4H), followed by removal of four protons from the nickel atom would yield a bis(η^5-pentadienyl)chromium "open chromocene" complex. This transformation could be regarded as a translocation of four protons from the nickel core to the cyclopentadienyl ligands. Bis(η^5-2,4-dimethylpentadienyl)chromium (**42**) is an example of just such a product. Therefore, when only the total number of valence electrons is considered (excluding the methyl substituents), this compound is isoelectronic with nickelocene.

Comprehensive reviews on the structures of and bonding in the chromium (*106,107*) and nickel compounds (*108*) above have appeared. In **42**, the methyl groups and the hydrogen on the central carbon atom are bent toward the metal atom while the framework carbon atoms are coplanar (*109*). In addition, the terminal π bonds are twisted such that the endo hydrogens are rotated away and the exo hydrogens toward the Cr atom. The average Cr—C bonds are 2.163 Å (2.169 Å in chromocene), remarkably similar to that found in nickelocene (2.164 Å) (*110*). Photoelectron spectral data are available for the nickelocene and cobaltacene but not for the open chromocene. Cp_2Ni has the ground-state electronic structure $(a_{1g})^2$, $(e_{2g})^4$, and $(e_{1g})^2$ and is paramagnetic (*111,112*).

Nickelocene reacts with a wide variety of reagents and is used as a starting material for many other nickel complexes. Bis(η^5-2,4-dimethylpentadienyl)chromium (**42**) shows some similarities in reactivity. For example, reaction with donor ligands such as phosphanes reductively removes both cyclopentadienyl ligands in nickelocene to give the tetrasubstituted $Ni(PR_3)_4$. Compound **42** reacts similarly with bis(dimethylphosphino)ethane (dmpe). On the other hand, the molybdenum and tungsten congeners react with PEt_3 by coordination of one ligand on the intact complex. Other differences do exist. For example, nickelocene is dimerized by isonitriles to give $[(Cp)Ni]_2(\mu\text{-}CNR)_2$ (*113,114*), but **42** is exhaustively substituted to give $(CNR)_6Cr$ (*115*).

3. Triple-Decker Sandwich Complexes

Until recently, the triple-decker tris(η^5-cyclopentadienyl)dinickel cation first made by Werner was the only example of this type of complex with all three rings composed of only carbon and hydrogen. Then in 1983 Jonas and co-workers detailed the novel $[(\eta^5\text{-}C_5H_5)V]_2(\mu,\eta^6\text{-}C_6H_6)$ (**43**), prepared from reaction of $(\eta^5\text{-}C_5H_5)V(\eta^3\text{-}C_3H_5)_2$ and 1,3-cyclohexadiene (*116*). A close analog is the triple-decker compound containing a hexaphosphabenzene as the central ligand, i.e., $[(\eta^5\text{-}Me_5C_5)Mo]_2(\mu,\eta^6\text{-}P_6)$ (**44**) (*117*).

The X-ray structures of **43** and **44** reveal three planar–parallel rings with the benzene or hexaphosphabenzene ring in the central position (Fig. 16) (*116,117*). In **43** the cyclopentadienyl rings are 1.922 Å from the vanadium atom, and the benzene ring is 1.702 Å from each of the two metal atoms (1.254 Å for the hexaphosphabenzene in **44**). The vanadium–vanadium distance of 3.403 Å is too long for a metal–metal bond. In **44** the Mo—C distances average 2.331 Å while the Mo—P average is 2.541 Å. This places the metal atoms much closer to the central phosphorus ring, and the Mo—Mo distance of 2.647 Å indicates bonding between the two atoms through the middle of the P_6 ring. The P—P bond lengths of 2.17 Å are about 0.17 Å longer than typical P=P double bonds.

The vanadium triple-decker compound **43** exchanges the benzene ring for other substituted benzenes, such as toluene and mesitylene, with retention of structure. It reacts with alkyl halides such as 1,2-dichloroethane in THF to give $(\eta^5\text{-}C_5H_5)V(Cl)(THF)$, which is a starting material for the preparation of other sandwich compounds such as $(\eta^5\text{-}C_5H_5)V(\eta^5\text{-}Me_5C_5)$. The triple decker is also oxidized by iodine to give $(\eta^5\text{-}C_5H_5)V(I)(THF)$. No reactions have been reported for the compound containing hexaphosphabenzene (**44**).

The nickel triple-decker tris(η^5-cyclopentadienyl)dinickel cation can similarly be compared with $[(\eta^5\text{-}Me_5C_5)Cr]_2(\mu,\eta^5\text{-}P_5)$ (*118*) and $[(\eta^5\text{-}C_5H_5)Mo]_2(\mu,\eta^4\text{-}As_5)$ (**45**) (Fig. 16) (*119a–c*). Here the central cyclopentadienyl ring is isoelectronic with P_5 and As_5. The metal–metal distances are Cr—Cr = 2.727 and Mo—Mo = 2.764 Å, respectively. Like the bridging cyclopentadienyl ring, the P_5 ring is a regular pentagon with P—P distances of 2.18 Å, however, the As_5 ring has two As—As bond

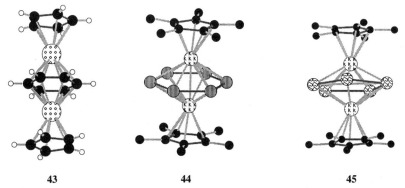

43   **44**   **45**

FIG. 16. Structures of $[(\eta^5\text{-}Cp)V]_2(\eta^6\text{-}C_6H_6)$ (**43**) (*116*), $[(\eta^5\text{-}Cp^*)Mo]_2(\eta^6\text{-}P_6)$ (**44**) (*117*), and $[(\eta^5\text{-}Cp)Mo]_2(\eta^4\text{-}As_5)$ (**45**) (*119a*) from X-ray parameters, showing triple-decker complexes containing isoelectronic units CH, P, and As. Similar complexes containing boron have been prepared by Siebert and colleagues (*119b,c*).

lengths of 2.393 Å and three of 2.753 Å. Effectively, one As atom has a reduced interaction with the metal atoms. Again, variation in the core size leads to significant changes in structure and bonding. Presumably, differences in the reactivity of these compounds also exist.

V

CLUSTERS

With homonuclear clusters we come to systems where an essential aspect of the bonding is delocalized. Because of this, the net bonding interactions become less easy to visualize than those of simple complexes or even those containing polyhapto ligands. Hence, the development of the skeletal electron counting rules (often referred to as Wade's rules) constituted an amazingly productive simplification of the problem (120). By dividing the total number of valence electrons into those associated with cluster skeletal bonding and all others (exo cluster ligand bonding, exo lone pairs, or nonbonding electrons), connections between apparently dissimilar species become evident (121). In addition, these rules give the experimental chemist a method for the generation of a reasonable cluster structure from an empirical molecular formula. Often this permits structural verification based on spectroscopic information. More recently, the application of the principles of Stone's tensor surface harmonic theory (122–124) and King's graph theory analysis (125) to the problem has provided additional insight into the origin and significance of the electron counting rules. An excellent review of cluster theories in general has been presented by Mingos and Johnson (126).

The skeletal electron counting rules are an integral part of cluster chemistry, and it is natural to assume that they will be useful in comparing, and ultimately in predicting, properties such as reactivity. However, as already indicated above, the electron counting rules separate the valence electrons into two types, those involved in skeletal bonding and those that are not involved. The ambiguities of this separation prevent any simple comparison of the type sought here. Hence, we emphasize how main group–metal interactions within a cluster bonding network in strictly isoelectronic species are perturbed by varying cluster atom identity.

First some simple calculational examples are presented to illustrate the changes in skeletal bonding as protons are moved about as well as to make connections with the electron counting rules. This is followed by a contemporary example illustrating how attempts to reconcile differences between molecules that are "isoelectronic" only in the sense of the electron counting rules can actually confuse the issue of electronic structure. These exercises lead to real examples of how properties of isoelectronic main group–metal clusters vary in a systematic and understandable fashion.

A. *Model Compounds*

The homonuclear cluster $B_6H_6^{2-}$ (Fig. 17a) constitutes the prototypical cluster and has been extensively discussed previously (*127*). As illustrated in Fig. 18, in the MO description of the electronic structure of this anion, there are 13 filled MOs. According to the electron counting rules, 6 are associated with exo cluster BH bonding and 7 are associated with endo cluster bonding. Although the proposed separation is analogous to the σ–π separation in benzene, there is a significant difference. In the case of benzene, the separation results from symmetry and there is no mixing between σ and π systems. In the case of $[B_6H_6]^{2-}$, even though it possesses very high (O_h) symmetry, the separation of exo and endo cluster MOs is not required by symmetry. Only the highest lying filled MOs of t_{2g} symmetry (see Fig. 19a) are required to be purely cluster bonding by symmetry. The expected t_{1u} and a_{1g} cluster bonding MOs are significantly mixed with exo cluster bonding MOs of the same symmetry types. Indeed, in the pioneering studies of Hoffmann and Lipscomb (*128*), the exo–endo separation was only one of several that were explored. As a cluster is substituted

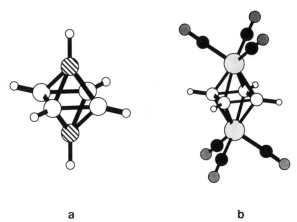

a   b

FIG. 17. Schematic drawings of the structures of (a) $[B_6H_6]^{2-}$ (striped atoms = B), 1,6-$C_2B_4H_6$ (striped atoms = C), 1,6-$N_2B_4H_4$ (hypothetical, striped atoms and attached balls = N), 1-NB_5H_6 (hypothetical, one striped atom = B, one striped atom = N), 1-PB_5H_6 (hypothetical, one striped atom = B, one striped atom = P), 1-SB_5H_5 (hypothetical, one striped atom = B, one striped atom and attached ball = S), and (b) $\{1,6-[(OC)_3Fe]_2B_4H_4\}^{2-}$ (hypothetical).

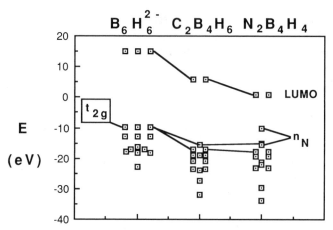

FIG. 18.   Plot of MO energies (Fenske–Hall) for $[B_6H_6]^{2-}$, 1,6-$C_2B_4H_6$, and 1,6-$N_2B_4H_4$. The energies of the MOs of $[B_6H_6]^{2-}$ have all been decreased by 10 eV, and t_{2g} refers to the set of "surface" MOs shown in Fig. 17a, while n_N refers to MOs containing the "lone pairs" on the nitrogen atoms.

with groups other than hydrogen or isolobal substitution is carried out, the exo–endo MOs become even more mixed. This provides ample scope for variation (or violation of the rules) and, consequently, generates "discussions" of exactly how the valence electrons should be divided up.

For these reasons, we present a brief overview of how heteroatoms perturb cluster bonding in strictly isoelectronic systems as well as a com-

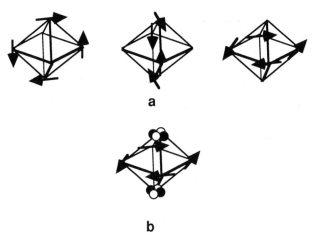

FIG. 19.   Sketch of (a) the "t_{2g} surface" skeletal MOs of $[B_6H_6]^{2-}$ and (b) the lowest lying "surface" MO of $\{1,6-[(OC)_3Fe]_2B_4H_4\}^{2-}$. Arrows represent the vectorial character of the $2p$ functions on the individual boron atoms.

parison of a pair of clusters "isoelectronic" only in terms of the number of electrons assigned to cluster bonding. We begin with a borane cage to illustrate the changes in cluster bonding MOs as electron-precise atoms, e.g., carbon in $1,6\text{-}C_2B_4H_6$, and electron-rich atoms, e.g., nitrogen in $1,6\text{-}N_2B_4H_4$ (hypothetical), are added to the cluster skeleton. Note that although most of the species investigated quantum chemically have yet to be prepared, the heteroatom fragments have been characterized in larger borane clusters (129–131). The six-atom closo cage constitutes a more transparent model system than the larger clusters.

Despite the large decrease in energy of all the MOs owing to the loss of anionic charge, there are still close similarites between the homo- and heteroatom cluster systems. In fact, the differences are exactly those expected from the considerations presented in earlier sections. For example, as shown in Fig. 18, the highest three filled MOs of the carborane remain exclusively cluster bonding "surface" orbitals, but the 3-fold degeneracy is removed as two now contain carbon character; in other words, two of the "surface" MOs (the e_g pair) pick up carbon character and are found at lower energy while the remaining "surface" MO (b_{2g} symmetry) remains purely boron in character and is found at higher energy (compare Figs. 17a and 19). For the nitraborane, the pattern of the "surface" orbitals is the same, but the splitting has increased (2.32 to 2.57 eV) as expected in going to the nitrogen derivative. The biggest difference results from the conversion of two exo-CH MOs to two exo-N "lone pairs," one combination of which constitutes the HOMO of the molecule. As $1,6\text{-}C_2B_4H_6$ reacts with Me_3N to form $5\text{-}Me_3N\text{-}2,4\text{-}C_2B_4H_6$, a $nido$-carborane, one might expect $1,6\text{-}N_2B_4H_4$ to self-associate in a rather interesting fashion (132).

By restricting our attention to the "surface" orbitals, we get a clear indication of how a set of cluster bonding orbitals are perturbed in an isoelectronic series. Keep in mind, however, that even these MOs can be made to pick up exo character by attaching substituents having filled MOs of π symmetry with respect to a radial direction of the cluster. The extent of this perturbation has been measured by photoelectron spectroscopy (133–135). The large exo–endo interactions in highly halogenated boron clusters result in B_nCl_n being the observed form rather than $[B_nCl_n]^{2-}$ (136).

In going from $[B_6H_6]^{2-}$ to the hypothetical dianion $\{1,6\text{-}[(OC)_3\text{-}Fe]_2B_4H_4\}^{2-}$ (Figs. 17 and 20) two BH fragments are replaced with their isolobal $Fe(CO)_3$ equivalents (136). In doing so, more complexity is introduced into the MO diagram, i.e., the metal compound possesses 47 filled MOs versus 13 for $[B_6H_6]^{2-}$. Nonetheless, considerable simplification is possible if we adopt the approach taken in earlier sections and separate the MOs associated primarily with the Fe $3d$ nonbonding electrons (6) and CO ligands (30) from the rest (see Fig. 20). In terms of the

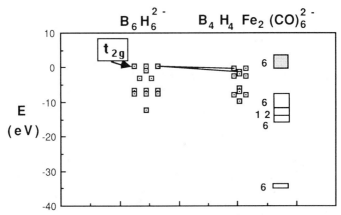

FIG. 20.   Plot of MO energies (Fenske–Hall) for $[B_6H_6]^{2-}$ and $\{1,6-[(OC)_3Fe]_2B_4H_4\}^{2-}$. Small squares correspond to single MOs, whereas for the metallaborane the larger blocks correspond to the CO-derived MOs (e.g., see Fig. 4) and the stippled block to the metal "nonbonding" MOs.

skeletal electron counting separation, 4 of the remaining 11 MOs are associated with BH bonding and 7 with skeletal bonding. In fact, the highest lying filled orbitals constitute analogs of the t_{2g} "surface" orbitals of $[B_6H_6]^{2-}$, but, just in the case of the carborane and nitraborane, the triple degeneracy is removed. Note, however, that the splitting is in the opposite direction, i.e., the analog of the b_{2g} MO in $C_2B_4H_6$ is the one found at lower energy. This can be ascribed to the differing energies of the Fe $3d$ and C $2p$ AOs. On the other hand, note that the splitting has been reduced (2.32 versus 0.75 eV). Because of the d functions on the cluster atoms the higher lying pair of "surface" orbitals contains boron character from all four boron atoms, and the lower lying "surface" MO has acquired significant Fe $3d$ character (126). In the latter case, Fig. 19b shows schematically how this takes place.

This feature of the metal-containing systems is not present in the main group only clusters and lets one appreciate how some of the "nonbonding" metal d electrons (see the block of six filled MOs in Fig. 20) can become involved in cluster bonding. An instructive example, $Fe_4(CO)_{12}(PH)_2$, has recently appeared (137). This cluster, which has an octahedral M_4E_2 core, is able to accommodate seven or eight skeletal bonding pairs while retaining a closo structure. In essence, a MO analogous to one of the high-lying t_{2u} antibonding surface orbitals of $[B_6H_6]^{2-}$ is sufficiently stabilized by interactions with the $3d$ δ orbitals of the PH fragments that it lies near the center of the HOMO–LUMO gap. Hence, this MO is low enough to be filled without causing any major structural change in the cluster core.

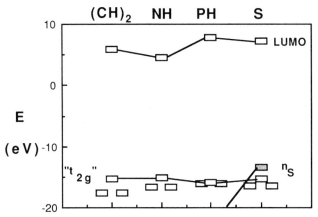

FIG. 21. Plot of the frontier MO energies (Fenske–Hall) for 1,6-$C_2B_4H_6$, i.e., $(CH)_2$, 1-NB_5H_6, i.e., NH, 1-PB_5H_6, i.e., PH, and 1-SB_5H_5, i.e., S. The stippled level (n_N) is the sulfur "lone pair."

Finally, we extend our analysis of models to second row main group atoms by comparing energies of frontier orbitals as cluster composition is varied. This is done in Fig. 21 for 1,6-$C_2B_4H_6$, 1-NB_5H_5, 1-PB_5H_5, and 1-SB_5H_4 (structures in Fig. 17a). Again, the differences follow the trends established in earlier sections. One nitrogen atom is not as effective as two carbon atoms in splitting the "t_{2g}" set, and one phosphorus atom is even less effective. The higher nuclear charge of S does a better job, but now the HOMO is a sulfur "lone pair." Bare sulfur atoms in a cluster environment should be reasonable bases, and, as seen below, this is indeed the case.

B. Electron Counting: A Case Study

Having the electron counting rules available, one might conclude that new E–M clusters will be easily classified and compared. To demonstrate that this is not always true, we present a brief synopsis of a controversy that recently surfaced in the literature concerning whether geometry fixes electron count or electron count geometry. We do not presume to act in judgment but do act as observers in this review simply because this tale nicely underlines one of our major points, namely, that unambiguous comparison, particularly if one is interested in reactivity and other useful properties, requires the comparison of isoelectronic compounds.

Kennedy and Greenwood described n-vertex metallaboranes possessing closed deltahedral structures which differ from the structure of the analogous ("isoelectronic") n-vertex *closo*-borane anion as isocloso. For example, the structure of $H(PPh_3)(Ph_2PC_6H_4)Ir(B_9H_8)$ exhibits a closed deltahedral structure which is not a bicapped square antiprism (*138*). In $HRh(PEt_3)_2(C_2B_7H_9)$, which exhibits the expected bicapped square antiprismatic structure, a closo count of $10 + 1 = 11$ is easily obtained by treating the metal fragment as contributing the normal three orbitals and two electrons to cluster bonding (*139*). In order to obtain a cluster count of 11 for the iridium compound, the metal fragment must contribute four electrons to cluster bonding and can do so by contributing four rather than the expected three orbitals to skeletal bonding (*140*). In essence, they assume any 10-atom closed deltahedral structure requires a closo electron count.

In challenging this view, Baker takes the position that geometry follows electron count rather than the reverse, i.e., the iridium metal fragment utilizes three orbitals and two electrons in cluster bonding (*141*). Thus, $H(PPh_3)(Ph_2PC_6H_4)Ir(B_9H_8)$ comes up one pair of skeletal bonding electrons short of those required for a closo cluster. Hence, Baker would describe this type of cluster as hypercloso to emphasize that there are n cluster pairs rather than the $n + 1$ required for a closo system. This is rationalized by invoking the capping principle (*126*).

Johnson and Mingos (*142*) have carried out Hückel calculations in an attempt to further explore the problem. They noted that when one goes from a spherical deltahedron [symmetrical distribution of vertices with the same valencies: observed structure of $HRh(PEt_3)_2(C_2B_7H_9)$] to a polar deltahedron [unique vertex of high connectivity on the principal rotational ʌis: observed structure of $H(PPh_3)(Ph_2PC_6H_4)Ir(B_9H_8)$], Stone's tensor surface harmonics approach shows that one formerly cluster bonding MO becomes nonbonding. Hence, in going from $[B_{10}H_{10}]^{2-}$ to $B_{10}H_{10}$, the polar deltahedral structure is favored relative to the spherical deltahedral structure (bicapped square antiprism). Because of the relatively high positive charge at the vertex of high connectivity, Johnson and Mingos conclude that the polar deltahedral structure is "resisted" for main group atoms but can be "further stabilized by a metal atom at this site." An analysis of Mulliken populations suggests that the metal atom in the polar deltahedral structure utilizes only three orbitals in cluster bonding interactions. Hence, these authors come down on the side of Baker's view.

However, we might well wonder how the polar deltahedral structure is "further stabilized" by placing a metal atom at the vertex of high connectivity. Is it by virtue of the fact that a positive site can be generated at the metal with little expenditure of energy? If so, where does the electron density come from, a fourth orbital? One might view the situation as analogous to the B_nCl_n versus $[B_nCl_n]^{2-}$ problem with a twist. In the case of the metallaborane, the extra electron density is not coming from exo ligands but exo nonbonding filled orbitals based on the metal itself.

Structural chemists are obsessed with determining the lowest energy structure in a given system. For a given set of atoms there are always alternative structures, but for classical main group systems these alternative structures tend to be at much higher energy or even unbonded, e.g., the only alternate structure for H_2 is repulsive! For main group atom clusters, these alternate structures are energetically much more accessible, and for clusters containing transition metals, the existence of facile isomeric equilibria is a fact of life (*143*). Rules such as the electron counting rules work best when energy differences are large, i.e., coloring the cluster bonding electrons red and the rest of the valence electrons blue is relatively easy to do for a small main group atom cluster. However, when the energy differences are very small, as they are in transition metal clusters, there may well be a significant number of purple electrons in the system. Whenever one takes an artificial categorization of valence electrons, even though extremely valuable and useful, and attempts to force reality into these preconstructed boxes, discussions such as that between Kennedy versus Baker are bound to arise. For close comparisons, it is much wiser to stick to strictly isoelectronic species.

C. *Selected Examples of EM₃ Clusters*

Clusters containing a equilateral triangular array of metal fragments symmetrically capped with a main group fragment (Fig. 22a) constitute the EM cluster systems presently characterized containing the greatest variety of E and M atoms. Hence, they provide the best opportunity of illustrating

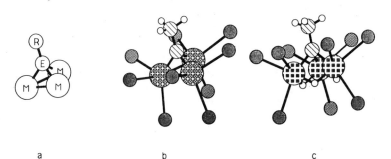

a   b   c

FIG. 22. Schematic structures of (a) EM_3, (b) $Co_3(CO)_9CMe$, and (c) $H_3Fe_3(CO)_9CMe$. Only the oxygen atoms of the CO ligands are shown for clarity.

TABLE I

EXAMPLES OF EM_3 CLUSTERS

Group	Compound	Reference
13	$HFe_3(CO)_9(H_3BH)$	144
	$Co_3(CO)_9Al$	145
	$Co_3(CO)_9Ga$	146
	$[Pt_3(CO)_3(PCy_3)_3Tl]^+$	147
14	$H_3Fe_3(CO)_9CR$	148
	$Co_3(CO)_9SiCo(CO)_4$	149
	$Co_3(CO)_9GeFe(CO)_2Cp$	150
15	$H_2Fe_3(CO)_9NR$	151
	$H_2Fe_3(CO)_9PR$	152
	$[Fe_3(CO)_{10}SbFe(CO)_4]^-$	153
	$H_3Fe_3(CO)_9Bi$	154
16	$[Fe_3(CO)_9O]^{2-}$	155
	$H_2Os_3(CO)_9S$	156
	$FeCo_2(CO)_9Se$	157
	$FeCo_2(CO)_9Te$	157
17	$[Ru_3(CO)_9I]^-$	158

the variation in behavior possible as the identity of both E and M atoms is changed. Table I (*144–158*) contains a selection of reported clusters organized according to the group and period of the main group atom. Not all known compounds are shown, nor are all those shown fully characterized in a structural sense. Table I, however, does give an indication of the breadth of possibilities for EM clusters. Generally the clusters have six skeletal electron pairs (or a cluster electron count of 48), but there are notable exceptions which contain an odd number of electrons (see below). The EM_3 cluster system is related isolobally to E_4 cages on the one hand, e.g., C_4R_4, and to M_4 tetrahedral clusters on the other, e.g., $Co_4(CO)_{12}$. On the main group atom side lie the closely related E_2M_2 and E_3M systems. The former could equally well have been chosen as a vehicle for the following discussion. Most of the types of behavior discussed below are found in EM clusters of various sizes and ratios of E to M. The reader is reminded of the reviews in which comparative structural and reaction chemistry EM clusters of various types are found (Section I).

1. *Electronic Structure*

As a mutually bonded four-atom array is the smallest entity we consider to be a cluster, the EM_3 system constitutes an example of the simplest cluster. As it is electron precise (the number of skeletal electron pairs is equal to the number of M–M and E–M edges), it also possesses the simplest, if not simple, electronic structure. For these reasons, a number of experimental and theoretical studies devoted to developing a detailed picture of the nature of the bonding in such clusters have appeared (159–163). For example, Fig. 23 presents the results of Fenske–Hall calculations that have been used in conjunction with photoelectron spectra to compare the electronic structures of $(OC)_9Co_3CMe$ and $H_3Fe_3(CO)_9CMe$ (Fig. 22b and c) (164). These two molecules are strictly isoelectronic. Using the approach of Section III,B, we can categorize the 64 filled MOs of $(OC)_9Co_3CMe$ as follows: 45 associated primarily with the CO ligands (9 with significant metal character), 12 primarily with the metals (with 3 having significant capping carbon character), and 7 with the CMe fragment (3 with significant metal character).

In a thought experiment, going from $(OC)_9Co_3CMe$ to $H_3Fe_3(CO)_9CMe$ requires three well-shielded protons to be moved from the cobalt nuclei to bridging positions between the iron atoms formed. The result (Fig. 23)

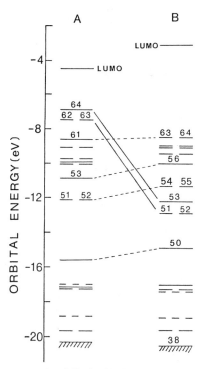

Fig. 23. Eigenvalue spectra for (A) $Co_3(CO)_9CMe$ and (B) $H_3Fe_3(CO)_9CMe$ from Fenske–Hall calculations (159).

closely parallels the changes in the simple molecules discussed in Sections III,A and III,B (Figs. 2 and 8), i.e., three MOs of $(OC)_9Co_3CMe$ associated with metal–metal bonding are stabilized by several electron volts in energy in going to $H_3Fe_3(CO)_9CMe$. Also consistent with the considerations of the mixed ligand complex in Fig. 4 or the heteroatom borane cages in Fig. 19, the LUMO for the cobalt cluster is lower than that for the iron cluster. Hence, even though these clusters are complex bonding systems, the perturbations caused by shifting protons in an isoelectronic series described earlier for simple molecules and complexes hold true as E and M are varied between isoelectronic clusters (*165*). Thus, as one goes from $H_3Fe_3(CO)_9CMe$ to $H_2Fe_3(CO)_9NMe$, one expects the MOs associated with E–M bonding to go to lower energy. On the other hand, in going from $H_2Os_3(CO)_9CCO$ to $H_3Os_3(CO)_9BCO$ the analogous E–M bonding MOs rise in energy (*166*). We expect these changes in the electronic structure to modulate the chemical behavior of the clusters.

Before discussing cluster properties, we emphasize that our understanding of the electronic structure of EM_3 systems is not based solely on calculations. The results from UV–photoelectron spectroscopic investigations of $(OC)_9Co_3CMe$ and $H_3Fe_3(CO)_9CMe$ clusters in the gas phase provide fully independent justification for the connection between the M–M bonding MOs of $(OC)_9Co_3CMe$ and the M–H–M bonding MOs of $H_3Fe_3(CO)_9CMe$ (*159,160,167*). Likewise, the "experimental quantum mechanics" of Dahl and co-workers, in which the effect of increasing or decreasing the number of electrons in a cluster on the atom–atom distances is used to define the character of the MO accepting or losing electrons, constitutes an unambiguous source of information on EM clusters (*168,169*). ESR studies of cluster radicals also provide particularly precise information on the MOs lying at the HOMO–LUMO gap. A classic example is the single-crystal ESR of $Co_3(CO)_9S$ diluted into $FeCo_2(CO)_9S$ (*170*). This study demonstrated that the unpaired electron is residing in an M–M antibonding MO with predominately metal $3d$ character. There is no contribution from the S atom and small contributions from the terminal ligands. Hence, in the normal 48-electron cluster $[FeCo_2(CO)_9S]$ the LUMO must be M–M antibonding and associated with the trimetal fragment. This approach has been extended to radical anions, and a variety of metal and main group systems including mixed metals have been studied (*171,172*). In the case of mixed metal clusters, the spin density is not evenly distributed over the metal centers. Consistent with Fig. 19, the LUMO contains a higher percentage of the metal or metals with higher nuclear charge; e.g., for $[FeCo_2(CO)_9S]^-$ the MO containing the single electron has 60% cobalt and 15% iron character.

2. EHM and MHM Hydrogens as Sites of Brönsted Acidity

One of the general properties of EM_3 clusters possessing endo hydrogen atoms is the existence of substantial Brönsted acidity. Relatively weak bases, amines, or silica gel can remove an M–H–M or E–H–M proton from the cluster skeleton. When E is B the E–M edge is generally the thermodynamically most stable site for proton location, whereas for elements to the right of B the M–M edge is usually, but not always, the most stable site (6). However, it has been shown that when E is C, kinetically controlled protonation (and presumably deprotonation) takes place at an E–M edge even though the most stable location of the proton added or lost is at an M–M edge (173). These observations are a consequence of three facts. Structures with E–H–M interactions are energetically accessible at ambient temperatures, there is only a small difference in the acidities of E–H–M and M–H–M protons, and the steric demands of the associated acid or base restricts attack to the more accessible E–M edge.

The kinetics of deprotonation have been studied for $H_3Fe_3(CO)_9CMe$ and $HFe_3(CO)_9S(t$-Bu$)$ (174,175). In both cases, the reaction is first order in the cluster and the amine base. Despite differences in E and the number of endo hydrogens, the second-order rate constants for deprotonation by small amines are comparable (E = C, $k = 0.33$ for NEt_3; E = S, $k = 2.8$ M^{-1} second^{-1} for n-BuNH$_2$). Steric effects were shown to be important for both clusters, but this is most dramatically shown in the latter study where the rate constant decreases four orders of magnitude in going from n-BuNH$_2$ to 2,2,6,6-tetramethylpiperidine (175).

The relative acidity appears to increase in the order B < C < N < O. $HFe_3(CO)_9(H_3BH)$ survives a silica gel column, but $Fe_3(CO)_9(CH)_4$ does not. $H_2Fe_3(CO)_9NR$ is known, but only the anion $[Fe_3(CO)_9O]^{2-}$ has been reported (151,155). The high acidity of $H_2Fe_3(CO)_9PR$ has received comment (176). In support of the trend N < O, note that $[Fe_4(CO)_{12}N]^-$ protonates readily with strong acid but that, thus far, no conditions have been found whereby $[MnFe_3(CO)_{12}O]^-$ can be protonated (159,160,177). On the other hand, $[Fe_3(CO)_9S]^{2-}$ is known (178), and the ruthenium and osmium analogs can be singly and doubly protonated (179,180). Hence, for E = S at least, it appears that the acidity decreases as the metal is changed from Fe to Os.

3. E and M Atoms as Sites of Lewis Acidity

In the EM_3 clusters in Table I, both the E and M atoms are saturated, i.e., they individually satisfy the 18- and 8-electron rules, respectively.

Hence, disregarding the removal of protons from the cluster framework by bases, one might expect attack by Lewis bases to be controlled by ligand dissociation. On the contrary, the EM_3 clusters are readily attacked by Lewis bases. Although the rate depends on the nature of E and M, susceptibility to attack is an intrinsic feature of the electronic structure of a cluster.

The electron might be considered the simplest base, and addition of electrons to M_3E clusters results in radical anions. A variety of E caps (S, PR, CR, GeR) and metals (Cr, Fe, Co, Ni, Mo, W) have been studied (171). The greatest accessibility to more than one oxidation state is observed when E is S, but many more capping atom types remain to be explored (e.g., E = B) (81). The ligands on the metal have a significant effect on the reduction behavior; for example, CP ligands are much more favorable than CO (82). The first reduction occurs over a fairly narrow range of potential as one would expect if the HOMO and LUMO have high metal d character. Ligand and capping atom influences decrease in the order $Cp/(CO)_3 > PR/S > CMe/CPh$, where the ligand or capping atom listed second in each pair makes the reduction easier (171). The effect of the change in metal is difficult to discern because the exo ligands on the metals are not constant; however, the overall effect of metal change is small.

The reaction of transition metal clusters with more classic Lewis bases has been well studied. In the case of EM_3 clusters, addition of another ligand results initially in the rupture of an M—M bond and cluster opening (Fig. 24) (83,84). Loss of the ligand results in reclosure of the cage. Note, however, that there is indirect evidence in the case where E is B or S for rupture of an E—M rather than M—M interaction (180,185). In clusters with several endo hydrogens, loss of H_2 can occur competitively with L. For example, when E is B or C the complete reaction shown in Scheme 1 takes place. Indeed, when E is C the reaction is reversible such that $H_3Fe_3(CO)_9CR$ and $HFe_3(CO)_{10}CR$ can be interconverted by thermal activation in the presence of CO and H_2, respectively (175). Careful mechanistic studies have shown that the addition or loss of H_2 to form or close the intermediate shown in Scheme 1 constitutes the rate-determining step (86). Similar reactivity is exhibited when E is N or P (87). However, although thermal activation of $H_2M_3(CO)_9NR$ (M = Fe, Ru) in the presence of CO results in $M_3(CO)_{10}NR$, the reverse reaction requires photochemical activation (88). In the case where E is P, chemical methods are used to effect the displacement of dihydrogen by a ligand (16,189).

The open cluster intermediate is susceptible to further attack by Lewis bases. If addition of base is competitive with H_2 loss, then degradation of the cluster ensues by cleavage of either M or E fragments from the cluster.

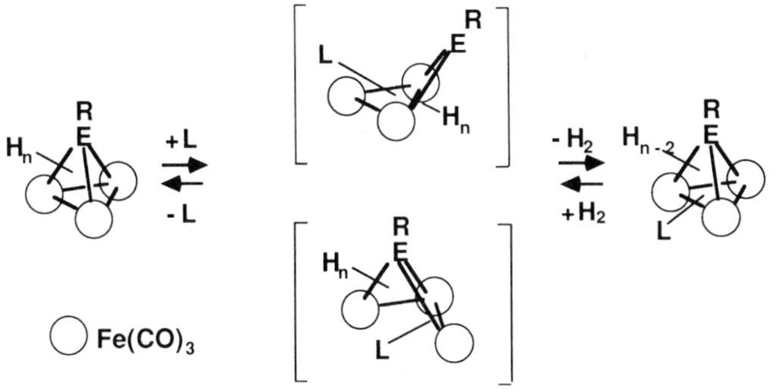

SCHEME 1. Generalized pathway for interconversion of the clusters $Fe_3(CO)_9EH_xR$ and $Fe_3(CO)_{10}EH_{x-2}R$ ($E = B$, $x = 4$; $E = C$, $x = 3$; and $E = N$, $x = 2$).

For example, in the presence of excess phosphine, $[HFe_3(CO)_9(H_2BH)]^-$ cluster cleavage is observed exclusively, whereas with 1 mol equiv or less $[Fe_3(CO)_9(PR_3)(HBH)]^-$ is a significant product (190). The reactions of $H_3M_3(CO)_9CH$ ($M = Fe$, Ru, Os) and $[HFe_3(CO)_9(H_2BH)]^-$ with excess base L are similar but not identical. For the former, CH_4 and stable $M_3(CO)_9L_3$ molecules are formed, while for the latter, one of the principal pair of products is $BH_3 \cdot L$ and $Fe_3(CO)_9L_2^-$ (172,186,191). The susceptibility to cluster degradation appears to decrease in the order $B > C > P(N)$. For example, the addition of 2 mol of ligand to $Fe_3(CO)_{10}PR$ results in the rupture of two M–M bonds. The reaction is reversible, and the addition–elimination sequence has been well studied (192). Indeed, Huttner and colleagues have emphasized the fact that substitution takes place by reversible addition–elimination reactions rather than by a dissociative pathway by likening the process to breathing: ligands are taken up by the cluster with rupture of M–M interactions and "expired" from the cluster with the formation of M–M interactions (192). If the ligand released differs from that taken up, the net reaction is substitution.

4. E and M Atoms as Sites of Lewis Basicity

As detailed in the discussion of Fig. 21, the formal conversion of an E'H fragment in a cluster to an E fragment results in a high-lying "lone pair" MO that should function as a Lewis base. Perhaps the most striking evidence for the basicity of a bare E atom is the spontaneous cyclic trimerization of $Co_3(CO)_9E$ ($E = P$, As) to $[M_3(CO)_8E]_3$ (183,193,194). In the case where E is As, the trimer can be converted back to the monomer with 15–20 atm overpressure of CO. This behavior depends on the nature of the metal fragment; for example, $Cp_3Mo_3(CO)_6As$ has been isolated and characterized as a monomer (*195*). Likewise, the nature of the E atom itself is important in that $H_2Ru_3(CO)_9S$ requires heating or photolysis to effect the same trimerization reaction (*196*). Corroborative evidence for the basicity of the bare E atom is the existence of compounds, such as $Co_3(CO)_9PMn(CO)_2Cp$ and $Co_3(CO)_9AsCr(CO)_5$, in which the EM_3 cluster acts as a normal two-electron donor to the single metal center (*197*).

For bare group 16 capping atoms, the Lewis basicity has synthetic consequences (*198*). The basicity of the capping atom promotes coupled clusters which can then undergo further transformations. For example, in $Os_3(CO)_{10}S:PtL_2$ coordination of the capping sulfur atom to platinium provides the initial link that, on loss of CO, leads to the closed cluster $L_2PtOs_3(CO)_9S$, which contains four metal atoms (*199*). Further, it has been suggested that the formation of a bicapped triosmium cluster from $Os_3(CO)_{10}S$ proceeds via a coupled intermediate (*200*). Finally, the capping oxygen atom in the raft cluster $Os_6(CO)_{19}O$ can be removed with $P(OMe)_3$ (*201*).

Although the basicity of bare group 15 and 16 capping atoms is not surprising, one wonders what circumstances would permit E atoms from groups 13 and 14 to behave as Lewis bases. Compounds such as $[Fe_3(CO)_{10}GeFe(CO)_4]^{2-}$ provide no help. The exo Ge—Fe bond might be considered as a "lone pair" on Ge being donated to the $Fe(CO)_4$ acid, but it can also be viewed as a lone pair on the $[Fe(CO)_4]^{2-}$ base being donated to an acid site on Ge or as a normal covalent bond with one electron coming from Ge and one from $[Fe(CO)_4]^-$ (*202*). The fact that $Co_3(CO)_9AsCr(CO)_5$ is known suggests that the last explanation is probably the most appropriate one (*203*). Efforts aimed at the generation of an M_3E cluster with a bare carbon atom have resulted in the characterization of $[M_3(CO)_9CCO]^{2-}$ (M = Fe, Ru, Os) in an elegant series of studies by Shriver and co-workers (*204–207*). These compounds have the properties of an exposed carbide but the structure of a so-called ketenylidene. When M is Fe, reaction with H^+ and Me^+ lead to $[Fe_3(CO)_{10}CH]^-$ and $[Fe_3(CO)_{10}CMe]^-$, respectively, i.e., the CO bound to the capping carbon reverts to the metal framework and the incoming electrophile is attached to the capping carbon atom. This behavior is suppressed in going from Fe to Ru and Os as the completely protonated $[M_3(CO)_9CCO]^{2-}$ anions have different forms, i.e., $HFe_3(CO)_{10}CH$ and $H_2M_3(CO)_9$-CCO (M = Ru, Os) (*204,208–210*).

The isoelectronic boron analog of $H_2Os_3(CO)_9CCO$, $H_3Os_3(CO)_9BCO$, has been characterized, and there is no tendency for the CO bound to the boron atom to migrate to the metal framework (*211*). In fact, the apically bound CO can be displaced by PMe_3 such that the boron atom behaves very much like an electropositive metal center (211). Similar reactivity is observed for $[FeCo_2(CO)_9CCO]^-$ in that reaction with PMe_3 results in displacement of the CO coordinated to the capping carbon atom, yielding $[FeCo_2(CO)_9CPMe_3]^-$ (212). The latter can be protonated to yield $HFeCo_2(CO)_9CPMe_3$, which can also be formed by the treatment of $HFeCo_2(CO)_9CCO$ with PMe_3. Spectroscopic information shows that initial attack of the phosphine is on a metal center, probably cobalt, and the $[FeCo_2(CO)_8(PMe_3)CCO]^-$ intermediate then rearranges to the final product. As the LUMO should have more cobalt than iron character (Fig. 23), attack of the phosphine at cobalt is reasonable.

Although the compound has not been crystallographically characterized, spectroscopic data suggest that the iron analog of $H_3Os_3(CO)_9BCO$, $HFe_3(CO)_9(HBH)$, has a structure related to that of $HFe_3(CO)_{10}CH$ (190). Hence, there is a possibility that the trianion would behave analogously to $[M_3(CO)_9CCO]^{2-}$.

The differing behavior of the isoelectronic clusters with bare capping atoms is consistent with the general principle discussed above. As one goes from group 16 to 13, there will be a tendency for the "lone pair" MO to rise in energy. Hence, EM_3 clusters where E is P are better donors than those where E is S. In the $H_2M_3(CO)_9S$ or $M_3(CO)_{10}S$ (M = group 18) or $M_3(CO)_9E$ (M = group 19; E = P, As) clusters, the E atom has a "lone pair" and is a heavy contributor to three M–E bonding MOs. With relatively high effective nuclear charges, the S and P atoms can handle the buildup of electronic charge. Possibly, when E is C, the form with bare carbon, $[M_3(CO)_{10}C]^{2-}$, has too high a buildup of electronic charge on the capping carbon, whereas the form in which the capping carbon acts as a formal acid, $[M_3(CO)_9CCO]^{2-}$, is more stable since the carbon shares only the pair from the CO bound to it. However, the reader will recall (Figs. 2 and 8) that protonation stabilizes a "lone pair" MO. In the case where M is Fe, this stabilization is sufficient to cause rearrangement, whereas when M is Ru or Os, it is not. The same arguments apply when E is B, but one expects the $[M_3(CO)_{10}B]^{3-}$ form (presently unknown) to be even less stable.

5. E and M Atoms as Hydrogen Atom Sources and Sinks

One of the fascinating aspects of EM clusters is the fact that, depending on the natures of M and E, endo hydrogen atoms can associate mainly with E or M. For $M_3(CO)_9EH_xR$ clusters with x endo hydrogen atoms, the dependence on E and M is known. When M is Fe, the endo hydrogens are found mainly as E–H–M bridges when E is B, and mainly as M–H–M for C and N (6). For Ru, there is a greater tendency for M–H–M bridges when E is B but the endo hydrogens are strictly found as M–H–M bridges for C and N (155,213,215).

The balance between E–H–M and M–H–M bridges is most precisely defined by tautomeric equilibria of the type first described by Calvert and Shapley (215) for the noncapped $Os_3(CO)_{10}CH_4$. This cluster exists as an equilibrium mixture of two forms, i.e., $H_2Os_3(CO)_{10}CH_2$ and $HOs_3(CO)_{10}(HCH_2)$. In the case of the capped EM_3 clusters, when E is B and M is Fe, $HFe_3(CO)_9(H_3BH)$ is the only form observed. However, when E is B and M is Ru, an equilibrium mixture of two forms, HRu_3-$(CO)_9(H_3BH)$ and $H_2Fe_3(CO)_9(H_2BH)$, is observed (13). Further, when E is C, an equilibrium mixture of three forms, $H_3Fe_3(CO)_9(CH)$, H_2-$Fe_3(CO)_9(HCH)$, and $HFe_3(CO)_9(H_2CH)$, is observed for Fe, but only one form, $H_3Ru_3(CO)_9(CH)$, is observed for Ru (172). The distribution can be more finely tuned by appropriate adjustment of the substituent on E as well as the metal. For example, the mixed-metal cluster $FeCo_2(CO)_9(CHR)$ also exists as an equilibrium mixture of two forms, but, in contrast to $Fe_3(CO)_9(CH_4)$, the form with a C–H–M interaction, $FeCo_2(CO)_9(HCR)$, predominates over the form with a M–H–M interaction, $HFeCo_2(CO)_9(CR)$ (16). Hence, by appropriate adjustment of E and M, the balance between the equilibrium distribution of endo hydrogens between E–M and M–M edges can be varied systematically.

As noted above (Section V,C,3), a number of examples of the direct addition of H_2 to or elimination of H_2 from EM_3 clusters are known. In the examples given, the H_2 added became endo hydrogens while the H_2 formed originated from endo hydrogens. The process is completed by the displacement or addition of a two-electron donor such as CO. The substituent on E can also get involved when it contains hydrogens. When the substituent is a simple alkyl group, the reaction is particularly interesting. Scheme 2 contains a partial summary of the information on the isoelectronic N, C, and B systems which differ solely in the number of endo hydrogen (155,175,190,217). At the present time, the most detailed information is known for the N and C clusters, and, although the close relationship is obvious, there are interesting and characteristic differences. For example, although both $H_2Fe_3(CO)_9NCH_2R$ and $HFe_3(CO)_9(H_3BMe)$ can be deprotonated and the anion reprotonated, the anion $[H_2Fe_3(CO)_9CMe]^-$ rapidly loses H_2 to form a vinylidene cluster anion. The mechanism for the formation of $[H_2Fe_3(CO)_9CMe]^-$ and its reconversion to $[H_2Fe_3(CO)_9CMe]^-$ has been studied in detail. Labeling studies show that the protonation–deprotonation reactions are centered on the carbon hydride fragment, while addition and loss of H_2 involve the metal sites. Protonation of the latter yields a neutral cluster, $HFe_3(CO)_9CCH_2$, identical to that produced thermally from $HFe_3(CO)_{10}CMe$ and analogous to that produced thermally from $H_2Fe_3(CO)_9NCH_2R$. Although the boron

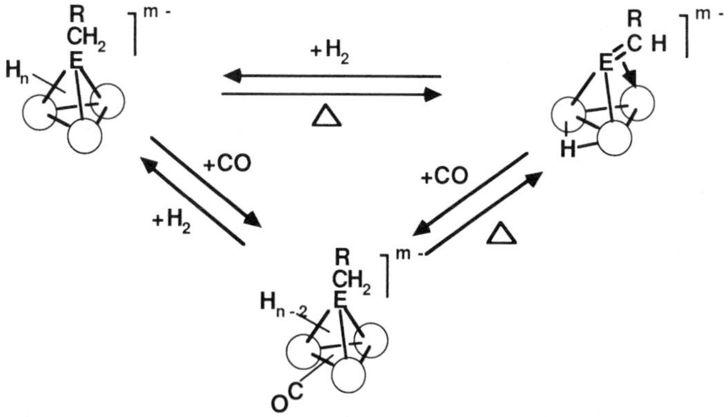

SCHEME 2. Involvement of the R substituent in the conversion of $Fe_3(CO)_9EH_xR$ and $Fe_3(CO)_{10}EH_{x-2}R$ ($E = B$, $x = 4$; $E = C$, $x = 3$; and $E = N$, $x = 2$) to generalized vinylidene clusters.

analog of the vinylidene cluster is unknown for iron, the osmium analog has been characterized by Jan and Shore (218). Likewise, there is an extensive, known, analogous chemistry of the triosmium alkylidene cluster in the literature (219).

VI

CONCLUSIONS

Space does not permit the analysis of other systems related to those presented above, and our discussions are merely a sampler of the kinds of information available from a comparison of isoelectronic main group–metal species. For example, the EM_4 "butterfly" cluster anions $[Fe_4(CO)_{12}B]^{3-}$, $[Fe_4(CO)_{12}C]^{2-}$, $[Fe_4(CO)_2N]^-$, and $[MnFe_3(CO)_{12}O]^-$ constitute another set of interesting clusters, and some calculational studies have already been reported (220). Indeed, butterfly clusters are the sole subject of a recent review (221). As more detailed studies reveal the systematic differences in new series of isoelectronic main group–metal compounds, we will gain even greater understanding, and ultimately full control, of the properties of these interesting materials.

ACKNOWLEDGMENTS

The continued support of our work in this area by the National Science Foundation is gratefully acknowledged. One of us (T. P. Fehlner) appreciates the support of the Guggenheim Foundation and the hospitality of the Chemistry Department at the University of Wisconsin, where much of this article was written.

REFERENCES

1. T. A. Albright, J. K. Burdett, and M.-H. Wangbo, "Orbital Interactions in Chemistry." Wiley, New York, 1985.
2. J. K. Burdett, "Molecular Shapes." Wiley (Interscience), New York, 1980.
3. R. L. DeKock and H. B. Gray, "Chemical Structure and Bonding." Benjamin/ Cummings, Menlo Park, California, 1980.
4. H. A. Bent, in "Molecular Structure and Energetics" (J. F. Liebman and A. Greenberg, eds.), Vol. 1, p. 17. VCH, Deerfield Beach, Florida, 1986.
5. N. N. Greenwood and A. Earnshaw, "Chemistry of the Elements." Pergamon, New York, 1984.
6. T. P. Fehlner, *Comments Inorg. Chem.* **7,** 307 (1988); T. P. Fehlner, *New J. Chem.* **12,** 307 (1988); C. E. Housecroft and T. P. Fehlner, *Adv. Organomet. Chem.* **21,** 57 (1982).
7. R. Hoffmann, *Science* **211,** 995 (1981).
8. D. M. P. Mingos, *Acc. Chem. Res.* **17,** 311 (1984).
9. K. Wade, *Adv. Inorg. Chem. Radiochem.* **18,** 1 (1976).
10. G. Schmid, *Angew. Chem., Int. Ed. Engl.* **17,** 392 (1978); G. Schmid, *Angew. Chem., Int. Ed. Engl.* **9,** 819 (1970).
11. K. H. Whitmire, *J. Coord. Chem.* **17,** 95 (1988).
12. W. A. Herrmann, *Angew. Chem., Int. Ed. Engl.* **25,** 56 (1986).
13. R. D. Adams and I. T. Horvath, *Prog. Inorg. Chem.* **33,** 127 (1985).
14. F. Bottomley and L. Sutin, *Adv. Organomet. Chem.* **28,** 339 (1988).
15. H. Vahrenkamp, *Adv. Organomet. Chem.* **22,** 169 (1983).
16. G. Huttner and K. Knoll, *Angew. Chem.* **99,** 765 (1987); *Angew. Chem. Int. Ed. Engl.* **26,** 743 (1987).
17. C. E. Housecroft, *Polyhedron* **6,** 1935 (1987)..
18. J. D. Kennedy, *Prog. Inorg. Chem.* **34,** 211 (1986); J. D. Kennedy, *Prog. Inorg. Chem.* **32,** 519 (1984).
19. B. F. G. Johnson and J. Lewis, *Adv. Inorg. Chem. Radiochem.* **24,** 225 (1981).
20. R. N. Grimes, ed., "Metal Interactions with Boron Clusters." Plenum, New York, 1982.
21. G. Wilkinson, F. G. A. Stone, and E. W. Abel, eds., "Comprehensive Organometallic Chemistry." Pergamon, New York, 1982.
22. J. G. Bentsen and M. S. Wrighton, *J. Am. Chem. Soc.* **109,** 4530 (1987).
23. U. Kölle, F. Khouzami, and B. Fuss, *Angew. Chem., Int. Ed. Engl.* **21,** 131 (1982).
24. R. L. Middaugh, in "Boron Hydride Chemistry" (E. L. Muetterties, ed.), p. 383. Academic Press, New York, 1975.
25. M. R. Churchill and J. Wormald, *J. Am. Chem. Soc.* **93,** 5670 (1971).
26. J. W. Rabalais, "Principles of Ultraviolet Photoelectron Spectroscopy." Wiley, New York, 1977.
27. T. P. Fehlner and J. R. Bowser, *J. Chem. Educ.* **65,** 976 (1988).
28. H. Fujimoto and K. Fukui, *Adv. Quantum Chem.* **6,** 177 (1972).
29. M. B. Hall and R. F. Fenske, *Inorg. Chem.* **11,** 768 (1972).
30. C. E. Housecroft and T. P. Fehlner, *Inorg. Chem.* **21,** 1739 (1982).
31. D. L. Lichtenberger and G. E. Kellogg, *Acc. Chem. Res.* **20,** 379 (1987).
32. J. C. Green, *Struct. Bonding (Berlin)* **43,** 37 (1981).
33. R. L. DeKock, in "Electron Spectroscopy: Theory, Techniques, and Applications" (C. R. Brundle and A. D. Baker, eds.), Vol. 1, p. 293. Academic Press, New York, 1977.
34. T. Koopmans, *Physica (Amsterdam)* **1,** 104 (1934).
35. M. Wu and T. P. Fehlner, *Chem. Phys. Lett.* **36,** 114 (1975).

36. S. Lacombe, D. Gonbeau, J.-L. Cabioch, B. Pellerin, J.-M. Denis, and G. Pfister-Guillouzo, *J. Am. Chem. Soc.* **110**, 6964 (1988).
37. N. Jonathan, A. Morris, M. Okuda, D. J. Smith, and K. J. Ross, *Chem. Phys. Lett.* **13**, 334 (1972).
38. T. P. Fehlner and D. W. Turner, *J. Am. Chem. Soc.* **95**, 7175 (1973).
39. T. J. Marks and J. R. Kolb, *Chem. Rev.* **77**, 263 (1977).
40. G. Parkin and J. E. Bercaw, *Polyhedron* **7**, 2053 (1988), and references therein; G. Parkin, E. Bunel, B. J. Burger, M. S. Trimmer, A. van Asselt, and J. E. Bercaw, *J. Mol. Catal.* **41**, 21 (1987).
41. F. R. Fronczek, E. C. Baker, P. R. Sharp, K. N. Raymond, H. G. Alt, and M. D. Rausch, *Inorg Chem,* **15**, 2284 (1976).
42. J. C. Green, M. L. H. Green, and C. K. Prout, *J. Chem. Soc., Chem. Commun.*, 421 (1972); J. L. Peterson, D. L. Lichtenberger, R. F. Fenske, and L. F. Dahl, *J. Am. Chem. Soc.* **97**, 6433 (1975).
43. J. W. Lauher and R. Hoffmann, *J. Am. Chem. Soc.* **98**, 1729 (1976).
44. J. C. Green, S. E. Jackson, and B. Higginson, *J. Chem. Soc., Dalton Trans.*, 403 (1975).
45. J. E. Bercaw, R. H. Marvich, L. G. Bell, and H. H. Brintzinger, *J. Am. Chem. Soc.* **94**, 1219 (1972); J. E. Bercaw, *J. Am. Chem. Soc.* **96**, 5087 (1974).
46. L. J. Guggenberger and R. R. Schrock, *J. Am. Chem. Soc.* **97**, 6578 (1975).
47. A. H. Klazinga and J. H. Teuben, *J. Organomet. Chem.* **157**, 413 (1978).
48. F. Bottomley, D. F. Drummond, G. O. Egharevba, and P. S. White, *Organometallics* **5**, 1620 (1986).
49. G. L. Hillhouse, A. R. Bulls, B. D. Santarsiero, and J. E. Bercaw, *Organometallics* **7**, 1309 (1988).
50. G. L. Hillhouse and J. E. Bercaw, *J. Am. Chem. Soc.* **106**, 5472 (1984).
51. D. M. Roddick, M. D. Fryzuk, P. F. Seidler, G. L. Hillhouse, and J. E. Bercaw, *Organometallics* **4**, 97 (1985).
52. D. J. Cardin, M. F. Lappert, C. L. Ruston, and P. I. Riley, *Compr. Organomet. Chem.* **3**, 583 (1982).
53. G. L. Hillhouse and J. E. Bercaw, *J. Am. Chem. Soc.* **106**, 5472 (1984).
54. A Shaver and J. A. McCall, *Organometallics* **3**, 1823 (1984).
55. R. R. Schrock, *J. Am. Chem. Soc.* **96**, 6796 (1974); R. R. Schrock, *J. Am. Chem. Soc.* **97**, 6577 (1975).
56. M. L. Green, *Pure Appl. Chem.* **50**, 27 (1978).
57. A. van Asselt, B. J. Burger, V. C. Gibson, and J. E. Bercaw, *J. Am. Chem. Soc.* **108**, 5347 (1986).
58. W. A. Herrmann, *Angew. Chem., Int. Ed. Engl.* **25**, 56 (1986), and references therein,
59. W. A. Herrmann, E. Voss, E. Guggolz, and M. L. Ziegler, *J. Organomet. Chem.* **284**, 47 (1985).
60. C. Hecht, E. Herdtweck, J. Rohrmann, W. A. Herrmann, W. Beck, and P. M. Fritz, *J. Organomet. Chem.* **330**, 389 (1987).
61. W. A. Herrmann, *J. Organomet. Chem.* **300**, 111 (1986).
62. W. A. Herrmann, C. Hecht, E. Herdtweck, and H.-J. Kneuper, *Angew. Chem., Int. Ed. Engl.* **26**, 132 (1987).
63a. W. A. Herrmann, B. Koumbouris, A. Schäfer, T. Zahn, and M. L. Ziegler, *Chem. Ber.* **118**, 2472 (1985).
63b. W. A. Herrmann, *Angew. Chem., Int. Ed. Engl.* **25**, 56 (1986).
64. W. A. Herrmann, C. Hecht, M. L. Ziegler, and B. Balbach, *J. Chem. Soc., Chem. Commun.*, 686 (1984).
65. M. Creswick, I. Bernal, and W. A. Herrmann, *J. Organomet. Chem.* **172**, C39 (1979).

66. B. E. R. Schilling, R. Hoffmann, and D. L. Lichtenberger, *J. Am. Chem. Soc.* **101**, 585 (1979); D. C. Calabro and D. L. Lichtenberger, *J. Am. Chem. Soc.* **103**, 6846 (1981).
67. D. C. Calabro, D. L. Lichtenberger, and W. A. Herrmann, *J. Am. Chem. Soc.* **103**, 6852 (1981).
68. W. A. Herrman, *Angew. Chem., Int. Ed. Engl.* **25**, 72 (1986).
69. W. A. Herrmann, C. Hecht, M. L. Ziegler, and T. Zahn, *J. Organomet. Chem.* **273**, 323 (1984).
70. W. A. Herrmann, B. Koumbouris, T. Zahn, and M. L. Ziegler, *Angew. Chem., Int. Ed. Engl.* **23**, 812 (1984).
71. P. J. Sullivan and A. L. Rheingold, *Organometallics* **1**, 1547 (1982).
72. W. I. Bailey, Jr., M. H. Chisholm, F. A. Cotton, and L. A. Rankel, *J. Am. Chem. Soc.* **100**, 5764 (1978).
73. C. F. Campana, A. Vizi-Orosz, G. Palyi, L. Marko, and L. F. Dahl, *Inorg. Chem.* **18**, 3055 (1979).
74. A. S. Faust, M. S. Foster, and L. F. Dahl, *J. Am. Chem. Soc.* **91**, 5633 (1969).
75. C. H. Wei and L. F. Dahl, *Inorg. Chem.* **4**, 1 (1965).
76. C. S. Cundy, B. M. Kingston, and M. F. Lapppert, *Adv. Organomet. Chem.* **11**, 253 (1973).
77. U. Schubert, G. Kraft, and E. Walther, *Z. Anorg. Allg. Chem.* **519**, 96 (1984).
78. W. Hönle and H. G. von Schnering, *Z. Anorg. Allg. Chem.* **464**, 139 (1980).
79. E. Colomer, R. J. P. Corriu, and A. Vioux, *Inorg. Chem.* **18**, 695 (1979).
80. G. Kraft, C. Kalbas, U. Schubert, *J. Organomet. Chem.* **289**, 247 (1985); G. Cerveau, E. Colomer, R. Corriu, and W. E. Douglas, *J. Chem. Soc., Chem. Commun.*, 410 (1975).
81. M. J. S. Dewar, *Bull. Soc. Chim. Fr.* **18**, C71 (1951); J. Chatt and L. A. Duncanson, *J. Chem. Soc.*, 2939 (1953).
82a. G. Medford and S. G. Shore, *J. Am. Chem. Soc.* **100**, 3953 (1978); J. S. Plotkin and S. G. Shore, *J. Organomet. Chem.* **182**, C15 (1979).
82b. R. L. DeKock, P. Deshmukh, T. P. Fehlner, C. E. Housecroft, J. S. Plotkin, and S. G. Shore, *J. Am. Chem. Soc.* **105**, 815 (1983).
83. G. Becker, G. Gresser, and W. Uhl, *Z. Naturforsch. B: Anorg. Chem. Org. Chem.* **36**, 16 (1981); B. Solouki, H. Bock, R. Appel, A. Westerhaus, G. Becker, and G. Uhl, *Chem. Ber.* **115**, 3747 (1982).
84. M. Regitz and P. Binger, *Angew. Chem., Int. Ed Engl.* **27**, 1484 (1988).
85. J. F. Nixon, *Chem. Rev.* **88**, 1327 (1988).
86. C. Krüger and K. Angermund, unpublished observations (1987).
87. G. Fachinetti, C. Floriani, F. Marchetti, and M. Mellini, *J. Chem. Soc., Dalton Trans.*, 1398 (1978).
88. P. Binger and B. Biedenbach, unpublished observations (1987); B. Biedenbach, Dissertation. Universität Kaiserslautern, 1988.
89. A. R. Barron, A. H. Cowley, S. W. Hall, C. M. Nunn, and J. M. Power, *Angew Chem., Int. Ed. Engl.* **27**, 837 (1988).
90. A. Nakamura and S. Otsuka, *J. Am. Chem. Soc.* **95**, 7262 (1973).
91. P. Binger, B. Biedenbach, C. Krüger, and M. Regitz, *Angew. Chem., Int. Ed. Engl.* **26**, 764 (1987).
92. W. E. Hunter, J. L. Atwood, G. Fachinetti, and C. Floriani, *J. Organomet. Chem.* **204**, 67 (1981).
93. G. Erker, C. Krüger, and G. Müller, *Adv. Organomet. Chem.* **24**, 1 (1985).
94. P. B. Hitchcock, M. J. Maah, and J. F. Nixon, *J. Chem. Soc., Chem. Commun.*, 737 (1986).

95. P. Binger, R. Milczarek, R. Mynott, C. Krüger, Y.-H. Tsay, E. Raabe, and M. Regitz, *Chem. Ber.* **121**, 637 (1988).
96. M. D. Rausch, G. F. Westover, E. Mintz, G. M. Reisner, I. Bernal, A. Clearfield, and J. M. Troup, *Inorg. Chem.* **18**, 2605 (1979).
97. K. Jonas, *Angew. Chem., Int. Ed. Engl.* **24**, 295 (1985).
98. P. M. Maitlis, *Adv. Organomet. Chem.* **4**, 95 (1966).
99. E. Efraty, *Chem. Rev.* **77**, 691 (1977).
100. R. P. Dodge and V. Schomaker, *Acta Crystallogr.* **18**, 614 (1965).
101. J. D. Dunitz, H. C. Mez, O. S. Mills, and H. M. M. Shearer, *Helv. Chim. Acta* **45**, 647 (1962).
102. J. C. T. R. Burkett-St. Laurent, P. B. Hitchcock, H. W. Kroto, and J. F. Nixon, *J. Chem. Soc., Chem. Commun.*, 1141 (1981).
103. S. Jagner, R. G. Hazell, and S. E. Rasmussen, *J. Chem. Soc., Dalton Trans.*, 337 (1976).
104. S. I. Al-Resayes, P. B. Hitchcock, M. F. Meidine, and J. F. Nixon, *J. Chem. Soc., Chem. Commun.*, 1080 (1984).
105. S. I. Al-Resayes, P. B. Hitchcock, and J. F. Nixon, *J. Chem. Soc., Chem. Commun.*, 928 (1987).
106. R. D. Ernst, *Chem. Rev.* **88**, 1255 (1988).
107. A. Haaland, *Acc. Chem. Res.* **12**, 415 (1979), and references therein.
108. P. W. Jolly, *Compr. Organomet. Chem.* **6**, 189 (1982).
109. T. D. Newbound, J. W. Freeman, D. R. Wilson, M. S. Kralik, A. T. Patton, C. F. Campana, and R. D. Ernst, *Organometallics* **6**, 2432 (1987).
110. P. Seiler and J. D. Dunitz, *Acta Crystallogr., Sect. B* **36**, 2255 (1980).
111. F. H. Koehler, *J. Organomet. Chem.* **110**, 235 (1976).
112. E. Gard, A. Haaland, D. P. Novak, and R. Seip, *J. Organomet. Chem.* **88**, 181 (1975).
113. R. D. Adams, F. A. Cotton, and G. A. Rusholme, *J. Coord. Chem.* **1**, 275 (1971).
114. L. R. Byers and L. F. Dahl, *Inorg. Chem.* **19**, 680 (1971).
115. T. D. Newbound, J. W. Freeman, D. R. Wilson, M. S. Kralik, A. T. Patton, C. F. Campana, and R. D. Ernst, *Organometallics* **6**, 2432 (1987).
116. A. W. Duff, K. Jonas, R. Goddard, H.-J. Kraus, and C. Krüger, *J. Am. Chem. Soc.* **105**, 5479 (1983).
117. O. J. Scherer, H. Sitzmann, and G. Wolmershäuser, *Angew. Chem., Int. Ed. Engl.* **24**, 351 (1985).
118. O. J. Scherer, J. Schwalb, G. Wolmershäuser, W. Kaim, and R. Gross, *Angew. Chem., Int. Ed. Engl.* **25**, 363 (1986), and references cited therein.
119a. A. L. Rheingold, M. J. Foley, and P. J. Sullivan, *J. Am. Chem. Soc.* **104**, 4727 (1982).
119b. T. Kuhlmanm, S. Roth, J. Roziere, and W. Siebert, *Angew. Chem., Int. Ed. Engl.* **25**, 105 (1986).
119c. W. Siebert, *Adv. Organomet. Chem.* **18**, 301 (1980).
120. K. Wade, *Inorg. Nucl. Chem. Lett.* **8**, 559 (1972).
121. K. Wade, *Inorg. Nucl. Chem. Lett.* **8**, 563 (1972).
122. A. J. Stone, *Mol. Phys.* **41**, 1339 (1980).
123. A. J. Stone, *Inorg. Chem.* **20**, 563 (1981).
124. A. J. Stone, *Polyhedron* **3**, 1299 (1984).
125. R. B. King and D. H. Rouvray, *J. Am. Chem. Soc.* **99**, 7834 (1977); R. B. King, *Inorg. Chim. Acta* **116**, 99, 109, 119, and 125 (1986).
126. D. M. P. Mingos and R. L. Johnson, *Struct. Bonding (Berlin)* **68**, 29 (1987).
127. K. Wade, "Electron Deficient Compounds." Nelson, London, 1971.
128. R. Hoffmann and W. N. Lipscomb, *J. Chem. Phys.* **36**, 2179 (1962).

129. A. Arafat, J. Baer, J. C. Huffman, and L. J. Todd, *Inorg. Chem.* **25**, 3757 (1986).
130. W. R. Pretzer and R. W. Rudolph, *J. Am. Chem. Soc.* **98**, 446 (1976).
131. G. B. Dunks and M. F. Hawthorne, in "Boron Hydride Chemistry" (E. L. Muetterties, ed.), p. 383. Academic Press, New York, 1975.
132. C. L. Bramlett and R. N. Grimes, *J. Am. Chem. Soc.* **88**, 4269 (1966).
133. J. A. Ulman and T. P. Fehlner, *J. Am. Chem. Soc.* **98**, 1119 (1976).
134. G. A. Beltram and T. P. Fehlner, *J. Am. Chem. Soc.* **101**, 6237 (1979).
135. E. L. Andersen, R. L. DeKock, and T. P. Fehlner, *J. Am. Chem. Soc.* **102**, 2644 (1980).
136. P. R. LeBreton, S. Urano, M. Shahbaz, S. L. Emery, and J. A. Morrison, *J. Am. Chem. Soc.* **108**, 3937 (1986).
137. J.-F. Halet and J.-Y. Saillard, *New J. Chem.* **11**, 315 (1987); J.-F. Halet, R. Hoffmanns, and J.-Y. Saillard, *Inorg. Chem.* **24**, 1695 (1985).
138. J. Bould, N. N. Greenwood, J. D. Kennedy, and W. S. McDonald, *J. Chem. Soc. Chem. Commun.*, 465 (1982).
139. G. K. Barker, M. P. Garcia, M. Green, F. G. A. Stone, J. M. Bassett, and A. J. Welch, *J. Chem. Soc., Chem. Commun.*, 653 (1981).
140. J. D. Kennedy, *Inorg. Chem.* **25**, 111 (1986).
141. R. T. Baker, *Inorg. Chem.* **25**, 109 (1986).
142. R. L. Johnson and D. M. P. Mingos, *Inorg. Chem.* **25**, 3321 (1986).
143. C. P. Horwitz and D. F. Shriver, *J. Am. Chem. Soc.* **107**, 8147 (1985).
144. J. Vites, C. E. Housecroft, C. Eigenbrot, M. L. Buhl, G. J. Long, and T. P. Fehlner, *J. Am. Chem. Soc.* **108**, 3304 (1986).
145. K. E. Schwarzhans and H. Steiger, *Angew. Chem., Int. Ed. Engl.* **11**, 535 (1972).
146. W. Kalbfus, J. Kiefer, and K. E. Schwarzhans, *Z. Naturforsch. B: Anorg. Chem. Org. Chem.* **28**, 503 (1973).
147. O. J. Ezomo, D. M. P. Mingos, and I. D. Williams, *J. Chem. Soc., Chem. Commun.*, 924 (1987).
148. K. S. Wong, K. J. Haller, T. K. Dutta, D. M. Chipman, and T. P. Fehlner, *Inorg. Chem.* **21**, 3197 (1982).
149. G. Schmid, V. Bätzel, and G. Etzrodt, *J. Organomet. Chem.* **112**, 345 (1976).
150. P. Gusbeth and H. Vahrenkamp, *Chem. Ber.* **118**, 1746 (1985).
151. M. A. Andrews and H. D. Kaesz, *J. Am. Chem. Soc.* **101**, 7255 (1979).
152. G. Huttner, J. Schneider, G. Mohr, and J. von Seyerl, *J. Organomet. Chem.* **191**, 161 (1980).
153. A. L. Rheingold, K. H. Whitmire, M. D. Fabiano, and J. S. Leigh, as referenced in Ref. 11.
154. K. H. Whitmire, C. B. Lagrone, and A. L. Rheingold, *Inorg. Chem.* **25**, 2472 (1986).
155. A. Ceriotti, L. Resconi, F. Demartin, G. Longoni, M. Manassero, and M. Sansoni, *J. Organomet. Chem.* **249**, C35 (1983).
156. A. J. Deeming and M. Underhill, *J. Organomet. Chem.* **42**, C60 (1972).
157. C. E. Strouse and L. F. Dahl, *J. Am. Chem. Soc.* **93**, 6032 (1971).
158. S.-H. Han, G. L. Geoffroy, B. D. Dombek, and A. L. Rheingold, *Inorg. Chem.* **27**, 4355 (1988).
159. B. E. R. Schilling and R. Hoffmann, *J. Am. Chem. Soc.* **101**, 3456 (1979).
160. P. T. Chesky and M. B. Hall, *Inorg. Chem.* **20**, 4419 (1981).
161. N. C. V. Costa, D. R. Lloyd, P. Brint, T. R. Spalding, and W. K. Pelin, *Surf. Sci.* **107**, L379 (1981).
162. G. Granozzi, S. Agnolin, M. Casarin, and D. Osella, *J. Organomet. Chem.* **208**, C6 (1981).

163. J. Evans, *J. Chem. Soc., Dalton Trans.,* 1005 (1980).
164. R. L. DeKock, K.-S. Wong, and T. P. Fehlner, *Inorg. Chem.* **21,** 3203 (1982).
165. J. C. Green, D. M.P. Mingos, and E. A. Seddon, *Inorg. Chem.* **20,** 2595 (1981).
166. R. D. Barreto, T. P. Fehlner, L.-Y. Hsu, D.-Y. Jan, and S. G. Shore, *Inorg. Chem.* **25,** 3572 (1986).
167. K. S. Wong, T. K. Dutta, and T. P. Fehlner, *J. Organomet. Chem.* **215,** C48 (1981).
168. P. D. Frisch and L. F. Dahl, *J. Am. Chem. Soc.* **94,** 5082 (1972).
169. L. R. Byers, V. A. Uchtman, and L. F. Dahl, *J. Am. Chem. Soc.* **103,** 1942 (1981).
170. C. E. Strouse and L. F. Dahl, *Discuss Faraday Soc.* **47,** 93 (1969).
171. B. M. Peake, P. H. Rieger, B. H. Robinson, and J. Simpson, *Inorg. Chem.* **20,** 2540 (1981); A. M. Bond, P. A. Dawson, B. M. Peake, P. H. Rieger, B. H. Robinson, and J. Simpson, *Inorg. Chem.* **18,** 1413 (1979).
172. P. N. Lindsay, B. M. Peake, B. H. Robinson, J. Simpson, U. Honrath, H. Vahrenkamp, and A. M. Bond, *Organometallics* **3,** 413 (1984).
173. T. K. Dutta, J. C. Vites, G. B. Jacobsen, and T. P. Fehlner, *Organometallics* **6,** 842 (1987).
174. L.-R. Frank, A. Winter, and G. Huttner, *J. Organomet. Chem.* **335,** 249 (1987).
175. T. K. Dutta, X. Meng, J. C. Vites, and T. P. Fehlner, *Organometallics* **6,** 2191 (1987).
176. W. Deck, M. Schwarz, and H. Vahrenkamp, *Chem. Ber.* **120,** 1515 (1987).
177. H. Vahrenkamp, V. A. Uchtman, and L. F. Dahl, *J. Am. Chem. Soc.* **90,** 3272 (1968).
178. R. D. Adams, I. T. Horvath, and H.-S. Kim, *Organometallics* **3,** 548 (1984).
179. B. F. G. Johnson, J. Lewis, D. Pippard, and P. R. Raithby, *J. Chem. Soc., Chem. Commun.,* 551 (1978).
180. B. F. G. John, J. Lewis, D. Pippard, P. R. Raithby, G. M. Sheldrick, and K. D. Rouse, *J. Chem. Soc., Dalton Trans.,* 616 (1979).
181. H. Vahrenkamp, *Adv. Organomet. Chem.* **22,** 169 (1983).
182. A. Vizi-Orosz, *J. Organomet. Chem.* **111,** 61 (1976).
183. J. Schneider and G. Huttner, *Chem. Ber.* **116,** 917 (1983).
184. H. Vahrenkamp, *Angew. Chem., Int. Ed. Engl.* **17,** 379 (1978).
185. C. E. Housecroft and T. P. Fehlner, *Inorg. Chem.* **25,** 404 (1986).
186. D. M. Dalton, D. J. Barnett, T. P. Duggan, J. B. Keister, P. T. Malik, S. P. Modi, M. R. Shaffer, and S. A. Smesko, *Organometallics* **4,** 1854 (1985).
187. G. Huttner and K. Evertz, *Acc. Chem. Res.* **19,** 406 (1986).
188. I. Fischler, R. Wagner, and E. A. Koerner von Gustorf, *J. Organomet. Chem.* **112,** 155 (1976).
189. K. Knoll, G. Huttner, L. Zsolnai, O. Orama, and M. Wasiucionek, *J. Organomet. Chem.* **310,** 225 (1986).
190. C. E. Housecroft and T. P. Fehlner, *J. Am. Chem. Soc.* **108,** 4867 (1986).
191. T. P. Duggan, D. J. Barnett, M. J. Muscatella, and J. B. Keister, *J. Am. Chem. Soc.* **108,** 6076 (1986); R. B. Calvert, Ph.D thesis. University of Illinois, Urbana, Illinois, 1978, as referenced by E. L. Muetterties, *Chem. Soc. Rev.* **11,** 283 (1982).
192. K. Knoll, G. Huttner, L. Zsolnai, I. Jibril, and M. Wasiucionek, *J. Organomet. Chem.* **294,** 91 (1985).
193. A. Vizi-Orosz, V. Galamb, G. Palyi, L. Marko, G. Bor, and G. Natile, *J. Organomet. Chem.* **107,** 235 (1976).
194. L. F. Dahl, *Abstr. Am. Crystallogr. Meet.* (1974).
195. K. Blechschmitt, H. Pfisterer, T. Zahn, and M. L. Ziegler, *Angew. Chem., Int. Ed. Engl.* **24,** 66 (1985).
196. R. D. Adams and D. A. Katahira, *Organometallics* **1,** 53 (1982).

197. H. Lang, G. Huttner, B. Sigwarth, I. Jibril, L. Zsolnai, and O. Orama, *J. Organomet. Chem.* **304,** 137 (1986).
198. R. D. Adams, *Polyhedron* **4,** 2003 (1985).
199. R. D. Adams, I. T. Horvath, P. Mathur, B. E. Segmuller, and L. W. Yang, *Organometallics* **2,** 1078 (1983).
200. R. D. Adams and T. S. A. Hor, *Inorg. Chem.* **23,** 4723 (1984).
201. R. J. Goudsmit, B. F. G. Johnson, J. Lewis, P. R. Raithby, and K. H. Whitmire, *J. Chem. Soc., Chem. Commun.,* 246 (1983).
202. K. H. Whitmire, C. B. Lagrone, M. R. Churchill, J. C. Fettinger, and B. H. Robinson, *Inorg. Chem.* **26,** 3491 (1987).
203. P. Gusbeth and H. Vahrenkamp, *Chem. Ber.* **118,** 1746 (1985).
204. J. W. Kolis, E. M. Holt, and D. F. Shriver, *J. Am. Chem. Soc.* **105,** 7307 (1983).
205. J. A. Hriljac and D. F. Shriver, *J. Am. Chem. Soc.* **109,** 6010 (1987).
206. M. J. Went, M. J. Sailor, P. L. Bogdan, C. P. Brock, and D. F. Shriver, *J. Am. Chem. Soc.* **109,** 6023 (1987).
207. M. J. Sailor, C. P. Brock, and D. F. Shriver, *J. Am. Chem. Soc.* **109,** 6015 (1987).
208. M. J. Sailor and D. F. Shriver, *Organometallics* **4,** 1476 (1985).
209. J. R. Shapley, D. S. Strickland, G. M. St. George, M. R. Churchill, and C. Bueno, *Organometallics* **2,** 185 (1983).
210. J. S. Holmgren and J. R. Shapley, *Organometallics* **3,** 1322 (1984).
211. S. G. Shore, D.-Y. Jan, L.-Y. Hsu, and W.-L. Hsu, *J. Am. Chem. Soc.* **105,** 5923 (1983).
212. S. Ching, M. Sabat, and D. F. Shriver, *J. Am. Chem. Soc.* **109,** 4722 (1987).
213. A. K. Chipperfield and C. E. Housecroft, *J. Organomet. Chem.* **349,** C17 (1988).
214. D. K. Bower and J. B. Keister, *J. Organomet. Chem.* **312,** C33 (1986).
215. R. B. Calvert and J. R. Shapley, *J. Am. Chem. Soc.* **99,** 5225 (1977).
216. R. D. Barreto and T. P. Fehlner, *J. Am. Chem. Soc.* **110,** 4471 (1988).
217. M. Lourdichi and R. Mathieu, *Organometallics* **5,** 2067 (1986).
218. D.-Y. Jan and S. G. Shore, *Organometallics* **6,** 428 (1987).
219. A. J. Deeming, *in* "Transition Metal Clusters" (B. F. G. Johnson, ed.), p. 391. Wiley, New York, 1980.
220. C. E. Housecroft and S. M. Owen, *Abstr. Am. Chem. Soc. Natl. Meet. 96th,* INORG., 248 (1988).
221. E. Sappa, A. Tiripicchio, A. J. Carty, and G. E. Toogood, *Prog. Inorg. Chem.* **35,** 437 (1987).

ADVANCES IN ORGANOMETALLIC CHEMISTRY, VOL. 30

Dissociative Pathways in Substitution at Silicon in Solution: Silicon Cations R_3Si^+, $R_3Si^+ \leftarrow Nu$, and Silene-Type Species $R_2Si{=}X$ as Intermediates

JULIAN CHOJNOWSKI and WŁODZIMIERZ STAŃCZYK

Centre of Molecular and Macromolecular Studies
The Polish Academy of Sciences
90-363 Łódź, Poland

I

INTRODUCTION

Silicon chemistry is one of the most dynamic areas of science. Silicon compounds have become increasingly important in the creation of many new materials including polymers, ceramics, glasses, and diverse composites. The use of silicon reagents in synthesis has also been constantly growing. Such reagents are applied in a number of synthetic approaches, including synthesis of natural products, models for biological studies, bioactive substances, and polymers. Paralleling this is increasing demand for basic knowledge of mechanisms of reactions of silicon compounds. A better understanding of reaction mechanisms is necessary for designing new synthetic routes and new products. Interest in basic knowledge on substitution at a silicon atom overruns the boundary of organosilicon chemistry since it proves to be very important for the understanding of transformations at heavier atom centers (*1–6*). Silicon occupies an exceptional position in the periodic table, being the closest neighbor to carbon among the heavier elements.

Mechanisms for the substitution at silicon in solution have been a subject of several reviews (*7–15*). Since expansion of the coordination number of silicon is a common feature, particular interest is directed to mechanisms involving intermediates or transition states with silicon having a coordination number of 5 or 6. So far, little attention has been devoted in the review literature to mechanistic pathways that do not invoke extracoordination. Knowledge of these mechanisms, however, has often become necessary for the understanding of chemical behavior of organosilicon compounds. In this article we discuss mechanistic pathways involving heterolytic cleavage

243

of a bond to silicon leading to the transient existence of tri- or tetracoordinate silicon cations or tricoordinate doubly bonded silicon species. The theory surrounding such intermediates in solution has recently drawn the attention of many chemists.

II

SILYLENIUM ION INTERMEDIATES

This section is devoted to displacement at the silicon atom involving transient formation of a tricoordinate silicon cation, e.g., according to Eq. (1). Generation of carbenium ions as intermediates is a common

$$R_3SiX \xrightarrow{\text{slow}} R_3Si^+ + X^-$$

$$R_3Si^+ + Nu^- \xrightarrow{\text{fast}} R_3SiNu$$

$$(1)$$

feature of substitution processes at carbon. The role of the silicon analog of the carbenium ion has been of interest to many organosilicon chemists (for reviews see Refs. 10 and 15). Numerous quantum mechanical calculations have pointed to the stability of various trivalent Si^+ ions, thus giving a solid foundation for experimental search. These ions are very common species in the gas phase. They comprise a major fraction of the ions formed in mass spectroscopy of a variety of organosilicon compounds, and they are often detected as intermediates in gas-phase reactions. Tricoordinate Si^+ ions have been postulated as intermediates in solution, and some kinetic and chemical evidence of their transient existence has been provided. The occurrence of stable long-lived ions of this type in solution is still the subject of controversy (16) although some evidence, recognized as significant, has recently been provided (17,18).

There is some divergence in the nomenclature: various names of these species are used, including silicenium, siliconium, and silylenium ion. According to Barton et al. (19), silylenium ion is the proper name since it is derived in a logical way from the name silylene in analogy to the carbenium ion originating from carbene.

A. Theoretical Studies

Quantum mechanical calculations of silylenium ion structures have been made mostly by *ab initio* methods. Although the results seem to depend to some extent on the procedure used, they provide important information concerning the structure and reactivity of silylenium ions.

The comparative stability of a silylenium ion and the corresponding carbenium ion has been evaluated as the energy difference (ΔE) between products and substrates in the isodesmic Eq. (2) (20–22).

$$SiH_2X^+ + CH_3X \rightarrow CH_2X^+ + SiH_3X \tag{2}$$

Calculations revealed that the parent structure of the silylenium ion H_3Si^+ is more stable than that of H_3C^+ by 41–73.5 kcal/mol. However ΔE tends to decrease strongly on replacement of the hydrogen atom, indicating that substituents, in particular π donors, are decidedly more effective in stabilizing carbenium ions than in stabilizing silylenium ions. Thus, the methyl cation is stabilized by an amino substituent by 93.8 kcal/mol but the silyl cation by only 38.3 kcal/mol (21).

The *ab initio* calculations also predict α-silylcarbenium ions $R_3SiC\overset{+}{R}_2$ to be generally less stable than the isomeric silylenium ions $R_2Si\overset{+}{C}R_2$ (23–25) and the vinyl cation to be less stable than its silicon analog $H_2C{=}Si\overset{+}{H}$ (22). Also the β-silyl substituted carbenium ion $H_3SiCH_2C\overset{+}{H}_2$ is higher in energy than its silylenium ion isomer $H_2\overset{+}{Si}CH_2CH_3$ (25). The higher thermodynamic stability of the silylenium ion might have been expected since silicon is a larger and more electropositive atom than carbon and should therefore better accommodate a positive charge.

Calculations of ΔE of the isodesmic Eq. (3) were executed to study the effect of substituents at silicon atom on the stability of the silylenium ion.

$$SiH_2X^+ + SiH_4 \rightarrow SiH_3^+ + SiH_3X \tag{3}$$

Only the strongly electron-withdrawing substituents Cl, F (26), CF$_3$, and CN (22) were found to destabilize the silylenium ion structure. In most cases, substitution at silicon increases the stability of H_3Si^+. Strongly electropositive groups like BeH or Li have a marked stabilizing effect, but the CH$_3$ group is also stabilizing (21). π donors like NH$_2$ and OH are effective stabilizers, too. One of the largest effects is calculated for the planar NH$_2$ group (21). The ethynyl group stabilizes the cation by 11 kcal/mol (22). According to Truong *et al.* (27), the vinyl group appears to be quite effective in stabilizing a positive charge on the silylenium ion, which may be considered as an analog of stable allyl cation. The calculated ΔE value is almost the same for vinyl as for the planar NH$_2$ group.

The greatly diminished stabilization of the silylenium compared to the carbenium structure by π substituents is connected with the known low effectiveness of $2p$–$3p$ (π) conjugation. Calculations were made on the $\equiv Si^+$ cation substituted with a phosphorus group with the hope of showing a superior $3p$–$3p$ (π) conjugation of P—Si$^+$ moiety (26,28). A smaller inductive destabilization by phosphorus than by electronegative substituents of the second period (N, O) was also expected. However, all calculation methods predicted PH_2 to be definitely inferior to NH_2 in stabilizing the $\equiv Si^+$ ion (26). The reason was that the planarization at phosphorus required for effective π donation was very expensive in terms of energy expended in $PH_2SiH_2^+$. It was further shown that the introduction of appropriate substituents at phosphorus may effectively decrease the planarization barrier (28). For example, $[(CH_3Si)_2P]_3Si^+$ is among the most stable silylenium ions yet evaluated.

Optimization of geometry for some silylenium cations was made (22,24). Such calculations predict that silylenium ions will adopt a planar structure in contrast to silyl anions, which are predicted to be pyramidal. It should be noted that theoretical studies also indicate a high ability of silicon to accommodate negative charge. The parent silyl anion H_3Si^- was calculated to be more stable than its carbon analog by about 50 kcal/mol (24). This implies a remarkable affinity of silylenium ions toward electron-rich species.

B. Silylenium Ions in the Gas Phase

Silylenium ions are common in gas-phase organosilicon chemistry, where they may be generated by various techniques including electron impact (29–33), photoionization (34–36), chemical ionization (37–45), collision-induced dissociation (25,46), and chemical-nuclear methods (15). Although this article is concerned with reactions in solution, a short account of gas-phase studies cannot be omitted, since they provide important information about chemical and physical properties of silylenium ions, which are so elusive in condensed phases.

Various mass spectroscopic techniques have been used. Fragmentation of organosilicon compounds under electron impact in a conventional mass spectrometer provides variety of silylenium ions. Intensities of peaks of the $\equiv Si^+$ ions exceed those of corresponding carbenium ions formed under analogous conditions by 1–2 orders of magnitude, giving an ample evidence of lower energies of formation of these ions (15). Advanced mass spectroscopic methods including ion cyclotron resonance (ICR), Fourier transform (FT), and tandem mass spectrometries have been widely applied for the generation and study of silylenium ions (for review, see Ref. 42).

1. Generation of Silylenium Ions in the Gas Phase

Electron-impact ionization of trimethylchlorosilane in an ICR mass spectrometer resulted in the formation of the trimethylsilyl cation, which could be detected for as long as a 800 mseconds (33). On collision it underwent further transformation to 1,1-dimethylsilaethylene with proton elimination [Eq. (4)].

$$(CH_3)_3SiCl \xrightarrow{e^-} Me_3Si^+ + e^- + Cl^- \xrightarrow{B}$$
$$BH^+ + Me_2Si{=}CH_2 \qquad (4)$$

ICR mass spectrometry has been employed to study the kinetics and equilibria of the generation of silylenium ions by hydride transfer from silanes to carbenium ions (43,47). The kinetics of the formation of silylenium ions from fluoromethylsilanes [$(CH_3)_nSiF_{4-n}$, $n = 1$–3] have also been investigated by this technique (48). Fluoride transfer was the dominant reaction in this system.

The extensive chemical ionization mass spectrometry studies of volatile silanes by Lampe and co-workers were comprehensively reviewed (42). They used tandem mass spectrometers, permitting investigation of the ion–molecule reactions over a broad range of collisional energies and pressures (37,39,42). Some selected kinetic and thermochemical results of studies of reactions of alkylium ions with series of silanes Me_nSiH_{4-n} (38,40) and Et_nSiH_{4-n} (39,40) ($n = 1$–4) are presented in Table I. Dominant reactions were hydride transfer, methide (ethide) transfer, and electron transfer. The contribution from each process depended appreciably on the structure of reactants. All simple reactions of the above type leading to the silylenium ions are strongly exothermic, reflecting the high stability of the silylenium ions.

Kinetic studies of silylenium ion formation by the fragmentation of a metastable parent ion produced by photoionization [Eq. (5)] have been

$$Me_3SiSiMe_3 \xrightarrow[-e^-]{h\nu} Me_6Si_2^+ \rightarrow Me_3Si^+ + Me_3Si \qquad (5)$$

performed (36). The special technique of photoelectron–photoion coincidence (PEPICO) used for this purpose makes possible the preparation of parent ions of well-defined internal energy.

Silylenium ions are also produced as a result of combined chemical-nuclear decomposition of tritiosilanes [Eq. (6)] (15). An analogous method

$$R_{4-n}SiT_n \rightarrow R_{4-n}Si^+T_{n-1} + He^0 \qquad (6)$$

has been used for the generation of carbenium ions (49). This method permits generation of the $\equiv Si^+$ species in the gas phase at higher pressures as well as in liquids and solids. The ions emerge in a free state, i.e., without a counterion or solvation shell.

TABLE I

KINETIC AND THERMOCHEMICAL DATA OF SELECTED GAS-PHASE REACTIONS OF TRIALKYLSILANES WITH CARBENIUM IONS

Reaction	ΔH (kcal/mol)	Relative cross section (Å^2)	$10^9\ k$ (cm^3/second)	Reference
$CH_3^+ + (CH_3)_3SiH \rightarrow (CH_3)_3Si^+ + CH_4$	-88	30	1.1	38
$\rightarrow (CH_3)_2HSi^+ + C_2H_6$	-64	66	2.4	38
$\rightarrow (CH_3)_3Si^{\cdot}{-}H^+ + CH_3^{\cdot}$	1	14.8	—	38
$C_2H_5^+ + (CH_3)_3SiH \rightarrow (CH_3)_3Si^+ + C_2H_6$	-49	22	0.57	42
$\rightarrow (CH_3)_2HSi^+ + C_3H_8$	-27	91	2	42
$(CH_3)_2CH^+ + (CH_3)_3SiH \rightarrow (CH_3)_3Si^+ + C_3H_8$	-27	36.0	—	42
$\rightarrow (CH_3)_2HSi^+ + i\text{-}C_4H_{10}$	-7.0	10.0	—	42
$CH_3^+ + (C_2H_5)_3SiH \rightarrow (C_2H_5)_3Si^+ + CH_4$	-90.0	3.9	0.14	39
$\rightarrow (C_2H_5)_2HSi^+ + C_3H_8$	—	74.0	2.7	39
$\rightarrow (C_2H_5)_2Si^{\cdot}{-}H^+ + CH_3^{\cdot}$	-8	35.0	1.3	39

2. Gas-Phase Reactions of Silylenium Ions

Mass spectroscopy has been often used to study the chemical behavior of silylenium ions. Intramolecular rearrangement with elimination of a neutral molecule has often been observed (e.g., Ref. 50 and references cited therein). For example, 1-phenyl-2-trimethylsilylethane transforms into the more stable phenyldimethylsilylenium ion [Eq. (7)] (51). Intermolecular rearrangements may occur as well (52).

$$ \qquad (7) $$

Silylenium ions are very reactive toward many organic and inorganic compounds. They readily form various addition complexes with compounds having nucleophilic centers like amines, ketones, alcohols, olefins, and aromatic rings (45,53–58). Long-lived complexes can be observed in chemical ionization mass spectra taken at relatively high pressures, i.e., 10^{-3}–1 Torr, and low collision energies. Reactions of SiH_3^+ with some unsaturated or aromatic compounds lead to adducts with lifetimes exceeding 10^{-5} second (45,53,54). The complex with benzene decomposes to give the phenylsilylenium ion as the major product. Study of the fully deuterated silylenium ion showed that no exchange of hydrogen occurs between silicon and carbon [Eq. (8)]. Thus, the reaction most probably proceeds

$$ \qquad (8) $$

through a complex analogous to that formed in the classical electrophilic aromatic substitution with a carbenium ion in solution. The SiH_3^+ ion–ethylene complex may add one or two additional ethylene molecules. Most probably, these adducts have the ultimate form of alkylsilylenium ions: $(C_2H_5)H_2Si^+$, $(C_2H_5)_2HSi^+$, $(C_2H_5)_3Si^+$ (53).

Kinetics of the reactions of Me_3Si^+ with water (57) and alcohol (31,32) were investigated by ICR mass spectrometry. Proposed mechanisms involve formation of an oxonium ion as the primary product, which in the case of the methanol adduct is eventually decomposed to methoxysilylenium ion according to Eq. (9). Reactions of halide transfer to silylenium ion [Eqs. (10) and (11)] have been studied by FT mass spectrometry (59) and by tandem mass spectrometry (44). Hydride transfer from silanes and alkanes to silylenium centers has also been investigated using both ICR ($47,60$) and tandem ($30,42$) mass spectroscopy.

$$Me_3Si^+ + MeOH \longrightarrow \underset{\underset{Me\ \ H}{|\ \ |}}{Me_2Si}\!\!-\!\!\overset{+}{O}\!\!-\!\!Me \xrightarrow{-CH_4} [Me_2\overset{+}{Si}OMe \leftrightarrow Me_2Si\!\!=\!\!\overset{+}{O}Me] \quad (9)$$

$$F_3Si^+ + CHCl_3 \rightarrow CHCl_2^+ + SiF_3Cl \quad (59) \quad\quad\quad (10)$$

$$H_3Si^+ + CF_4 \rightarrow F_3C^+ + SiH_3F \quad (44) \quad\quad\quad (11)$$

It should be noted that conditions for the generation and reactions of silylenium ions in the gas phase are somewhat different from those in solution. Energies of collision leading to chemical transformations are often much larger than in solution where colliding molecules have thermal energy showing the Boltzmann distribution. Unless the pressure is high enough, intermediates formed in the gas phase are not able to transmit the excess energy to surrounding molecules. Thus, the kinetic parameters and mechanisms of reactions determined depend on the techniques and conditions used. Reactions in the gas phase must therefore be compared with those in solution with caution. Nevertheless, experiments in the gas phase have provided a great deal of evidence for the high stability of isolated silylenium ion species and their ability to react with nucleophilic reagents. In particular, similarities in behavior to carbenium ions were disclosed.

Gas-phase methods also constitute a source of important information on basic physical properties of silylenium ions. In particular, the thermochemical behavior is well characterized ($30,33,34,47,61$). Thermochemical data are applied for the evaluation of relative thermodynamic reactivities of silylenium ions in some systems. For example, affinities of R_3Si^+ and R_3C^+ toward various bases may be compared as the heterolytic dissociation energies of corresponding bonds [Eq. (12)] ($47,61$). It was shown that

$$D(R_3M^+\!\!-\!\!X^-) = \Delta H_f(R_3M^+) + \Delta H_f(X^-) - \Delta H_f^0(R_3MX)$$

$$\Delta H_f = \text{heat of formation; M = Si, C} \quad (12)$$

the respective hydride affinities of $SiMe_3^+$, $SiMe_2H^+$, and $SiMeH_2^+$ are 220.5, 230.1, and 245.9 kcal/mol. These ions are more stable (with H^- as a reference base) than their carbenium analogs by 13.1, 21.4, and 24.6 kcal/mol, respectively (47). On the other hand, the Me_3C^+ ion is much more stable than its silicon analog toward F^-. The calculated fluoride affinity is 37.5 kcal/mol higher for Me_3Si^+ (61).

C. Search for Stable Silylenium Ions in Solution

Organosilicon chemists have long searched for a solution system in which silylenium ions are stable. Analogy to the carbenium ion has been a general strategy of studies aiming in this direction. It is usually assumed that silicon analogs of stable carbenium ions should also exhibit relative stability in solutions; however, many early attempts were unsuccessful. It was found that Ph_3SiCl does not ionize in SO_2, DMF, or nitrobenzene–$AlBr_3$ (62), although the carbon analog is well known to be ionic in these systems. Also (p-$Me_2NPh)_3SiCl$ (63) and silyl perchlorates (64,65) were initially found to be covalent. Similarly, ionization of Ph_3SiOH in liquid HCl (66) or (p-$NMe_2Ph)_3SiOH$ in acidic systems (67) were not found. The application of superacids for the generation of the $\equiv Si^+$ ion also failed (68–70). All these and other unsuccessful attempts up to 1973 were comprehensively reviewed by Corriu and Henner (10).

Two reactions were explored in continuation of these attempts: hydride transfer from silanes to a carbenium center and halide transfer from silanes to a Lewis acid. Media of low nucleophilicity and high ionization power were employed. Stable silylenium ions A and B were postulated to be formed by H^- transfer from corresponding silyl hydride to triphenylmethylium perchlorate in CH_2Cl_2 (71,72). However, soon after, and in view of results of additional experiments, this evidence was shown to be insufficient (19,73).

Further attempts were made by Lambert and Schulz (17), who claimed to be able to generate stable silylenium ions as a result of H^- transfer under conditions similar to those used elsewhere (71,72). Trisopropylmercaptosilyl hydride was used as the precursor [Eq. (13)]. Maximization of

$$(i\text{-}PrS)_3SiH + Ph_3C^+ClO_4^- \rightarrow Ph_3CH + (i\text{-}PrS)_3Si^+ClO_4^- \qquad (13)$$

positive charge stabilization through polarization and $3p$–$3p$ overlap was expected. The most important supporting evidence for the silylenium ion was as follows: (1) The hydride transfer products in dilute methylene chloride solution showed a high specific conductance comparable to that of fully ionic species such as trityl perchlorate. (2) Cryoscopically determined molecular weights corresponded well to dissociated species. (3) ^1H- and

^{13}C-NMR spectra were in accord with the formation of only one silicon-containing species having only one type of isopropyl group. According to the authors, the isomeric sulfonium ion structure and the dimer structure can be eliminated with a high degree of confidence. (4) Treatment of the product with diisobutylaluminum hydride led only to the recovered silane precursor.

Later Lambert *et al.* reconsidered the nature of triphenylsilyl perchlorate (*18,74,75*) and trimethylsilyl perchlorate (*76*) produced in an analogous way. These authors postulated full ionization and dissociation of these species in acetonitrile and sulfolane solutions, as evidenced by a high molar conductance in these solvents (for Ph_3SiClO_4, 179.5 and 12.0 mho cm^2/mol in acetonitrile and sulfolane[1], respectively) and the i factor of 2 from cryoscopically measured molecular weight. According to these authors, the silyl perchlorates are also ionized in methylene chloride solution but appear as a strong ion pair, which leads to a low molar conductance for Ph_3SiClO_4 (1.13 mho cm^2/mol). This statement is at variance with the earlier view concerning the nature of triphenylsilyl perchlorate, which has been generally considered as a covalent species (*63–65,77*).

Lambert *et al.* (*18,75*) consider that the observation of a ^{13}C chemical shift of the ipso-aromatic carbon in Ph_3SiClO_4 eliminates the possibility of coordination of the solvent molecule to the silylenium ion. The value of this shift is almost the same in CH_3CN (128.73 ppm) as in CH_2Cl_2 (128.70 ppm), and addition of sulfolane has no effect. On the other hand, addition of the stronger nucleophiles pyridine or imidazole, known to coordinate easily to the $\equiv Si^+$ center (*78*), shifted the signal in CH_2Cl_2 about 3 ppm downfield. Another test was provided by studies of the ^{15}N-NMR spectra. The coordination of the nitrile group to the $\equiv Si^+$ center is expected to move the ^{15}N resonance about 100 ppm upfield. However, the additional shift observed after introduction of 1 mol equiv of CH_3CN to the solution of triphenylsilyl perchlorate in CD_2Cl_2 (concentrations not quoted) was only about 1% of this value (*18,74*). In contrast, the introduction of an equivalent amount of pyridine shifted the signal by almost 100 ppm upfield.

The [35]Cl-NMR resonance may be an important tool for the differentiation of ionic and covalent forms (*16,18*). [35]Cl is a quadrupolar nucleus having nuclear spin 3/2 and natural abundance 75.5%. The perchlorate anion has spherical symmetry, leading to a sharp line with half-width less than 20 Hz (*79*). In contrast, the [35]Cl signal of covalently bonded unsymmetrical perchlorate is very broad. Lambert *et al.* observed that the [35]Cl resonance of triphenylsilyl perchlorate in sulfolane at 0.002 mol/dm^3 gave a signal

---

[1] The relatively low molar conductance is associated with the high viscosity of this solvent.

at 4.5 ppm (from external standard $HClO_4$—H_2O) with a half-width of 18 Hz, which they attributed to silylenium perchlorate (18,74). On raising the concentration, the signal broadened and moved to higher field, which was interpreted in terms of a dynamic equilibrium with the ion pair or covalent form of the perchlorate.

Lambert's interpretation in terms of the ionic nature of the silyl perchlorate species in solution met criticism (16,80). The main objections are as follows. (1) ^{29}Si-NMR spectroscopy has not shown any signal whatsoever which could be attributed to the silylenium ion. Olah and Field (81) found that a correlation exists between the ^{29}Si chemical shift of organosilicon compounds and the ^{13}C chemical shift of their carbon analogs. This correlation, which was demonstrated for 35 silicon compounds, allows for prediction of δ ^{29}Si values for these silicon compounds for which ^{13}C chemical shifts of their carbon analogs are known. Corresponding values for the Me_3Si^+ and Ph_3Si^+ ions are expected to lie within the ranges 225–275 and 100–150 ppm, respectively. Values observed for $Me_3SiOClO_3$ and $Ph_3SiOClO_3$ [46 (76) and 2.0 ppm (75), respectively], are within the range predicted for a covalent form of these compounds (80). (2) ^{1}H- and ^{13}C-NMR data obtained by Lambert are not diagnostic of a differentiation of the ionic structure from a covalent one (16). (3) X-Ray crystallographic study indicated that in the solid state triphenylsilyl perchlorate showed a covalent nature (16). (4) Studies of the ^{35}Cl-NMR spectrum of Ph_3SiClO_4 proved that at the concentration level 10^{-1} M there is no sharp ^{35}Cl signal expected for ClO_4^-. Such a signal is observed at a concentration level about 10^{-3} M. However, it is not reliable as diagnostic for the $\equiv Si^+$ ion since the water concentration in the system may exceed the concentration of silyl perchlorate, causing its hydrolysis (16). It should be noted that Olah and colleagues (16) do not exclude the possibility that in highly diluted solutions at sufficiently high dielectric constant and low nucleophilicity the covalent silyl perchlorates would undergo ionization to the silylenium ion.

The other approach in attempts to synthesize stable long-lived silylenium ions explored the reaction of Lewis acids with silyl halides. The reaction of tris(trimethylamino)silyl chloride and fluoride with $AlCl_3$ was carried out in an effort to prepare the $(Me_2N)_3Si^+$ ion (82). Three amine groups substituted to silicon were expected to stabilize strongly the positive charge on the silicon atom. The attempt, however, was futile as only a donor–acceptor complex involving one of the Me_2N groups was formed.

The reaction of lithium salts of tert-butylaminodiorganylfluorosilanes with $AlCl_3$ was shown to lead to silicon ylides [Eq. (14)] (83). The partial zwitterionic character of these species was deduced from ^{29}Si- and ^{27}Al-NMR data.

$$R-\underset{\underset{F}{|}}{\overset{\overset{R'}{|}}{Si}}-N\underset{Li}{\overset{CMe_3}{<}} \xrightarrow[-LiF]{+\frac{1}{2}Al_2Cl_6} \underset{R}{\overset{R'}{>}}\overset{+}{Si}\underset{Cl}{\overset{N\underset{>}{\overset{CMe_3}{|}}}{<}}\overset{-}{Al}Cl_2$$

R, R' = *t*-Bu, *t*-Bu or *t*-Bu, Ph   (14)

Olah and Field (*81*) showed that the interaction of Lewis acids, such as BI_3, BBr_3, $TaBr_5$, $AlBr_3$, and $AlCl_3$, with trimethylsilyl halides in weakly nucleophilic solvents, methylene bromide, chloride, and iodide as well as CS_2 and PBr_3 leads to strongly polarized donor–acceptor complexes, but no free silylenium ions were observed. The largest deshielding of the silicon nucleus [$\delta(^{29}Si)$ 62.7 ppm] was observed for $Me_3Si^{\delta+}$---Br-->$^{\delta-}$ $AlBr_3$. The line width of the ^{27}Al-NMR resonance was reduced from 1250 Hz for $AlBr_3$ in CH_2Br_2 to 950 Hz in the complex, while tetrahedral $AlBr_4^-$ in CH_2Br_2 has a line width of only about 20 Hz. Complex formation is accompanied by ligand exchange at the metal center. The reaction of trialkylsilyl trifluoromethanesulfonate with BCl_3 and BBr_3 also leads to polarized donor–acceptor complexes, giving rise to considerable deshielding of the ^{29}Si-NMR shift and marked sharpening of ^{27}Al-NMR resonance (*80*).

To briefly summarize, it should be pointed out that, although Lambert's results are generally considered to be interesting and significant, there is still controversy about the true silylenium nature of the conducting species involved in the systems (*16*). Further studies are welcome. In particular, the divergence between expected and experimental values of ^{29}Si chemical shifts should be explained. Additionally, criticism concerning possible perturbations of some experiments by fortuitous water cannot be ignored.

In light of the great affinity of silylenium ions for electron-rich species, one could be also skeptical about the possibility of the existence of the tricoordinate Si^+ species in solvents like acetonitrile (AN) and sulfolane. These solvents are known to have nucleophilic coordination ability. Gutmann's donicity number (*84*), 14.1 and 14.8 for AN and sulfolane, respectively, is comparable to that for acetone, 17.0. Coordination of acetonitrile to silicon has been considered in some systems (*85,86*). The small equivalent conductance of triphenylsilyl perchlorate in CH_2Cl_2 may be explained by the domination of the covalent form (*18*). Consequently, the absence of the coordination of acetonitrile and sulfolane with the tricoordinate Si^+ observed in CH_2Cl_2 (with 1 or 6 equiv of acetonitrile) may simply indicate that ClO_4^- coordinates to Si^+ under these conditions more readily than acetonitrile and sulfolane [Eq. (15)]. The reverse situation in acetonitrile

$$R_3SiOClO_4 + CH_3CN \rightleftharpoons (R_3Si\cdot CH_3CN)^+ + ClO_4^- \qquad (15)$$

[equilibrium (15) lies to the right] or sulfolane should not be a surprise as these solvents show considerably higher ionizing power (dielectric constant 38 and 42.0, respectively) and are present in higher concentrations.

D. Silylenium Ions as Intermediates in Reactions at the Silicon Center

1. Hydride Transfer

Hydride transfer from silicon to a carbenium center has also been explored as the main object of studies aiming to detect a silylenium ion intermediate. This reaction constitutes a part of many reduction processes with silyl hydride reagents, which are often utilized in organic chemistry (87–89). Early investigations were performed in systems containing fairly strong nucleophiles like Cl^- counterion or acetic acid solvent which were conducive to a one-step concerted hydride transfer. Corey and West (90) studied the reaction of triorganosilyl hydrides with trityl chloride in benzene and also in more polar solvents such as a nitromethane. They suggested that the hydride halide exchange proceeds through a four-center transition state in which nucleophilic attack by the chloride of the ion pair on silicon occurs while hydride is being transferred.

$$R_3Si\text{--------}H$$
$$Cl\text{--------}CPh_3$$

Such an explanation has received support from the observation that the reaction of trityl chloride with (+)-α-naphthylphenylmethylsilane in benzene leads to complete retention of configuration at the silicon atom (91).

The kinetics of the reaction of various silyl hydrides with tris(2,6-dimethoxyphenyl)methyl cation and some other stable carbocations in acetic acid have been systematically investigated by Carey and Wang-Hsu (92). The results were interpreted in terms of a four-center transition state involving a trigonal bipyramid at silicon with nucleophilic participation of the solvent.

The reaction shows a negative value of Hammett's ρ coefficient (-1.84) for the series

$$\underset{X}{\bigodot} \text{Me}_2\text{SiH}$$

Sommer and Bauman investigated the reaction of optically active α-naphthylphenylmethylsilyl hydride with trityl chloride and found that the stereochemical course was retention in benzene solution, inversion in methylene chloride, and full racemization in methylene chloride (93). The result in CH_2Cl_2 was explained as consistent with an exchange in which electrophilic attack on the silicon hydride by a triphenyl cation is of primary importance; however, the counterion appears in the transition state [Eq. (16)]. Racemization of the unreacted R_3SiH could arise from

$$Ph_3CX \rightleftharpoons Ph_3C^+X^-$$
$$Ph_3C^+X^- + R_3\overset{*}{S}iH \rightleftharpoons [R_3Si^{\delta+}\text{---}H\text{---}\overset{X^-}{\overset{\delta+}{C}}Ph_3] \xrightarrow{X^-} R_3Si^+ + Ph_3CH \tag{16}$$

partial breaking of the Si—H bond and rotation of the R_3Si moiety followed by return of the hydride.[2]

The first kinetic evidence of the stepwise course of hydride transfer involving a silicocation intermediate was provided in 1977 (94). Kinetic studies of reaction (17) were performed in methylene chloride with exclu-

$$Ph_3C^+SbF_6^- + R_3SiH \rightarrow Ph_3CH + R_3SiF \tag{17}$$

sion of any nucleophile that could be stronger than the solvent. Free ion and ion pair concentrations were controlled since the equilibrium constant of ion pair formation had been determined (95). Fortuitous water and other common nucleophiles were excluded by high vacuum experiments. It was shown that the final product, i.e., R_3SiF, appears with a considerable delay with regard to the Ph_3C^+ conversion. The rate of the Ph_3C^+ conversion is the same for free ions and ion pairs, which is equivalent to the kinetic Eq. (18) also being valid in the range of exclusive free ion existence.

$$\frac{d(Ph_3CH)}{dt} = \frac{-d(Ph_3C^+)}{dt} = \frac{-d(R_3SiH)}{dt} = k(Ph_3C^+)(R_3SiH) \tag{18}$$

The above results indicate that the counterion does not participate in the transition state, which contradicts the synchronous mechanism. This con-

---

[2] According to localized orbital theories of the chemical bond, during the rotation the system must pass through a structure with no overlap between p(Si) and s(H) orbitals.

TABLE II

RATE CONSTANTS FOR HYDRIDE TRANSFER OF Et_3SiH
WITH VARIOUS CARBENIUM SALTS

Carbenium ion	Anion	k (dm^3/mol second)
Ph_3C^+	SbF_6^-	139^a
	AsF_6^-	120^a
	PF_6^-	117^a
	BF_4^-	110^a
	$SbCl_6^-$	120^a
	$FeCl_4^-$	120^a
	ClO_4^-	146^b
(ring structure)$^+$	SbF_6^-	0.046^a
	PF_6^-	0.037^a
	BF_4^-	0.031^a
	ClO_4^-	0.052^b

^{a} From Chojnowski et al. (96).
^{b} From W. Fortuniak, Ph.D. thesis. Center of Molecular
and Macromolecular Studies, The Polish Academy of Sciences, Łódź, Poland, 1989.

clusion was confirmed (96) by the observation that the rate is independent
of the structure of various counterions (Table II). Moreover, the reaction
is not sensitive to steric effects unless a strongly sterically hindered substrate is considered (such at t-Bu_3SiH, for which the reaction is very slow).
Also, it shows a Taft's ρ^* value of approximately -3.5 and k_H/k_D values
for $PhMe_2SiL$ (L = H, D) with Ph_3C^+ and $C_7H_7^+$ (tropylium) of 1.49 and
1.45, respectively.

An important argument against any S_N2 mechanism in the process is the
observation that hydride transfer from a silicon atom in a strained ring
(k_1 1.63×10^2 at 25°C) proceeds at approximately the same rate as the
transfer from its acyclic analog (k_1 1.64×10^2 at 25°C) under the same
conditions.

The S_N2 substitutions are faster for the cyclic system by a factor of several
orders of magnitude (97–99) as there is a considerable release of ring strain
on going to a trigonal bipyramid structure.

The mechanism postulated involves formal hydride transfer leading to the formation of a silylenium ion intermediate [Eq. (19)]. It

is suggested

$$R_3SiH + Ph_3C^+ \xrightarrow{\text{slow}} R_3Si^+(solv) + Ph_3CH$$

$$R_3Si^+(solv) + SbF_6^- \xrightarrow{\text{fast}} R_3Si \underset{F}{\overset{F}{\diagdown}} \underset{F}{\overset{F}{\underset{F}{Sb}}} \xrightarrow{\text{slow}} \tag{19}$$

$$R_3SiF + SbF_5 \xrightarrow{\text{CH}_2\text{Cl}_2} \text{unspecified products}$$

that the solvation of this ion with CH_2Cl_2 may influence its character. A further step is formation of the neutral complex with the counterion which decomposes to the product. On the other hand, formation of a silylenium ion in a one-step hydride transfer process seems improbable in light of the rather high energy of heterolytic cleavage of the Si—H bond [220.5 kcal/mol for Me_3SiH (47)], and it is not compatible with the small values of energy of activation observed (5–7 kcal/mol). Certain features of the reaction seem to be best rationalized on the basis of the single electron transfer (SET) mechanism [Eq. (20)] (96). The rate-limiting step

$$R_3SiH + {}^+C\!\!\equiv\ \ \rightleftharpoons\ \ \underset{\substack{\text{charge transfer or encounter} \\ \text{complex}}}{[R_3SiH\ {}^+C\!\!\equiv]} \xrightarrow{\text{rate limiting}}$$

$$\underset{\substack{\text{free radical pair} \\ \text{in a solvent cage}}}{\{R_3\overset{+}{Si}\!\!-\!\!\overset{\cdot}{H}\ \overset{\cdot}{C}\!\!\equiv\}} \xrightarrow{\text{fast}} R_3Si^+ + HC\!\!\equiv \tag{20}$$

is electron transfer within a charge transfer or encounter complex with the formation of a free radical pair, which decomposes to the silylenium ion and triphenylmethane. The last step occurs presumably within the same solvent cage, as the whole process proceeds stoichiometrically to analytical precision (100). It should be mentioned that the SET mechanism has been considered in similar redox processes at the silicon center (101–103).

The concept of the generation of silylenium ion by hydride abstraction with a stable carbenium ion has been explored in designing the synthesis of saturated and unsaturated cyclic silaethers (104). For this purpose the incipient silylenium ion is intramolecularly trapped by a properly located ethereal nucleophile, which leads to cyclization [Eq. (21)]. If the triphenyl-

$$\tag{21}$$

methyl ethers are used, the reaction is catalytic through the continuous regeneration of the trityl ion. Synthesis according to Eq. (21) in many cases proved to be successful.

2. 1,2 Migration in α-Functional Silanes

The Lewis acid-induced 1,2 migration in (α-chloroalkyl)trialkylsilanes was first reported in 1947 (105). A mechanism involving the silylenium ion was proposed [Eq. (22)]. However, further studies did not confirm the

$$\text{Me}_3\text{SiCH}_2\text{Cl} \xrightarrow{\text{AlCl}_3} \text{Me}_3\text{SiCH}_2{}^+ + \text{AlCl}_4{}^- \longrightarrow \text{Me}_2\text{Si}^+\text{CH}_2\text{Me} + \text{AlCl}_4{}^- \longrightarrow$$

$$\text{EtMe}_2\text{SiCl} + \text{AlCl}_3 \tag{22}$$

stepwise pathway (106–110). A four-center transition state was proposed from the kinetic results (106).

Pathways involving full ionization of the carbon–chlorine bond followed by migration of the alkyl group synchronously with nucleophilic attack at silicon or preceded by the attack have also have considered (109,110).

The general opinion that the 1,2-alkyl migration from silicon to the carbenium center is a synchronous process was, however, at variance with calculations of Hopkinson and Lien showing that there is essentially no barrier for the strongly exothermic rearrangement of $\text{H}_3\text{SiCH}_2{}^+$ to $\text{H}_2\text{Si}^+\text{CH}_3$ (24). Referring to these calculations, Barton and co-workers suggested the formation of a silylenium ion intermediate in the reaction of (chloromethyl)vinylsilanes with AlCl_3 in CS_2 solution to afford cyclo-propylchlorosilane products [Eq. (23)]. According to these authors the

(23)

intermediate could be formed by β or γ closure of an initially formed carbocation followed by the rearrangement to the silacyclopropylsilylenium ion (*111*).

However, essential progress in proving silylenium ion intermediacy in the 1,2-alkyl migration from silicon to $^+C\equiv$ has been achieved only relatively recently (*112,113*). Apeloig and Stanger found a system in which a carbenium ion, generated in the α position to silicon in a process of spontaneous solvolysis, transforms to a silylenium ion by a 1,2 shift of the methyl group to the positively charged carbon atom. These authors performed successful experiments with chemical trapping of the transient silylenium ion by a nucleophilic solvent molecule according to Eq. (24). It

$$X = p\text{-}O_2NPhCOO^-, \; Cl^-; \; OS = \text{solvent moiety} \qquad (24)$$

was shown that solvolysis of substrates **A** may lead to two products: **C**, which is quenched with a solvent carbenium ion **B**, and **E**, which is quenched with the solvent silylenium ion **D**. The ratio of both products depends in a dramatic way on the nucleophilicity of the solvolytic system. In acetone containing 20% water **E**:**C** is 1:99, whereas in hexafluoroisopropanol the product derived from the silylenium ion is formed almost exclusively (>99%). The reactivity pattern argues against solvent-assisted methyl migration.

3. Nucleophilic Substitution in Trisyl-Type Silicon Species

A useful approach to tracing disociative pathways of substitution at silicon proved to be studies of sterically hindered compounds with bulky nonreactive ligands and relatively good leaving groups. The majority of experimental data in this area are concerned with a particular class of compounds, namely, organosilicon derivatives of tris(trimethylsilyl)methane. The $(Me_3Si)_3C$ ligand is commonly denoted as "trisyl" (Tsi). Synthetic aspects of this chemistry and some earlier results have been reviewed (114–116).

Trisyl and related groups exert a dramatic effect on the reactivity of silanes, leading in most cases to the inhibition of direct nucleophilic attack at the silicon atom. Eaborn and colleagues (117) found that solvolysis of trisyl perchlorate in methanol exhibits some classic features of a unimolecular S_N1 reaction [Eq. (25)]. In particular, no significant rate increase was found on addition of MeONa to the system, indicating that participation of the nucleophile is not essential for breaking the bond to the leaving group. On the other hand, the reaction is accelerated by strong electrophiles (114). The effect of addition of LiCl, $NaClO_4$, and $LiNO_3$ was as expected for a primary salt effect in the ionization of the perchlorate in the rate-determining step. In the presence of $LiNO_3$ a considerable amount of nitrate product is formed, indicating trapping of the silicocation by NO_3^-. The S_N1 pathway was postulated.

$$(Me_3Si)_3CSiMe_2ClO_4 \xrightarrow{\text{slow}} [(Me_3Si)_3CSiMe_2]^+ + ClO_4^-$$

$$[(Me_3Si)_3CSiMe_2]^+ + MeOH \xrightarrow{\text{fast}} (Me_3Si)_3CSiMe_2OMe$$

(25)

Studies of the reaction of trisyl-substituted silanes of the general formula $(Me_3Si)_3CSiMeZX$, in which X is a leaving group and Z is, for example, OMe or aryl (116–122), revealed that the solvolysis reaction is often accompanied by 1,3 migration of the nonreactive substituent Z. The bridged structure (**A**) of the silicocation intermediate has been proposed to account for this observation. It bears a close analogy to that well established for the organoaluminum dimer (**B**).

Strong evidence for the intermediacy of bridged silicocations was provided by the observation of the formation of a considerable amount of rearranged products in the presence of Ag^+ and Hg^+ or other electrophiles in reactions of trisyl species both in solvolytic systems and in aprotic media [Eq. (26)] (*120,123–126*). The ionization of the silane takes place as a result of electrophilic attack of the cation on the leaving group. The extent of the migration is governed by steric hindrance at the respective 1 and 3 silicon atoms, which are targets of the consecutive nucleophilic attack.

On the other hand, methanolysis or hydrolysis of $TsiSiR_2I$ species ($R_2 = Ph_2$, Et_2 or HPh), in the absence of electrophilic agents, gives

$$
\begin{array}{c}
Me_3Si \diagdown \quad \diagup SiMe_2CH=CH_2 \\
\qquad C \\
Me_3Si \diagup \quad \diagdown Si-I \\
\qquad\quad Et_2
\end{array}
+ Ag^+CF_3C(O)O^- \xrightarrow[-AgI]{}
\begin{array}{c}
\qquad\qquad Me_2 \\
Me_3Si \diagdown \quad \diagup Si \\
\qquad C+ \qquad CH=CH_2 + O_2CCF_3 \\
Me_3Si \diagup \quad \diagdown Si \\
\qquad\qquad Et_2
\end{array}
$$

$(Me_3Si)_2C(SiMe_2CH=CH_2)(SiEt_2O_2CCF_3)$ $(Me_3Si)_2C(SiMe_2O_2CCF_3)(SiEt_2CH=CH_2)$

33%   67%

(26)

exclusively unrearranged $TsiSiR_2OMe$ or $TsiSiR_2OH$ products (*127*). Consequently, the bridged cationic intermediate does not appear. Although these reactions are not catalyzed by lyate ion, some nucleophilic assistance by the solvent must be involved, and, in analogy to solvolysis of *tert*-butyl halides (*128*), the S_N2 intermediate mechanism was proposed (*127*) but recently disproved (*129*). In some cases, however, S_N2-type reactions were observed, e.g., for $TsiSiMe_2I$ with suitable nucleophiles (F^-, NCS^-, NCO^-) as well as for methanolysis of less hindered $TsiSiPhXH$ ($X = Br$, F, NO_3, NCO^-, NCS^-, $p\text{-}MeC_6H_4SO_3$) (*127,130*). Thus, the spectrum of mechanisms of nucleophilic substitution on the trisyl-bound silicon atom is broad.

It should be mentioned that rearranged product formation is often observed together with a strong rate enhancement by the γ-substituted electron-donating group Z, which is indicative of anchimeric assistance from this group (*121–123,126,131–133*). In these cases the interaction of Z with the silicon center already takes place in the rate-determining step, making departure of the leaving group easier.

$$
\left[
\begin{array}{c}
\qquad\qquad R_2 \\
\qquad\quad Si \\
(Me_3Si)_2C \diagdown \qquad Z \\
\qquad\quad \diagdown Si\cdots\cdots X^{\delta-} \\
\qquad\quad R_2'
\end{array}
\right]
$$

Some remarkable effects of the electronegative neighboring group such as OMe (*124*) or OAc (*134*) have been observed in these systems. For example, although tetrakis(trimethylsilyl)methane ($TsiSiMe_3$) is completely inert toward CF_3COOH, even under reflux, the methoxy derivative TsiSiMe$_2$OMe gives an impressive cleavage of a normally stable Si—Me bond at room temperature [Eq. (27)].

It appears that the measured ability of Z to supply anchimeric assistance decreases in the following order: $OMe > SO_3CF_3 > ClO_4 > F > Cl, I > Me$, in agreement with MNDO calculations of the relative stability for acyclic

$$(Me_2Si)_2C \begin{matrix} SiMe_2OMe \\ SiMe_2O_2CCF_3 \end{matrix} \xrightarrow{CF_3COOH} (Me_3Si)_2C(SiMe_2O_2CCF_3)_2 \qquad (27)$$

$[H_2C(SiH_2Z)(SiH_2{}^+)]$ and bridged

cations (135). Recently, anchimeric assistance of azide (123), aryl (122), and vinyl groups (133) has been also reported.

4. Other Systems

Observations considered as evidence in favor of silylenium ion intermediacy have been made in other systems. Most of them have already been reviewed (10).

Sommer et al. (136) found that optically active silyl halides readily undergo racemization induced by halide ions. The reaction is related to halide ion exchange discovered by Allen and Modena (137). The first tandem kinetic studies of the racemization and halide exchange reaction led to conclusion that the reaction proceeds via a silylenium ion pair [Eq. (28)],

$$(-)\text{-}R_3Si*Cl + Cl^- \underset{}{\overset{k_E, k_R}{\rightleftharpoons}} (\pm)\text{-}R_3Si^{36}Cl + *Cl^- \qquad (28)$$

where k_E and k_R are the specific rates of exchange and racemization, respectively. Later reinvestigation of these reactions (138,139) revealed

$$(-)\text{-}R_3SiCl \rightleftharpoons [R_3Si^+Cl^-] \overset{{}^{36}Cl^-}{\rightleftharpoons} [R_3Si^{+36}Cl^-] \rightleftharpoons (+)\text{-}R_3Si^{36}Cl \qquad (29)$$

that some features are in conflict with silylenium ion pair formation. Nevertheless, the value for k_E/k_R of 1.3 observed in chloroform may still be interpreted as partial ionization according to Eq. (29).

Attention has also been focused on reactions of β-functional silanes. Sommer et al. (140) first reported the β-elimination reaction of β-chloro-alkylsilanes catalyzed with Lewis acids [Eq. (30)]. Transient formation of

$$R_3SiCH_2CH_2Cl \xrightarrow{AlCl_3} R_3SiCH_2CH_2^+ + AlCl_4^- \longrightarrow R_3Si^+ + CH_2CH_2 + AlCl_4^- \longrightarrow$$
$$R_3SiCl + AlCl_3 \tag{30}$$

the silylenium ion was initially postulated. A synchronous mechanism, however, was preferred by Bott et al. (106).

Solvolysis of β-halogenosilanes occurs in an analogous way [Eq. (31)] (141). The reaction rate

$$R_3SiCH_2CH_2Cl \xrightarrow{EtOH/H_2O} R_3SiOEt + H_2C{=}CH_2 + HCl \tag{31}$$

is dependent on the solvent ionizing power but is not affected by the nucleophilic character of the medium. Base catalysis does not seem to be significant, and electron-attracting substituents slow down the reaction. A "limiting siliconium ion" mechanism was proposed [Eq. (32)], but this mechanism has been criticized (142).

$$\equiv SiX \xrightarrow{slow} [\equiv Si^+ \text{-----} X^-] \xrightarrow{fast} [S\text{-----}\overset{\overset{\delta+}{\diagup}}{\underset{|}{Si}}\text{-----}\equiv X] \longrightarrow S{-}Si\equiv + X^-$$
$$X = \beta\text{-chloroethyl}; \ S = \text{solvent moiety} \tag{32}$$

The stereochemical course of solvolysis of a diastereomeric β-bromo-silane, which was found to be trans elimination, as well as the observation of 1,2-silyl group migration in a process of substitution at the β carbon (143) are evidence of anchimeric assistance of the silyl group in ionization of the bond to the β carbon. It was further noticed (144) that after quenching the solvolysis reaction of β-bromo-β-dideuterioethylsilane the substrate recovered is a mixture of α and β deuterated isomers [Eq. (33)]. These

$$Me_3SiCH_2CD_2Br \xrightarrow[H_2O]{MeOH} Me_3SiCH_2CD_2Br + Me_3SiCD_2CH_2Br \tag{33}$$

observations are consistent with the intermediacy of a nonclassical silacy-clopropenium ion although an alternative explanation is also feasible [Eq. (34)] (145,146).

$$\text{Me}_3\text{SiCH}_2\text{CH}_2\text{X} \longrightarrow \underset{\overset{|}{\underset{\text{H}_2\text{C}\underline{\hspace{1.2cm}}\text{CH}_2}{}}}{\overset{\text{Me}_3}{\overset{\text{Si}}{\diagup\;\;\overset{+}{}\;\;\diagdown}}} + \text{X}^- \xrightarrow{\text{ROH}} \text{H}_2\text{C}{=}\text{CH}_2 + \text{Me}_3\text{SiOR} \quad (34)$$

The transient existence of the silylenium ion has been considered in other reactions of chlorosilanes in the presence of Lewis acids. Guyot (*147*) successfully used the silyl chloride–silver salt system for initiation of the cationic polymerization of vinyl ethers [Eq. (35)]. Gel permeation

$$\text{Ph}_3\text{SiCl} + \text{Ag}^+\text{PF}_6^- \longrightarrow \text{Ph}_3\text{Si}^+\text{PF}_6^- + \text{AgCl}$$

$$\text{Ph}_3\text{Si}^+\text{PF}_6^- + \text{H}_2\text{C}{=}\text{CHOR} \longrightarrow \underset{\overset{|}{\text{OR}}}{\text{Ph}_3\text{SiCH}_2\text{CH}^+\text{PF}_6^-} \xrightarrow{n\text{H}_2\text{C}{=}\text{CHOR}} \text{polymer} \quad (35)$$

chromatography analysis, using in tandem the refractive index and UV detectors, established the presence of aromatic groups built up at the end of the polymer chain. Attempts at electrophilic aromatic substitution performed using silyl triflate–boron trihalide complexes (*80*), however, failed. The silylenium ion intermediate has also been postulated in the disproportionation of alkylhalogenosilanes (*148,149*).

III

SILENE-TYPE SPECIES IN ELIMINATION–ADDITION REACTIONS

The elimination–addition route [Eq. (36)] constitutes a rare but important dissociative mechanism for displacement at silicon (*150*). Along the reaction coordinate one must assume the formation of unsaturated silicon species, which are now of interest to great number of organosilicon chemists (*151*).

Even in the late 1970s one could find the statement that stable compounds with double bonds from silicon to carbon, oxygen, and nitrogen were not known (*152*). In the 1980s knowledge about such systems has developed dramatically. Theoretical calculations and studies of gas-phase processes, though still numerous, are slowly giving way to the relatively new area of stable unsaturated molecules and their reactions in solution (*153–157*).

$$\underset{\overset{|}{\text{X}}}{\text{R}_2\text{Si}{-}\text{Y}^-} \xrightarrow{\text{slow}} [\text{R}_2\text{Si}{=}\text{Y}] + \text{X}^-$$

$$\text{(36)}$$

$$[\text{R}_2\text{Si}{=}\text{Y}] + \text{NuH} \xrightarrow{\text{fast}} \text{R}_2\text{Si}(\text{YH})(\text{Nu}) \qquad \text{Y} = \text{CR}_2, \text{O}, \text{S}, \text{NR}$$

A. *Theoretical Predictions and Stable π-Bonded Compounds*

There have been number of calculations concerning energy and structural features of the double-bonded silicon compounds, including silenes (Si=C), disilenes (Si=Si), silanones (Si=O), silathiones (Si=S), and silaimines (Si=N). Some earlier estimates concerning the π bond energy in silene (silaethylene) varied significantly (3–62 kcal/mol) (*155,158*). It is now generally regarded to be in the range of 35–37 kcal/mol (*159,160*), compared to that of 65 kcal/mol in ethylene (experimental value) (*161*) and, e.g., 31 kcal/mol of $H_2Ge=CH_2$ (*160*). *Ab initio* calculations predict a value for the length of the Si=C double bond in $Me_2Si=CH_2$ of 1.70 Å (*154,162*), being close to the experimental distances for stable silenes (see below). In light of this, the first electron diffraction result of 1.83 Å for 1,1-dimethyl-1-silaethylene (*163*) is generally regarded as unreliable.

Theoretical calculations estimate the respective π bond energy and double bond length as 24 kcal/mol (*164*) and 2.0–2.2 Å (*162,165,166*) for Si=Si in the simple parent disilene. For the two similar systems, silanone and silathione, the latter is expected to be thermodynamically more stable, with π bond energy of 33 and 42 kcal/mol, respectively, and approximately similar 10% bond length shortening on going from H_3Si—XH to $H_2Si=X$ (X = O, S). The respective distances between silicon and heteroatoms were calculated to be 1.498 Å in $H_2Si=O$ and 1.936 Å in $H_2Si=S$ (*167,168*). The calculated value of the Si=N bond length in silaimine (H_2SiNH, 1.53–1.57 Å) (*169,170*) is also about 10% shorter than for most Si—N single bonds (*7,151,171*).

The above *ab initio* and related calculations clearly reveal that unsaturated silicon compounds are less stable than their carbon analogs (*156,170*), but as gas-phase studies and the chemistry of sterically crowded silicon derivatives have proved (*153,154,156,158,162,172–175*) such compounds can no longer be labeled as "nonexistent" (*176,177*). A number of isolable double-bonded silicon derivatives have recently been synthesized and characterized in agreement with theoretical predictions (Table III) (*169,173,178–195*). The data illustrate well that the kinetic stability of multiply bonded silicon species can be greatly improved by substitution of sufficiently bulky groups across the p_π–p_π bond. The ease of addition reactions is then severely restricted. The new target for organosilicon chemists appears to be silicon triple-bonded compounds (*196,197*).

B. *Elimination in Gas Phase*

Elimination processes of the type depicted in Eq. (36) constitute an important pathway for generating unsaturated silicon compounds.

TABLE III

PHYSICAL PROPERTIES OF SELECTED STABLE π-BONDED SILICON COMPOUNDS[a]

Compound	Double bond length (Å)	^{29}Si NMR (ppm)	IR (cm^{-1})	Reference(s)

Silenes (Si=C)

$$\begin{array}{ccc} Me_3Si & & OSiMe_3 \\ & \diagdown\;/ & \\ & Si=C & \\ & \diagup\;\diagdown & \\ Me_3Si & & R \end{array}$$

Compound	Double bond length (Å)	^{29}Si NMR (ppm)	IR (cm^{-1})	Reference(s)
R = t-Bu	—	41.5	1130	*178–180*
R = CEt$_3$	—	54.3	1133	*180–182*
R = Methylcyclohexane	—	43.5	—	*180*
R = Adamantyl	1.764	41.4	1135	*179–181*

$$\begin{array}{ccc} Me_3SiO & & SiMe_2{}^tBu \\ & \diagdown\;/ & \\ & Si=C & \\ & \diagup\;\diagdown & \\ Me & & Adamantyl \end{array}$$

| | | 126.5 | — | *183* |

$$\begin{array}{ccc} THF & & SiMe_3 \\ \diagdown & & / \\ Me-Si=C & & \\ \diagup & & \diagdown \\ Me & & SiMe^tBu_2 \end{array}$$

| | 1.747 | — | — | *184* |

$$\begin{array}{ccc} Me & & SiMe_3 \\ \diagdown & & / \\ & Si=C & \\ \diagup & & \diagdown \\ Me & & SiMe^tBu_2 \end{array}$$

| | 1.702 | 144.2 | — | *185,186* |

Disilenes (Si=Si)

R(Mes)Si=Si(Mes)R

Compound	Double bond length (Å)	^{29}Si NMR (ppm)	IR (cm^{-1})	Reference(s)
R = Mes	2.160	63.6	—	*187–190*
R = t-Bu (trans)	2.143	90.3	—	*173,189*
R = t-Bu (cis)	—	94.7	—	
R = Me$_3$SiN (trans)	—	61.9	—	*173,189*
R = Me$_3$SiN	—	49.4	—	
(Ar')$_2$Si=Si(Ar')$_2$	—	64.06	—	*191,192*

Silaimines (Si=N)

$$\begin{array}{c} {}^tBu \\ \diagdown \\ \quad Si=NSi^tBu_3 \\ \diagup \\ {}^tBu \end{array}$$

| | 1.568 | 78.29 | — | *169* |

$$\begin{array}{c} THF \\ \diagdown \\ Me-Si=N-Si^tBu_3 \\ \diagup \\ Me \end{array}$$

| | 1.588 | 1.08 | — | *169,193* |

$$\begin{array}{c} {}^iPr \\ \diagdown \\ \quad Si=N-Ar \\ \diagup \\ {}^iPr \end{array}$$

| | — | 60.3 | — | *194* |

Silaphosphines (Si=P)

$$\begin{array}{c} Mes \\ \diagdown \\ \quad Si=P-Ar \\ \diagup \\ Is \end{array}$$

| | — | 148.7 | — | *195* |

$$\begin{array}{c} Is \\ \diagdown \\ \quad Si=P-Ar \\ \diagup \\ {}^tBu \end{array}$$

| | — | 175.9 | — | *195* |

[a] Mes, 2,4,6-Trimethylphenyl; Ar, 2,4,6-tri-*tert*-butylphenyl; Ar, 2,6-dimethylphenyl; Is, 2,4,6,-triisopropylphenyl.

Historically it was the gas-phase process which first unequivocally proved the existence of species with a π bond to silicon [Eq. (37)] (198). In the absence of nucleophiles, the dimethylsilene gives a head-to-tail dimerization product, 1,1,3,3-tetramethyl-1,3-disilacyclobutane. On the other hand, addition products across the double bond of the intermediate silene are formed in the presence of various trapping agents.

$$
\begin{array}{ccc}
\underset{\substack{\text{Me}}}{\overset{\text{H}_2\text{C}-\text{CH}_2}{\underset{|}{\overset{|\qquad|}{\text{H}_2\text{C}-\text{Si}-\text{Me}}}}} & \xrightarrow[-\text{H}_2\text{C}=\text{CH}_2]{400^\circ\text{C}} & \left[\text{H}_2\text{C}=\text{Si}\!\!\begin{array}{c}\nearrow\text{Me}\\\searrow\text{Me}\end{array}\right] & \longrightarrow & \underset{\substack{\text{Me}}}{\overset{\text{H}_2\text{C}-\overset{\text{Me}}{\overset{|}{\text{Si}}}-\text{Me}}{\underset{|}{\overset{|\qquad|}{\text{Me}-\text{Si}-\text{CH}_2}}}}
\end{array}
$$

$$
\underset{\text{CH}_3\text{SiMe}_2\text{OMe}}{\overset{\text{MeOH}}{\swarrow}} \quad \underset{\text{CH}_3\text{SiMe}_2\text{OH}}{\overset{\text{H}_2\text{O}}{\downarrow}} \quad \underset{\text{CH}_3\text{SiMe}_2\text{NH}_2}{\overset{\text{NH}_3}{\searrow}} \tag{37}
$$

Many of the results in this field have been already reviewed (153,154,158,162,199–201), and a few recent ones concerning thermal elimination of $R_2Si{=}S$ species are illustrated in Scheme 1 (202–205). Weber and co-workers described a transannular reaction of 6-oxa-3-silabicyclo[3.1.0]hexanes leading to transient silaoxetanes and elimination of dimethylsilanone, trapped by oxadisilacyclopentane (202). Barton and co-workers reported elimination of dimethylsilanone from alkoxysilanes (203) and silylketenes (204). Analogous silylthioketenes yielded silathione (205), and kinetics of gas-phase addition reactions of dimethylsilene and dimethylsilanone were also studied (206,207).

C. Elimination Processes in Condensed Phase

Apart from examples of thermal extrusion of unsaturated species (e.g., $Et_2Si{=}S$ from dithiocyclohexasilanes) (201,208) and photolytic elimination reactions (158,209,210), most of the processes utilizing the elimination route are induced by generation of an anion in the α position to silicon. The pioneering studies by Wiberg (156,211–213) and Jones (214,215) and colleagues have been continued (216–218) and extended as a preparative method for relatively stable π-bonded silicon compounds (195). Reactions of lithium silylamides with chlorosilanes in nonpolar solvents were also rationalized in terms of a mechanism involving formation of the unstable silaimine, $Me_2Si{=}NR$, and subsequent dimerization to account for a high yield of cyclodisilazanes in a mixture with expected substitution products (219). Examples are given in Scheme 2 (50–53).

Trisyl compounds (see Section II,D,3) of the form $(TsiSiR_2X)$ (Scheme 3) (54–56) provide a useful model for studies of elimination–addition pro-

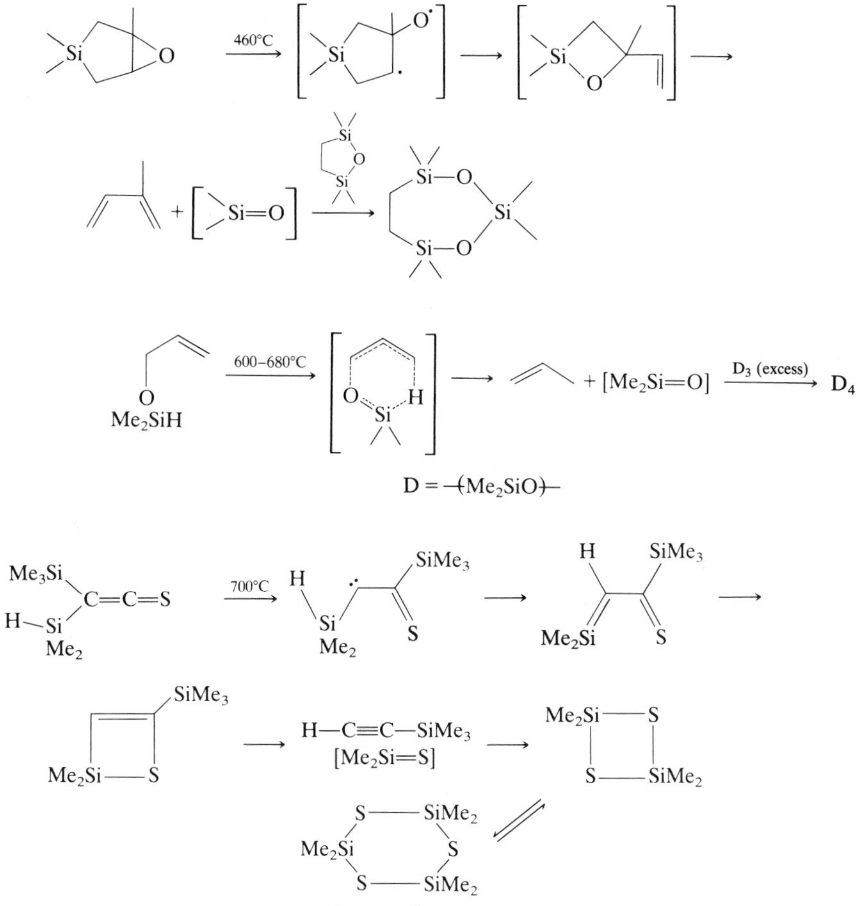

460°C

D_3 (excess)

D_4

$D = -(Me_2SiO)-$

600–680°C

Me_2SiH

$+ [Me_2Si=O]$

Me_3Si

$H-Si$

Me_2

$C=C=S$

700°C

$SiMe_3$

H $SiMe_3$

Me_2Si S

$SiMe_3$

Me_2Si S

$H-C\equiv C-SiMe_3$

$[Me_2Si=S]$

Me_2Si S

S $SiMe_2$

S $SiMe_2$

Me_2Si S

S $SiMe_2$

SCHEME 1

cesses. As a result of the steric bulk of the Tsi group, typical substitution reactions are made difficult, and the elimination–addition pathway is favored (220). The unusually high reactivity of trisyl silanolates, generated in the reaction medium, is explained by the unimolecular process involving transient silanone species. TsiSiPh(OH)I undergoes immediate complete conversion to TsiSiPh(OH)(OMe) in 0.17 M MeOH solution with 0.25 M MeONa, whereas there is no Si—I bond cleavage even in boiling MeOH (221). The "silanone mechanism" (115) was also suggested as accounting for the abnormally high reactivity of the Si—H bond and one-step formation of TsiSi(OH)$_3$ from TsiSiH(OH)I on hydrolysis (222).

Kinetic studies of the pathways shown in Scheme 3 were performed for base-catalyzed cleavage of diol RSiMe(OH)$_2$ [Eq. (38)] and triol

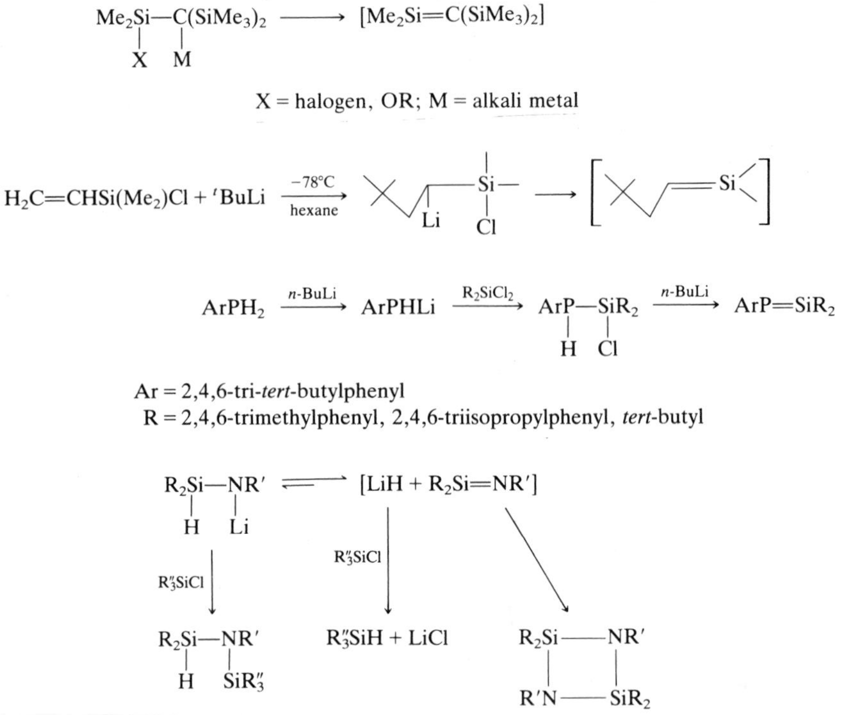

$$Me_2Si\overset{|}{\underset{X}{-}}\overset{|}{\underset{M}{C}}(SiMe_3)_2 \longrightarrow [Me_2Si{=}C(SiMe_3)_2]$$

X = halogen, OR; M = alkali metal

$$H_2C{=}CHSi(Me_2)Cl + {}^tBuLi \xrightarrow[\text{hexane}]{-78°C} \left. \bigtimes \overset{|}{\underset{Li}{\diagup}}\overset{|}{\underset{Cl}{Si}}{-} \right. \longrightarrow \left[\bigtimes {=}{Si}\diagup \right]$$

$$ArPH_2 \xrightarrow{n\text{-BuLi}} ArPHLi \xrightarrow{R_2SiCl_2} Ar\overset{|}{\underset{H}{P}}\overset{|}{\underset{Cl}{-}}SiR_2 \xrightarrow{n\text{-BuLi}} ArP{=}SiR_2$$

Ar = 2,4,6-tri-*tert*-butylphenyl
R = 2,4,6-trimethylphenyl, 2,4,6-triisopropylphenyl, *tert*-butyl

SCHEME 2

$$MeOSiMe_3 + [(Me_3Si)_2C{=}SiR_2] + X^-$$

$$[(Me_3Si)_2C{=}SiR_2] + MeOH \longrightarrow (Me_3Si)_2CH{-}SiR_2(OMe)$$

R = Me, Ph; X = F, Cl, Br, I

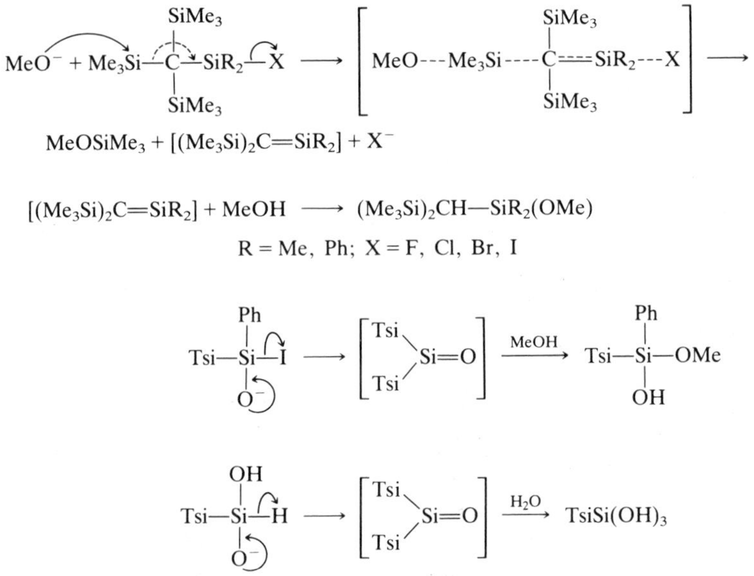

SCHEME 3

$RSi(OH)_3$ ($R = m\text{-}ClC_6H_4CH_2$) in $H_2O–MeOH$ and $H_2O–DMSO$ media.

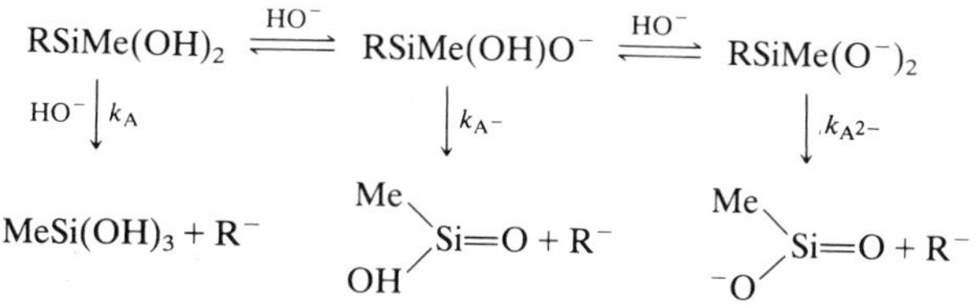

$$RSiMe(OH)_2 \xrightleftharpoons{HO^-} RSiMe(OH)O^- \xrightleftharpoons{HO^-} RSiMe(O^-)_2$$

$$HO^- \Big\downarrow k_A \qquad\qquad \Big\downarrow k_{A^-} \qquad\qquad \Big\downarrow k_{A^{2-}}$$

$$MeSi(OH)_3 + R^- \qquad \underset{OH}{\overset{Me}{\diagdown}}Si{=}O + R^- \qquad \underset{^-O}{\overset{Me}{\diagdown}}Si{=}O + R^-$$

(38)

It was shown that at a high base concentration the unimolecular processes of internal nucleophilic displacement (k_A^- and k_A^{2-}) dominate over S_N2 substitution, giving evidence for the generation of unstable π-bonded silaacetate (*223*) and monomeric metasilicate ions (*224*). Earlier, a contribution from the analogous elimination–addition process had been suggested for Si—C bond cleavage in silanols $RSiMe_2OH$ ($R = PhCH_2$, *m*-$ClC_6H_4CH_2$, $PhC{\equiv}C$) (*221,225*), involving transient formation of dimethylsilanone. The importance of this novel dissociative substitution pathway in organosilicon chemistry, often overshadowed by faster S_N2 reactions, is also being shown in other systems. Recently, it was proposed for Si—Si bond cleavage in polysilanes by aqueous–alcoholic base (*226*) (Scheme 4), and its features (partial silicon–oxygen double bond) (*150,227*) have been discussed for polysiloxanes (*228*) (Scheme 4).

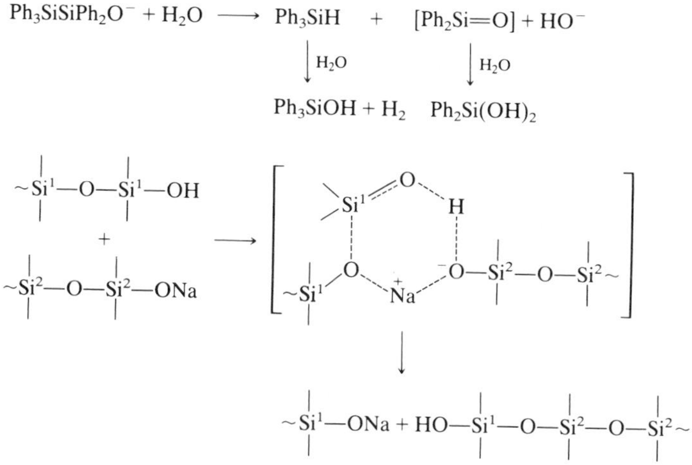

$$Ph_3SiSiPh_2O^- + H_2O \longrightarrow Ph_3SiH + [Ph_2Si{=}O] + HO^-$$

$$\Big\downarrow H_2O \qquad\qquad \Big\downarrow H_2O$$

$$Ph_3SiOH + H_2 \qquad Ph_2Si(OH)_2$$

SCHEME 4

IV

BORDERLINE S_N1–S_N2 MECHANISMS

Classification of nucleophilic displacement as a one-step S_N2 or a step-wise S_N1 process is not adequate as there is a continuous spectrum of nucleophilic substitution mechanisms, of which the classic S_N1 and S_N2 pathways can be considered as extremes. There is no clearly defined line between them but rather a region of borderline pathways for which a mechanistic diagnosis is difficult (229,230). When the classic S_N1 mechanism operates, the intermediate lives long enough to become diffusionally equilibrated in the solution. If, however, its lifetime is short, about 10^{-10} second or less, the intermediate has little chance to leave the solvent cage in which it is generated. The product may be formed either by trapping of the intermediate by a reagent molecule in the closest solvation shell or by preassociation with the reagent prior to intermediate formation, which enforces the reaction toward the product. In these cases, the reaction is an S_N1 process but shows some features of the S_N2 substitution. Its rate depends on the concentration of the nucleophile.

It may, however, also happen that the barrier to reaction of the intermediate toward the product is insignificant. The lifetime of the intermediate becomes shorter than the time of molecular vibration along the reaction coordinate (10^{-13} second), and so the intermediate does not exist. The reaction becomes a one-step concerted process, but it involves a loose "exploded" transition state with little bond making and highly advanced bond breaking (230). The reaction is S_N2 substitution with the intermediatelike transition state exhibiting features of the S_N1 processs. In particular, the rate is only slightly dependent on nucleophile basicity. A somewhat similar borderline mechanism is the S_N2 (intermediate) pathway when nucleophilic attack occurs on a preformed ion pair intermediate (128,231,232).

The borderline S_N1–S_N2 mechanisms play an important role in substitution at carbon and phosphorus, elements which are the closest neighbors of silicon. Most solvolyses of secondary alkyl halides have been described as belonging to this category (229). Even the solvolysis of tert-butyl halides, long thought to be a classic S_N1 process, was shown to involve considerable nucleophilic assistance (128). Many important phosphorylation reactions, like that in Eq. (39), have been postulated as occurring along the elimination–addition pathway via monomeric metaphosphate transient species (233), analogs of the monomeric metasilicates considered as intermediates in the substitution at silicon (224). However, Knowles and colleagues (234) found that alcoholysis of O^{17},O^{18}-labeled phenylphosphate monoanion proceeds with complete inversion of the configuration at

$$(39)$$

phosphorus. Evidently, a borderline mechanism is operating. The choice would be between S_N2 "exploded" transition state **A** and preassociation S_N1 mechanisms **B** (230,234).

   A   B

   Borderline mechanisms may play an important role in nucleophilic displacement at the silicon center. Actually, almost all arguments in favor of the transient formation in solution of silylenium ions and silicon double-bonded species advanced so far could be interpreted in terms of borderline mechanisms as well. These pathways have often been proposed in studies of the silylenium ion question. Good examples of this are concepts of the "limited silylenium ion" developed by Sommer and Baughman (141) and the "incipient silylenium ion" considered by Barton and co-workers (111). Eaborn discussed in broad terms the formation of a bridged silicon cation either by intramolecular trapping of the preformed silylenium ion or as a result of the anchimeric assistance of the neighboring group (114,115). Nucleophilic participation of the solvent has been also suggested (61,94), and extrapolation of the results of gas-phase studies to solutions pointed to a particular role of silylenium ion–solvent interaction (47,61). Borderline pathways are also consistent with some observations in studies of double-bonded silicon intermediates in solution (227) and could, possibly, be responsible for some unsuccessful attempts at trapping these species.

V

TETRACOORDINATE SILICON CATIONS AS INTERMEDIATES IN NUCLEOPHILE-ASSISTED SUBSTITUTION AT SILICON

   This section is devoted to processes beginning with the heterolytic cleavage of a silicon bond by a neutral nucleophile. Thus, some aspects of intermediacy of cations of silicon having a coordination number of 4 are discussed. The subject is somewhat related to the silylenium ion problem since the species discussed here may be considered as an $\equiv$Si$^+$ ion which coordinates a neutral molecule. Basic knowledge concerning the interaction of the trivalent positively charged silicon center with neutral species rich in electrons is essential for understanding the character of the solvation of silylenium ions around which the controversy arose (Section II).

The other justification of the extension of the scope of this article to tetracoordinate Si$^+$ intermediates is the notable interest in the nucleophilic activation to substitution at the silicon atom, which is closely related to the tetrahedral Si$^+$ intermediacy problem. Little attention has been paid so far to this aspect of organosilicon chemistry in other review articles. This discussion does not include, however, the vast area of tetrahedral Si$^+$ intermediates in reactions catalyzed by protic acids; thus, the protic cations, i.e., those having acidic hydrogen, are in principle excluded.

A. Relation to Problems of Catalysis in Silylation Processes

The tetracoordinate silicon cation is a rather common species in solution. It may be generated by heterolytic cleavage of a bond from silicon to a reactive ligand, as a result of interaction of the silicon center with an uncharged nucleophile like amine, imine, phosphine, phosphine oxide, and amide. Since these nucleophiles are also known to be effective catalysts for many displacements at silicon including important silylation processes (86,89,235–238), the cations of tetracoordinate silicon have received attention as possible intermediates in these reactions according to Eq. (40) (78,235,239–243).

$$\begin{aligned} Nu + R_3SiX &\rightarrow NuSiR_3{}^+ + X^- \\ NuSiR_3{}^+ + Y^- &\rightarrow R_3SiY + Nu \end{aligned} \tag{40}$$

Two other pathways, however, should be considered. First, the nucleophile may act as a Brönsted base, receiving proton from the substrate. This way the nucleophilicity of the substrate toward the silane is increased. Specific [Eq. (41)] or general [Eq. (42)] base catalysis is possible (7,137,244–246). The second possibility is a pathway involving an extra-

$$\begin{aligned} HY + Nu &\rightleftharpoons NuH^+ + Y^- \\ Y^- + R_3SiX &\rightarrow R_3SiY + X^- \end{aligned} \tag{41}$$

$$\begin{aligned} HY + Nu &\rightleftharpoons Nu^{\delta+} \cdots\cdots HY^{\delta-} \\ Nu^{\delta+} \cdots\cdots HY^{\delta-} + R_3SiX &\rightarrow R_3SiY + NuH^+ + X^- \end{aligned} \tag{42}$$

coordinate silicon intermediate formed as a result of direct coordination of the nucleophile (for reviews, see Refs. 10–13 and 247). A more general scheme of this mechanism is given in Eq. (43).

$$\text{Nu} + R_3SiX \longrightarrow Nu\overset{\displaystyle \overset{RR}{\diagdown\!\diagup}}{\underset{\displaystyle R}{Si}}X \xrightarrow{\;Y^-\;} R_3SiY + \text{Nu} + X^- \qquad (43)$$

The spectrum of reactions assisted by nucleophiles is broad, and Eqs. (40)–(43) are all generally regarded to be significant. There are, however, opposing views on the scope of their operation. Some details of the pathway shown in Eq. (43) concerning the explanation of how extra-coordination activates the silane toward nucleophilic displacement are also controversial. In particular, the mechanism of nucleophilic displacement of chlorine in triorganochlorosilanes has become the subject of continuing controversy. These reactions are of considerable importance: triorgano-chlorosilanes are among the most inexpensive and most common silylating agents (*89,235–238*) and are often used in mixtures with Lewis bases which efficiently promote the silylation. Corriu *et al.* found (*248–250*) that the reactions are strongly accelerated by nucleophiles like HMPA, DMSO, and DMF. These authors also observed that the nucleophiles dramtically change the stereochemistry of the reaction from inversion to retention (*251*), which together with the kinetics of the process can be well explained by the pathway involving pentacoordinate (and possibly hexacoordinate) Si intermediates [Eq. (44)] (*248–251*). It was, however, pointed out (*239*)

$$\underset{R^2}{\overset{R^1}{\diagdown}}\!\!\!\underset{Cl}{\overset{R^3}{Si}}\;\;\overset{\text{Nu}}{\underset{}{\rightleftharpoons}}\;\; \left[\underset{R^2}{\overset{R^1}{\diagdown}}\!\!\underset{Cl}{\overset{\overset{\displaystyle \text{Nu}}{|}}{Si}}\!-\!R^3\right] \xrightarrow{\text{ROH}} \left[\underset{R^2}{\overset{R^1}{\diagdown}}\!\!\underset{\overset{|}{Cl}}{\overset{\overset{\displaystyle \text{Nu}}{|}}{Si}}\!\!\!\underset{}{\overset{R^3}{\diagup}}\;\!\!\text{OH}\right] \xrightarrow{-\text{Nu}\cdot\text{HCl}} \underset{R^2}{\overset{R^1}{\diagdown}}\!\!\!\underset{OH}{\overset{R^3}{Si}} \qquad (44)$$

intermediate
or transition state

that both the kinetics and the stereochemistry of these reactions could also be understood on the basis of Eq. (45) involving a tetracoordinate silicon cation intermediate, and this view found support in other results (*240,242,243,252,253*).

$$\underset{R^2}{\overset{R^1}{\diagdown}}\!\!\!\underset{Cl}{\overset{R^3}{Si}}\;\;\underset{\text{inversion}}{\overset{\overset{\displaystyle \text{Nu}}{\text{fast}}}{\rightleftharpoons}}\;\; \left[\underset{R^2}{\overset{R^1}{\diagdown}}\!\!\!\underset{R^3}{\overset{\text{Nu}}{Si}}\right]^{+}\!\!\!Cl^- \;\;\xrightarrow[\text{inversion}]{\overset{\text{ROH}}{\text{(slow)}}}\;\; \underset{R^2}{\overset{R^1}{\diagdown}}\!\!\!\underset{OR}{\overset{R^3}{Si}} \qquad (45)$$

Arguments for the intermediacy of penta- or hexacoordinate silicon intermediates have been discussed in detail in earlier reviews (*10–13,247*), but little attention has so far been devoted to Eq. (40). There is, however, a body of evidence that nucleophilic displacement at silicon may follow pathway (40). Evidence for this mechanism includes the following: (1) the common existence of compounds postulated to be intermediates or modeling the intermediates i.e., positively charged ionic silane–nucleophile complexes containing tetracoordinate silicon; (2) the dynamic behavior of these compounds when mixed with their components, and the behavior of the mobile ligand; (3) a high reactivity of the complexes toward nucleophiles; and (4) agreement with stereochemical results, kinetic laws, activation parameters, structure–reactivity dependences, and medium effects of the nucleophile-induced silylation reactions.

B. *Formation of Stable Salts Having a Tetracoordinate Silicon Atom in the Cation*

Cations having a tetracoordinate silicon may appear by three general routes. Apart from the ionization of a silicon–reactive ligand bond, the cations can be formed by transformation of a group bound to silicon, in particular by the addition of positively charged ion. For example, the quaternization of trimethylsilylamine with methyl iodide leads to the same ionic complex as the reaction of trimethylsilyl iodide with trimethylamine [Eq. (46)] (*254*).

$$\begin{matrix} Me_3SiNMe_2 + MeI \\ \\ Me_3SiI + NMe_3 \end{matrix} \searrow_{\nearrow} [Me_3SiNMe_3]^+I^- \qquad (46)$$

Addition of an uncharged nucleophile to a silylenium ion is also possible, and some sterically hindered tetrahedral Si^+ products are likely to appear. Compounds of this type which have so far been generated could be classified according to the participating nucleophile as

1. Complexes of nitrogen-containing nucleophiles or silylammonium salts, $[R_3SiNR_3]^+X^-$
2. Complexes of phosphorus nucleophiles or silylphosphonium salts, $[R_3SiPR_3]^+X^-$
3. Complexes of oxygen nucleophiles or silyloxonium salts, for example,

$$[R_3P\!\!=\!\!O \cdots\cdots SiR_3]^+X^-$$

(silyloxyphosphonium or silylphosphoxonium salts) and

$$[RC\!\!=\!\!O \cdots\cdots SiR_3]^+X^-$$
$$\mid$$
$$NMe_2$$

(silyloxycarbenium or silylcarboxonium salts)

Since the formal positive charge is on the ligand, the proper names of these salts should originate from corresponding nucleophiles. Thus, they should be regarded as silylated ammonium, phosphonium, and oxonium salts. Therefore we use the name silylonium ion. Most often, the following counterions X appear: I^-, $CF_3SO_3^-$, Br^-, ClO_4^-, Cl^-. The positive charge is usually strongly delocalized over other atoms of the complex ion.

For example, the cation generated by the interaction of $(Me_3N)P$ with Me_3SiI could be represented by the mesomeric structures in Eq. (47) (255).

$$
Me_3Si^+ : P \overset{\displaystyle \diagup NMe_2}{\underset{\displaystyle \diagdown NMe_2}{\text{—}NMe_2}} \longleftrightarrow Me_3Si\overset{+}{\text{—}}\!\!P \overset{\displaystyle \diagup NMe_2}{\underset{\displaystyle \diagdown NMe_2}{\text{—}NMe_2}} \longleftrightarrow Me_3Si\text{—}P \overset{\displaystyle \diagup \overset{+}{N}Me_2}{\underset{\displaystyle \diagdown NMe_2}{\text{—}NMe_2}} \longleftrightarrow
$$

$$
Me_3Si\text{—}P \overset{\displaystyle \diagup \overset{+}{N}Me_2}{\underset{\displaystyle \diagdown NMe_2}{=\!\overset{+}{N}Me_2}} \longleftrightarrow Me_3Si\text{—}P \overset{\displaystyle \diagup NMe_2}{\underset{\displaystyle \diagdown \overset{+}{N}Me_2}{\text{—}NMe_2}} \tag{47}
$$

Some of the salts were isolated in pure crystalline form and showed fairly high thermal stability, but many of the cations were observed only in solution. In certain cases, however, solid complexes are readily formed, like $Me_3N \cdot Me_3SiOSO_2CF_3$, although at ambient temperature in solution only unchanged reactants are observed (78). They are often easily soluble in aprotic solvents of high or moderate polarity like acetonitrile and methylene chloride. Some of these complexes are unstable at room temperature, decomposing reversibly to components, and may be observed only at a low temperature (78,242,252,255,256). Sometimes irreversible decomposition to other products takes place. An example is shown in Eq. (48). The majority of these complexes are hydrolytically very unstable

$$
[Me_3SiP(NMe_2)_3]^+I^- \rightarrow Me_3SiNMe_2 + IP(NMe_2)_2 \tag{48}
$$

(239,242,257,258) and can be observed only in aprotic solvents. Some protic complexes having bulky substituents at silicon (259–261) or bridgehead structures (262) show a considerable resistance to hydrolysis, however.

Important evidence of the ionic structure of many silane–uncharged nucleophile complexes in solution is their high electrical conductance, characteristic of strong electrolytes (239,252,254–256,263). The 1:1 stoichiometry of certain complexes in solution was proved by conductometric titration (Fig. 1). The stoichiometry in solids was, in many cases, confirmed by elemental analysis (239,260).

X-Ray studies of crystals of pyridine complexes with trimethylsilyl bromide and iodide (264) and of N-methylimidazole adducts to trimethylchlorosilane (265) were performed. The tetracoordinate silicon structure of these complexes was proved. The distances between the halogen and silicon atoms in pyridine adducts are 4.359 and 4.559 Å for bromine and iodine, respectively, which is approximately 2 Å greater than the sum of the covalent radii and significantly longer (0.5 Å) than the sum of the van der Waals radii. This result is consistent with the ionic structures of the complexes in the solid state. The distance from silicon to nitrogen (1.86 Å) is evidently larger than the length of typical Si—N bonds (1.75 Å), which

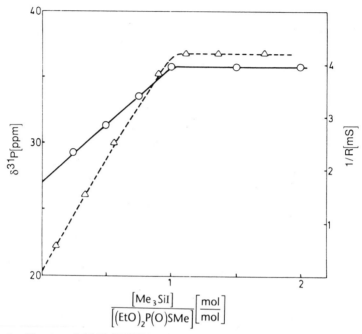

FIG. 1.   Titration of $(EtO)_2P(O)SMe$ with Me_3SiI in CH_2Cl_2 at $-70°C$ followed by conductance (Δ---Δ) and ^{31}P-NMR spectroscopy ($\bigcirc$—$\bigcirc$). $[(EtO)_2P(O)SMe]_0$, 0.1 mol/dm^3 (conductance) or 0.5 mol/dm^3 (NMR). (From M. Cypryk, Ph.D thesis. Centre of Molecular and Macromolecular Studies, The Polish Academy of Sciences, Łódź, Poland, 1982. See also Ref. 256.)

indicates that the nucleophile is only weakly bonded to the central silicon atom. The CSiC angles are around 6° larger and the CSiN angle about 6° smaller than the usual tetrahedral angle. Thus, considerable distortion of the trigonal pyramidal structure occurs, making nucleophilic attack on silicon easier from the opposite side to the nitrogen ligand. The ionic structure of a crystalline N-methylimidazole–Me_3SiCl complex was also unequivocally established (265).

NMR spectroscopy is a powerful tool for studies of silylonium complexes. The ^{29}Si-NMR chemical shift provides important structural information. Bassindale and Stout (252) showed that the ^{29}Si δ value for the same cation is virtually independent of the counterion in such solvents as acetonitrile, deuterochloroform, and methylene chloride, which may be used as evidence of the ionic structure. Since the range of the ^{29}Si chemical shift is large and highly dependent on coordination number, it may serve as a particularly useful probe for coordination at silicon. For example, the complex Me_3SiX (X = I, Br, CF_3SO_3) with N,N-dimethylacetamide in CD_2Cl_2 shows a δ value about 44 ppm (240), which is strongly indicative of

the tetracoordinate structure **A**, since the pentacoordinate silicon resonance appears in a quite different region (*266*). Species (**B**) of well-defined
pentacoordinate structure (*267*), which may serve as a good model for the
silicon pentacoordination in DMF–triorganohalosilane systems, shows the
^{29}Si resonance at -29.4 ppm in the same solvent, at a considerably higher
field (*240*).

A   B

The dependence of NMR chemical shifts on the molar ratio of complex
components may be used for studies of the stoichiometry of the complex
(*252,256,258,263*) (Fig. 1). The resonance of some nuclei of the electron
donors changes considerably on complexation with silanes. Particularly
large changes are observed for the ^{15}N resonance (~ 100 ppm upfield) since
the paramagnetic component of the ^{15}N-NMR chemical shift is eliminated
on quarternization of nitrogen. Mixing of equimolar amounts of pyridine
and triphenylsilyl perchlorate was demonstrated to lead to a shift of the
^{15}N-NMR signal from 314.0 to 216.8 ppm, which was taken as evidence for
the formation of a 1:1 ionic complex (*18*).

A marked change in the position of resonance on the complex formation
is also observed in the ^{31}P-NMR resonance. The additional shift caused by
complexing is, to a considerable extent, dependent on the structure of the
nucleophile. For example, in the interaction of PhEtMePO with Me$_3$SiI in
CH$_2$Cl$_2$, phosphorus is deshielded by 36 ppm (*238*), while in the case of
the analogous complexing of (Me$_2$N)$_3$PO deshielding is only 3 ppm. On
the other hand, the conversion of (Me$_2$N)$_3$P to the cation shown in Eq. (47)
is accompanied by considerable additional shielding of the ^{31}P nucleus
($\Delta\delta-56$ ppm), owing to charge delocalization to the nitrogen atom (*255*).

Proton- and ^{13}C-NMR spectroscopy have been also used as diagnostic
tools for the silane–nucleophile complex formation (*252,253*). In some
cases the coupling constant is very sensitive to complexing with a silane and
may be used as a probe of the formation of an ionic complex. A good
example is the selenium–phosphorus coupling in selenophosphoryl nucleophiles. Selenium bridging to phosphorus and another atom in a neutral
molecule gives a J_{P-Se} value of about 400–500 Hz, while J_{P-Se} for the
P=Se group is at least twice as large, usually within the range 800–
1000 Hz (*268*). The J_{P-Se} constant in the ionic silyloxyphosphonium salt
lies somewhat between these values because of the contribution from the

mesomeric structure involving a selenium–phosphorus π bond [Eq. (49)].

$$\left[\begin{array}{cc} \overset{\displaystyle SeMe}{\underset{\displaystyle OEt}{EtO-\overset{+}{P}-OSiMe_3}} & \longleftrightarrow & \overset{\displaystyle \overset{+}{SeMe}}{\underset{\displaystyle OEt}{EtO-P-O-SiMe_3}} \end{array} \right] \quad J_{P-Se} = 670 \text{ Hz} \quad (CH_2Cl_2, -90°C)$$

$$(49)$$

$$\underset{\underset{J_{P-Se} = 480 \text{ Hz}}{\displaystyle EtO}}{\overset{EtO}{\diagdown}} P \overset{SeMe}{\underset{O}{\diagup}} \qquad \underset{\underset{J_{P-Se} = 900 \text{ Hz}}{\displaystyle EtO}}{\overset{EtO}{\diagdown}} P \overset{Se}{\underset{OMe}{\diagup}}$$

The direct silicon–phosphorus bond coupling constant (J_{P-Si} 115–Hz, CH_2Cl_2, $-60°C$) was observed for the complex shown in Eq. (47) (225). Selected complexes for which the silylonium structure in solids or in solution was proved are collected in Table IV (see also Refs. 269 and 270).

C. Thermodynamics of Silylonium Ion Formation

Formation of an ionic tetracoordinate Si^+ complex from an uncharged nucleophile and a functional silane is an exothermic process accompanied by a marked drop in entropy. Many qualitative observations indicated that these complexes are generated more readily at lower temperatures (78,242,252,256). Unfortunately, there are few data on the thermodynamic parameters of complex formation. From the temperature variation of the ^{29}Si resonance position, Bassindale and Stout (252) determined the enthalpy and entropy of the formation of bis(N,N-trimethylsilyl)-imidazolium chloride (Table IV, entry 10). A similar procedure permitted Chaudhry and Kummer (242) to determine the enthalpy of formation of complexes of 2-trimethylsilyl-1,1,3,3,-tetramethylguanidine (Table IV, entries 6, 7).

The equilibrium and dynamic behavior of the silylonium complexes were extensively studied by Bassindale *et al.* (252,253), who used mostly bis(N,N-trimethylsilyl)imidazolium salts as model compounds [Eq. (50)].

$$Me_3SiX + N \diagup \diagdown N-SiMe_3 \rightleftharpoons Me_3SiN \diagup\overset{+}{\diagdown} NSiMe_3 X^- \qquad (50)$$

The position of the equilibrium depends to a considerable extent on the structure of the counterion and was found to lie well toward the side of the complex when X is Br^-, I^-, $CF_3SO_3^-$, or ClO_4^-, but to the side of components for Cl⁻. However, silylonium salts with the Cl counterion stable at room temperature have recently been obtained by complexing Me_3SiCl with N-methylimidazole (265) and 2-trimethylsilyl-1,1,3,3-tetramethylguanidine (242).

On the basis of NMR studies of equilibrium [Eq. (51)] in CD_2Cl_2 and CD_3CN, the following order of the ability of complex formation was established (253): $Me_3SiCl < Me_3SiBr < Me_3SiI < Me_3SiOSO_2CF_3 < Me_3SiOClO_3$.

$$Me_3SiN \underset{(+)}{\frown} NSiMe_3X^- + Me_3SiY \rightleftharpoons Me_3SiN \underset{(+)}{\frown} NSiMe_3Y^- + Me_3SiX \quad (51)$$

Ion pairing in the salt was observed in methylene chloride, which suggests that solvation of the counterion in organic solvents is an important factor determining the formation of the salt (253). The order for ion pairing in the N-trimethylsilylimidazolium salts is $Cl^- > Br^- > I^- > {}^-OSO_2CF_3 > {}^-OClO_3$, thus exactly reversing the order of complex formation constants. Investigations of competition reactions between nucleophiles according to Eq. (52) allowed establishment of the order of the effectiveness of nucleophiles in complex formation (78) (see Table V).

$$Nu + Nu'^+SiMe_3 \rightleftharpoons Nu^+SiMe_3 + Nu' \quad (52)$$

In contrast to the existence of numerous stable ionic triorganosilylonium complexes, there is no well-defined pentacoordinate molecular 1:1 adduct of acyclic structure of a triorganohalosilane with an uncharged nucleophile. This implies a higher thermodynamic stability of the ionic silylonium complexes than their molecular isomers for the open chain triorganosilyl structure [Eq. (53)]. Stable penta- or hexacoordinate 2:1 nucleophile–silane open chain complexes with three organic groups bound to

$$Nu\overset{\displaystyle R \ R}{\underset{\displaystyle R}{\overset{\displaystyle \diagdown \diagup}{—Si—}}}X \rightleftharpoons R_3Si^+NuX^- \quad R = alkyl\ or\ aryl,\ X = good\ leaving\ group \quad (53)$$

silicon are not known either. Instead, diorganosilyl complexes such as $[Me_2HSi(NMI)_2]^+Cl^-$ and $[Me_2Si(NMI)_3]^{2+} \cdot 2Cl^-$ have recently been well characterized (265,271). In general, the tendency for the triorganosilylonium ion to accept a fifth uncharged ligand does not seem to be high [Eq. (54)].

$$R_3SiNu^+ + Nu \rightleftharpoons \overset{\displaystyle R \ R}{\underset{\displaystyle R}{\overset{\displaystyle \diagdown \diagup +}{NuSiNu}}} \quad (54)$$

TABLE IV

Selected Tetrahedral Silicocation Complexes[a]

Entry	Complex	Form and method of identification	Properties	Reference(s)
1	PySiMe$_3$$^+I^-$	Cryst., X-ray, el. anal., Ir	mp 100°C	254,264
2	PySiMe$_3$$^+Br^-$	Cryst., X-ray	mp 27°C	264
3	Me$_3$NSiMe$_3$$^+I^-$	Cryst., el. anal., sol.,[b] IR, conduct.		254
4	Me$_3$NSiMe$_3$$^+ClO_4$$^-$	Cryst., el. anal., sol.,[b] conduct.		270
5	(Me$_2$N)$_2$C=N(SiMe$_3$)$_2$$^+Cl^-$	Cryst., el. anal., IR, sol.,[c] ^{1}H NMR	ΔH −23	242
6	(Me$_2$N)$_2$C=N(SiMe$_3$)$_2$$^+Br^-$	Cryst., el. anal., IR, sol.,[c] ^{29}Si, ^{1}H NMR	ΔH −30, $\delta(^{29}$Si) 14.84[c]	242
7	NMISiMe$_3$$^+Cl^-$	Cryst., X-ray, el. anal.		265
8	NMISiMe$_3$$^+CF_3SO_3$$^-$	Sol.,[d] ^{1}H, ^{29}Si NMR	$\delta(^{29}$Si) 26.8[d]	78
9	TMSISiMe$_3$$^+Cl^-$	Sol.[d] (−85°C), ^{29}Si NMR	$\delta(^{29}$Si) 25.34[d] (−85°C), ΔH −10, ΔS −41	252
10	TMSISiMe$_3$$^+Br^-$	Cryst., el. anal., sol.,[b,c,d] conduct., ^{29}Si NMR	$\delta(^{29}$Si) 26.3,[b] 26.5,[c] 26.3[d]	252
11	TMSISiMe$_3$$^+I^-$	Cryst., el. anal., sol.,[b,c,d] conduct., ^{29}Si NMR	$\delta(^{29}$Si) 26.4,[b] 26.5,[c] 26.9[d]	252
12	TMSISiMe$_3$$^+CF_3SO_3$$^-$	Cryst., el. anal., sol.,[b,c,d] conduct., ^{29}Si NMR	$\delta(^{29}$Si) 26.7,[b] 26.7,[c] 26.3[d]	252
13	TMSISiMe$_3$$^+ClO_4$$^-$	Cryst., el. anal., sol.,[b,c,d] conduct., ^{29}Si NMR	$\delta(^{29}$Si) 26.1,[b] 26.1,[c] 26.1[d]	252

14	HC(O)NMe$_2$SiMe$_3$$^+I^-$	Sol.,^{d} ^{29}Si NMR	δ(^{29}Si) 43.94^d	240
15	HC(O)NMe$_2$SiMe$_3$$^+CF_3SO_3$$^-$	Sol.,^{d} Si NMR	δ(^{29}Si) 44.27^d	240
16	(Me$_2$N)$_3$POSiMe$_3$$^+Br^-$	Cryst., el. anal., sol.,^{d} conduct., ^{31}P NMR	mp 40°C, δ(^{31}P) 26.2^d	239
17	(Me$_2$N)$_3$POSiMe$_3$$^+I^-$	Cryst., el. anal., sol.,^{d} conduct., ^{31}P NMR	mp 85.5°C, δ(^{31}P) 26.2^d	239
18	Ph$_3$POSiMe$_3$$^+I^-$	Cryst., el. anal., sol.,^{d} ^{31}P NMR	mp 161–164°C, δ(^{31}P) 51^d	239,257
19	Me$_3$POSiMe$_3$$^+I^-$	Cryst., IR, sol.,^{b} conduct., ^{31}P NMR		270
20	Me$_3$POSiMe$_3$$^+Br^-$	Sol.,^{b,d} conduct., ^{31}P NMR	δ(^{31}P) 83	271
21	(Me$_3$SiO)$_3$POSiMe$_3$$^+I^-$	Cryst., el. anal., sol.,^{d} conduct.	mp 151°C, subl. 120°C/0.5 Torr, δ(^{31}P) 34.40	258
22	(EtO)$_2$(MeS)POSiMe$_3$$^+I^-$	Sol.^{d} (−80°C), conduct., ^{31}P NMR	δ(^{31}P) 35.8^d (−80°C)	256
23	(Me$_2$N)$_3$PSiMe$_3$$^+I^-$	Sol.^{d} (−60°C), conduct., ^{31}P NMR	δ(^{31}P) 64^d (−60°C), δ(^{29}Si) −0.3d^d (−30°C), $J_{\text{P-Si}}$ 115 Hz	255
24	(Me$_2$N)$_3$PSiMe$_3$$^+Br^-$	Sol.^{d} (−60°C), conduct., ^{31}P NMR	δ(^{31}P) 64^d (−60°C)	255

[a] For review, see also MacDiarmid (269). δ(^{29}Si) and δ(^{31}P), Chemical shift in ppm; ΔH, enthalpy of complex formation in kcal/mol; ΔS, entropy of complex formation in cal/mol degree; Py, pyridine; NMI, N-methylimidazole; TMSI, trimethylsilylimidazole.
[b] In CH$_3$CN or CD$_3$CN.
[c] In CHCl$_3$ or CDCl$_3$.
[d] In CH$_2$Cl$_2$ or CD$_2$Cl$_2$.

TABLE V

EQUILIBRIUM CONSTANTS[a] FOR REACTIONS

$$Me_3Si^+PyCF_3SO_3^- + Nu \overset{K}{\rightleftharpoons} Me_3Si^+NuCF_3SO_3^- + Py$$

Nucleophile	K
Pyridine	1
3,5-Dimethylpyridine	5.5
Dimethylformamide (DMF)	8.1
N-Methylpyridine	10
Triphenylphosphine oxide	110
Pyridine N-oxide	660
1,2-Dimethylimidazole	54,300
Hexamethylphosphoroamide (HMPA)	99,000
N-Methylimidazole (NMI)	328,000

[a] Shown are approximate values deduced from Bassindale and Stout (78). Reactions were conducted in CH_2Cl_2.

D. *Dynamic Behavior of Tetracoordinate Si$^+$ Complexes*

Knowledge of the dynamic behavior of silylonium complexes is of great importance for understanding the role of these species as intermediates in processes of substitution at silicon. These complexes are very reactive toward nucleophiles. For example, observations of their hydrolytic unstability have been made (e.g., Refs. *239, 242, 254,* and *257*).

The energy barrier to ionization of functional electrophilic silanes by uncharged nucleophiles must be very low. It was shown (*256*), for example, that equilibrium [Eq. (57)] (X = Br, I) in methylene chloride is established quickly when measured on the ^{31}P-NMR scale, even in temperatures as low as $-90°C$. Considering the favorable thermodynamics for complex formation under these conditions, and accounting for the negative entropy change, the rate of ionization must be very high and the energy of activation very low. Similar behavior is exhibited by some other adducts including those of trimethylsilyl chloride (*242,252*). This behavior fits very well the role of these ionic adducts as intermediates formed in the fast preequilibrium step of nucleophile-assisted substitutions at the silicon atom.

The tetrahedral Si$^+$ complexes readily exchange the mobile ligand (*242,253*) of the nucleophile on the silicon center, which was proved for the system shown in Eq. (55) (*253*). The reaction is rapid as compared with the

$$Me_3SiN \overset{}{\underset{(+)}{\bigcirc}} NSiMe_3 + Me_3SiN \overset{*}{\underset{}{\bigcirc}} N \rightleftharpoons Me_3SiN \overset{*}{\underset{(+)}{\bigcirc}} NSiMe_3 + Me_3SiN \overset{}{\underset{}{\bigcirc}} N \quad (55)$$

NMR time scale for all counterions studied, i.e., Br^-, I^-, ClO_4^-, $CF_3SO_3^-$, including also Cl^- at low temperature. The rate constant was estimated to be at least of the order of 10^3 second^{-1} at 30°C. In light of this result it is reasonable to postulate a ligand exchange process for the nucleophile-induced racemization of optically active silanes (discussed in Section V,F).

Exchange of the silyl group between the ionic complex and the functional silane according to Eq. (56) (253) was also studied. It was proved

$$Me_3SiN \overset{\frown}{\underset{+}{\bigcirc}} NSiMe_3X^- + Me_3\overset{*}{Si}X \;\rightleftharpoons\; Me_3SiN \overset{\frown}{\underset{+}{\bigcirc}} N\overset{*}{Si}Me_3X^- + Me_3SiX \quad (56)$$

that the reaction proceeded by the dissociation–recombination pathway involving slow decomposition of the salt followed by fast recombination of the nucleophile and silane. The main evidence was the order of decreasing reactivity in the series of counterions $Cl^- > Br^- > CF_3SO_3^- > I^-$, ClO_4^- and solvents $CDCl_3 > CD_2Cl_2 > CD_3CN$, which parallels the order of increasing salt stability. The results of these studies gave, therefore, important information about the rate of ionic complex decomposition to components. For example, the rate constant of the decomposition of bis(N,N'-trimethylsilyl)imidazolium triflate in $CDCl_3$ at 300 K is 13.25 second^{-1}.

Tetracoordinate Si^+ complexes may undergo transformations in the mobile ligand, since nucleophilic attack of the counterion may be directed toward electrophilic centers located in the ligand group. Some reactions involving these transformations are of importance in synthesis (86,272–274). For example, the silylation of phosphorus alkyl esters has been broadly explored by bioorganic chemists as a convenient method of generation of phosphorus acids [Eq. (57)] (273). The mechanism of the silylation reaction [Eq. (57)] has been well documented for the cases where X is

$$\underset{/}{\overset{\backslash}{P}}\underset{OAlk}{\overset{O}{\diagup}} \underset{-Me_3SiX}{\overset{\overset{fast}{Me_3SiX}}{\rightleftharpoons}} \left[\underset{O-Alk}{\overset{|}{\underset{|}{-POSiMe_3}}} \right]^+ X^- \;\underset{-Alk-X}{\overset{slow}{\longrightarrow}}\; \underset{/}{\overset{\backslash}{P}}\underset{O}{\overset{OSiMe_3}{\diagup}} \quad (57)$$

Br or I by NMR, conductometric, stereochemical, and kinetic studies (256). In particular, it was demonstrated that silylation of optically active alkyl thiophosphonate with a stoichiometric amount of Me_3SiBr leads to racemic silyl esters and the rate of optical rotation decay is equal to the rate of silyl ester formation (256) in spite of the fact that no bond to phosphorus was cleaved. Thus, the reaction must involve an achiral transient species [Eq. (58)]. This result is also proof of fast exchange of silyl groups between the complex and the silane, which occurs via the dissociation–recombination mechanism.

$$
\begin{array}{c}
\underset{\text{EtS}}{\overset{\text{Me}}{>}}\!\!P\!\!\overset{O}{\underset{\text{OMe}}{<}} \xrightarrow[\text{Me}_3\text{SiBr}]{\text{slow}}
\end{array}
$$

$$
\underset{\text{EtS}}{\overset{\text{Me}}{>}}\!\!P\!\!\overset{\text{OSiMe}_3}{\underset{O}{<}} \underset{-\text{Me}_3\text{SiBr}}{\overset{\text{fast}}{\underset{\rightleftharpoons}{\text{Me}_3\text{SiBr}}}} \left[\underset{\text{EtS}}{\overset{\text{Me}}{>}}\!\!P\!\!\overset{\text{OSiMe}_3}{\underset{\text{OSiMe}_3}{<}}\right]^{+} \text{Br}^{-} \underset{+\text{Me}_3\text{SiBr}}{\overset{\text{fast}}{\underset{\rightleftharpoons}{-\text{Me}_3\text{SiBr}}}} \underset{\text{EtS}}{\overset{\text{Me}}{>}}\!\!P\!\!\overset{O}{\underset{\text{OSiMe}_3}{<}}
$$

racemic product

$$(58)$$

E. Features of Uncharged Nucleophile-Assisted Substitution at Silicon

Many important substitution reactions at silicon are known to be effectively accelerated by uncharged Lewis bases. Silylation processes catalyzed by amines and nitrogen heterocycles are the most common and important examples. Amines are often used in the mixture with halosilanes in the silylation of protic substrates to play the role of both catalyst and hydrohalogen acceptor. Earlier reports on the subject are quoted in Pierce's review (235). Other reviews on silylation and selected aspects of this process have appeared (86,89,236–238,273). Transient formation of amine–silyl halide silylonium complexes has been often suggested (see Refs. 76, 78, and 235 and references cited in Refs. 241–243, 252, 253, 275, and 276). Silylation with other functional silanes is also subject to nucleophilic catalysis. Good examples of this may be the silylation of silanols with acetoxysilanes (275), alcohols with silanethiols (276), or alcohols with silyl triflates (76). Activation of silylation reagents with oxygen nucleophiles like HMPA DMF, and DMSO has been pointed out and studied by Corriu et al. (248,249,251,263), and the possibility of participation of silylonium intermediates was also considered (78,239,240,252,253). The intermediacy of the ionic tetrahedral Si complex has been postulated in other transformations, e.g., ligand exchange at trivalent phosphorus (255) and synthesis of Mannich salts [Eq. (59)] (274).

$$
\text{Me}_2\text{NCH}_2\text{NMe}_2 \xrightarrow{\text{Me}_3\text{SiI}} \underset{\underset{\text{Me}}{|}}{\overset{\overset{\text{SiMe}_3}{|}}{\text{Me}-\overset{+}{\text{N}}-\text{CH}_2}}\overset{\text{I}^-}{-}\overset{\overset{\curvearrowleft}{\cdot\cdot}}{\text{N}}\overset{\text{Me}}{\underset{\text{Me}}{<}} \longrightarrow \text{H}_2\text{C}=\overset{+}{\text{N}}\overset{\text{Me}}{\underset{\text{Me}}{<}} + \text{I}^- + \text{Me}_3\text{SiNMe}_2
$$

$$(59)$$

Assistance with uncharged Lewis bases is particularly effective if a bulky silyl group is to be introduced to an organic compound (76,241,277,278) or the silylated object is sterically crowded (243,278). Chaudhry and Hernandez (241) demonstrated the promotion of silylation of alcohols with tert-butyldimethylchlorosilane by a mixture of triethylamine and 4-dimethylamino pyridine (DMAP). The silylation proceeded readily in the presence of catalytic amounts of DMAP but did not occur in its absence, although an equimolar amount of the stronger Brönsted base Et_3N was present. This result points to the nucleophilic character of the catalysis. Frye and co-workers showed that some nitrogen heterocycles of high nucleophilicity are unusually effective catalysts in the silylation of vinyl-dimethylcarbinol with Ph_2SiCl_2 and $Ph_2Si(OR)Cl$ (243). As pointed out by Bassindale and Stout (78), the catalytic constants for various promoters correlate well with the relative values of silylonium complex formation constants.

The activities of silanes Me_3SiX bearing various functional groups X as silylating agents used in mixture with Et_3N were compared in silyl ether formation studies (279). The reaction presumably takes place according to Eq. (60) (253). The rate of reaction decreased with varying X groups in the

$$Me_3SiX + Et_3N \rightleftharpoons Me_3Si\overset{+}{N}Et_3\,X^-$$

$$Me_3Si\overset{+}{N}Et_3 + \left[\!\!\begin{array}{c}\\\end{array}\!\!\right]\!\!=\!\!O \rightleftharpoons \left[\!\!\begin{array}{c}\\\end{array}\!\!\right]\!\!=\!\!\overset{+}{O}\!\!\diagup^{SiMe_3} + Et_3N \qquad (60)$$

$$\left[\!\!\begin{array}{c}\\\end{array}\!\!\right]\!\!=\!\!\overset{+}{O}\!\!\diagup^{SiMe_3} + Et_3N \longrightarrow \left[\!\!\begin{array}{c}\\\end{array}\!\!\right]\!\!-\!\!OSiMe_3 + Et_3\overset{+}{N}H$$

order $I > OSO_2CF_3 > Br > OSO_2CH_3 > Cl$, which is roughly parallel to the order of the decreasing ability of Me_3SiX to ionize as a result of interaction with nucleophiles (253) (Section V,C). The parallel behavior of nucleophiles and silanes in complex formation and in nucleophile-assisted silylation points strongly to the nucleophilic character of this catalysis and the participation of silylonium complexes.

The kinetics of nucleophilic substitution at the silicon atom assisted by uncharged nucleophiles have been studied by Corriu et al. (248–251). Hydrolysis of triorganochlorosilanes induced with HMPA, DMSO, and DMF was used as the model. The reaction proceeded according to the third-order kinetic law, first order with respect to the nucleophile, the silane, and the silylation substrate. Very low values of activation enthalpy and high negative entropy of activation were observed (Table VI). These results were taken as evidence for the intermediacy of silicon hypervalent species (249,251); however, they are also perfectly consistent with

$$Nu + R_3SiX \underset{k_{-1}}{\overset{k_1}{\rightleftharpoons}} [R_3Si^+Nu]X^- \overset{R'OH}{\underset{k_2}{\longrightarrow}} R'OSiR_3 + Nu \cdot HX \qquad (61)$$

TABLE VI

CATALYTIC RATE CONSTANTS k AND ACTIVATION PARAMETERS FOR
HYDROLYSIS OF CHLOROSILANES IN ANISOLE AT 20°C

Substrate	Nucleophile	k (mol^{-2} dm^6 second^{-1})	$\Delta H^{\ddagger}$ (kcal/mol)	$\Delta S^{\ddagger}$ (cal/mol degree)	Reference
Ph$_3$SiCl	HMPA	1200 ± 100	−3.4	−56	251
Ph$_3$SiCl	DMSO	50 ± 10			249
Ph$_3$SiCl	DMF	6 ± 1			249
α-NpPhMeSiCl	HMPA	3500 ± 400			249
α-NpPhMeSiCl	DMSO	180 ± 20	1.4	−43	251
α-NpPhMeSiCl	DMF	40 ± 5	2.6	−40	251
α-NpPh(MenO)SiCl	HMPA	220 ± 20	−2.9	−46	251

Eq. (61). If steady-state conditions are met[3] and the second step is rate limiting, Eq. (61) leads to the kinetic law represented by Eq. (62).

$$\text{Rate} = \frac{k_1 k_2}{k_{-1}} [\text{Nu}][\text{R}_3\text{SiX}][\text{R}'\text{OH}] \qquad (62)$$

$$\Delta H^{\ddagger} = \Delta H_1 + \Delta H_2^{\ddagger} \qquad \Delta S^{\ddagger} = \Delta S_1 + \Delta S_2^{\ddagger} \qquad (63)$$

The enthalpy and entropy of activation are calculated according to Eq. (63). Since ionic silylonium complex formation is exothermic, its enthalpy, ΔH_1, is negative. The activation enthalpy for exchange of the mobile ligand, $\Delta H_2^{\ddagger}$, is expected to be rather small since a considerable partial positive charge on the silicon center makes nucleophilic attack easier. Moreover, the mobile ligand X is an excellent leaving group. Consequently, $\Delta H^{\ddagger}$ of the overall process is expected to be small and may even be negative. ΔS_1 is strongly negative (252), mostly due to the organization of the weakly polar solvent by the ionic complex. The contribution from $\Delta S_2^{\ddagger}$ is expected to be negative as well, because of the bimolecularity of the ligand exchange process. Thus, a highly negative entropy of activation should be a general feature of the process proceeding according to pathway [Eq. (61)].

One of the most interesting results concerning features of nucleophile-assisted substitution at silicon is retention of configuration at the silicon center observed by Corriu et al. (248,251), even when the leaving group is

---

[3] The analysis is not applied to cases in which $k_{-1} < k_1[\text{Nu}]$, i.e., if the equilibrium of complex formation lies on the side of the complex.

chlorine which as a rule is substituted with inversion $(8,10,12,280)$. This result is in agreement with a silylonium ion intermediate. The ionization, most probably induced by the uncharged nucleophile, is an S_N2 reaction, which occurs with Walden inversion. The second step is displacement of the nucleophile bound relatively loosely to silicon. Obviously, it should also proceed with inversion. Thus, the expected net stereochemical result of the process is retention.

F. Nucleophile-Induced Inversion at Silicon

Nucleophile-induced racemization of chiral halosilanes has been extensively studied $(8,99,280–288)$ in connection with interest in the mechanism of nucleophilic catalysis in substitution at silicon. This reaction was first reported by Sommer, who found that optically active silanes readily undergo spontaneous racemization in some solvents (8). Corriu et al. $(280–283,285)$ investigated the kinetics of the racemization assisted by catalytic amounts of such uncharged nucleophiles as HMPA, DMSO, and DMF in an inert solvent, CCl_4. The reaction was found to be second order with respect to the nucleophile, which indicates that two nucleophile molecules participate in formation of the transition state. The reaction showed a low value for energy of activation and strongly negative entropy of activation. These results were interpreted in terms of a mechanism involving hypervalent achiral silicon intermediates A or B [Eq. (64)]. It was later pointed out

$$(64)$$

that the kinetics also fit well to the double-displacement pathway involving a silylonium intermediate [Eq. (65)] (239).

If steady-state conditions are imposed i.e. $k_{-1} > k_1[Nu]$, Eq. (65) leads to the general kinetic law represented by Eq. (66), where R_{rac} is the rate of racemization. When the second step is rate limiting, the equation adopts

$$
\begin{array}{ccc}
\underset{R^2}{\overset{R^1}{\diagdown}}\mathrm{Si}\underset{\diagdown Cl}{\diagup R^3} & \underset{\underset{(\mathrm{inversion})}{k_{-1}}}{\overset{\overset{(\mathrm{fast})}{\mathrm{Nu}}}{\underset{\rightleftharpoons}{k_1}}} & \left[\underset{R^2}{\overset{R^1}{\diagdown}}\mathrm{Si}\underset{\diagdown R^3}{\diagup \mathrm{Nu}}\right]^{+} \mathrm{Cl}^{-}
\end{array}
$$

$$
\mathrm{Nu}^*\,k_2 \big\Updownarrow (\text{slow}) (\text{inversion})
$$

$$
\underset{R^2}{\overset{R^1}{\diagdown}}\mathrm{Si}\underset{\diagdown R^3}{\diagup Cl} \quad \underset{\underset{(\mathrm{inversion})}{-\mathrm{Nu}^*}}{\overset{(\mathrm{fast})}{\rightleftharpoons}} \quad \left[\underset{R^2}{\overset{R^1}{\diagdown}}\mathrm{Si}\underset{\diagdown \mathrm{Nu}^*}{\diagup R^3}\right]^{+} \mathrm{Cl}^{-} \tag{65}
$$

the form of a third-order kinetic equation [Eq. (67)]. The reaction should thus be second order with respect to the nucleophile. This is perhaps a common case of nucleophile-induced racemization of silanes and related processes of nucleophile-catalyzed substitution at silicon taking into account the fact that the complex is formed very fast (Section V,D). If the first step is rate determining the reaction becomes first order with respect to the nucleophile [Eq. (68)]. The crossover from the kinetics described by Eq. (67) at lower concentrations of nucleophiles to the kinetics according to Eq. (68) at higher nucleophile concentrations was observed by McKinnie and Cartledge (99,286) for cis–trans isomerization of 1-chloro-1,2-dimethylsilacyclobutane. The authors interpreted this result as consistent with Eq. (64), structure **B**.

$$
R_{\mathrm{rac}} = \frac{n k_1 k_2 [\mathrm{Nu}]^2}{k_{-1} + k_2 [\mathrm{Nu}]} [\mathrm{R_3SiCl}] \tag{66}
$$

$$
R_{\mathrm{rac}} = \frac{2 k_1 k_2}{k_{-1}} [\mathrm{Nu}]^2 [\mathrm{R_3SiCl}] \tag{67}
$$

$$
R_{\mathrm{rac}} = k_1 [\mathrm{Nu}][\mathrm{R_3SiCl}] \tag{68}
$$

An analogous analysis to that in Section V,E leads to the conclusion that small $\Delta H^{\ddagger}$ and large negative $\Delta S^{\ddagger}$ values of racemization (Table VII) are fully compatible with the silylonium pathway. They can also be interpreted in terms of the mechanisms involving a hypervalent silicon intermediate (10,263,285).

It should be stressed that Eq. (66) is not the sole representation of kinetics of nucleophile-activated racemization of optically active silanes. Silylonium complexes often appear to be thermodynamically more stable than the substrate in the reaction systems, i.e., $k_1[\mathrm{Nu}] > k_{-1}$. The steady-state conditions are not met, which usually leads to complex kinetics. On the other hand, the different kinetic laws observed in these processes (287,288) are related to other types of mechanisms, which are discussed later.

TABLE VII

ACTIVATION PARAMETERS OF NUCLEOPHILE-INDUCED
RACEMIZATION OF TRIORGANOSILANES

Silane	Nucleophile	Solvent	$\Delta H^{\ddagger}$ (kcal/mol)	$\Delta S^{\ddagger}$ (cal/mol second degree)	Reference
NpFcPhSiCl	HMPA	Benzene	3.61	−54	285
NpFc-i-PrSiCl	HMPA	Benzene	2.69	−58	285
NpPh-i-PrSiCl	HMPA	CCl$_4$	3.15	−55	251
NpPhEtSiCl	HMPA	CCl$_4$	0.39	−57	251
NpPhEtSiCl	DMF	CCl$_4$	0	−70	251

Mechanisms involving ionic intermediates, i.e., Eqs. (64) pathway (through **B**) and (65), are strongly supported by the observation that the rate of racemization increases markedly with increasing dielectric constant of the medium. This was demonstrated by kinetic investigations of the racemization of NpPhMeSiCl induced by catalytic amounts of HMPA in CH$_2$Cl$_2$–CCl$_4$ binary solutions of various composition (284) (Fig. 2).

Important requirements for formation of hexacoordinate silicon complexes is small bulk and a strongly electron-withdrawing character of the substituents at the Si center (171). These conditions are not met in the chlorosilane racemization and solvolysis processes studied (251,285). Also the faster rate of halogen exchange relative to inversion on silicon observed by Cartledge et al. (286) remains in conflict with the hexacoordinate silicon intermediate (**A**) pathway [Eq. (64)].

An argument for pathway [Eq. (65)] was provided by kinetic investigation of the racemization of NpPhMeSiCl induced with (PhO)$_2$(Me$_3$SiO)-P=O in competition with silyl group exchange in the ester (transsilylation) (284). The transsilylation was found to be faster than racemization, and it also leads to an optically active product, in conflict with the hypervalent Si pathways if a reasonable assumption is made that both transsilylation and racemization occur via the same intermediate. On going to this intermediate [Eq. (64), **A** or **B**], chirality would be lost, and the racemization could not be faster than transsilylation.

Recently Bassindale et al. (288) discovered a new mechanism of halosilane racemization in the presence of nucleoophiles. They performed dynamic NMR studies of the inversion at silicon catalyzed by HMPA and N-trimethylsilylimidazole, using as a model a halosilane having diastereotopic methyl groups. Diastereotopic separation of the ^{1}H- and ^{13}C-NMR signals is apparent in both the substrate and the silylonium complex. Thus, inversion at the Si center leading to coalescence may be

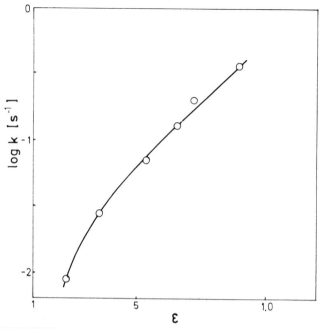

FIG. 2. Dependence of the specific rate of racemization of NpPhMeSiCl induced by HMPA in CCl₄–CH₂Cl₂ solvent systems at 25°C on the dielectric constant of the medium (variable CH₂Cl₂ to CCl₄ ratio). [HMPA], 0.0025 mol/dm³. (Deduced from M. Cypryk, Ph.D thesis. Centre of Molecular and Macromolecular Studies, The Polish Academy of Sciences, Łódź, Poland, 1982. See also Ref. *284.*)

$$\underset{\underset{\text{Me}\quad\text{Me}}{\vert\qquad\vert}}{\overset{\overset{\text{H}\quad\text{Me}}{\vert\qquad\vert}}{\text{Ph}-\text{C}-\text{Si}-\text{X}}} \qquad \text{X = Br, Cl}$$

observed simultaneously in the substrate and its complex.[4] It turned out that, over a certain temperature interval, diastereotopic separation in both the ¹³C- and ¹H-NMR spectra was evident only in the complex while the substrate showed one signal. This observation indicated that the halosilane was undergoing inversion of configuration while the complex remained stereochemically rigid; thus, nucleophilic displacement could not account for the racemization. This result was explained by a mechanism involving halide exchange. On increasing the temperature, the pair of the signals of

---

[4] Since for X=Br the substrate–complex equilibrium lies well to the side of the complex, the silane-to-nucleophile molar ratio 2 : 1 was used to ensure the 1 : 1 silane-to-complex molar ratio under the NMR experimental conditions.

the complex underwent coalescence before the averaging of signals of the complex and bromosilane took place. This indicates that both mechanisms, halide exchange and nucleophile displacement, operate in parallel, and the full scheme of the process is as in Eq. (69). For the model used (X = Br)

$$R_3SiX + Nu \underset{inv}{\overset{inv}{\rightleftharpoons}} [R_3SiNu]^+X^-$$

$$R_3SiX + Nu \underset{inv}{\overset{inv}{\rightleftharpoons}} [R_3SiNu^*]^+X^- \qquad (69)$$

and under the experimental conditions applied, halide exchange dominates. The authors suggest that this mechanism is also more important in other systems including chlorosilanes. This, however, would be in conflict with the second-order kinetics with respect to the nucleophile reported by Corriu *et al.* (*283,285*). The authors assume that the only role of the nucleophile in the halide exchange pathway is to generate the halogen ion.

There is also a well-documented mechanism of nucleophile-catalyzed racemization of silane that does not involve an ionic intermediate. This mechanism operates in at least some peculiar classes of cyclic silanes, having particular ability to form complexes with nucleophiles. Stevenson and Martin (*287*) studied carefully the kinetics of the inversion of a bicyclic spirosilane **A** [Eq. (70)]. The reaction was catalyzed by a weak nucleophile, a substituted benzaldehyde. The mechanism consistent with

$$Nu = p\text{-}NMe_2PhC(O)H \qquad (70)$$

experimental observations assumed the formation of a 1:1 molecular complex **B** having a pentacoordinate silicon, followed by a rate-determined sequence of five pseudorotations inverting the chirality of **B**. Loss of the nucleophile gives the inverted silane. The most important evidence was the first-order kinetics of the reaction in nucleophile; acceleration of the process by electrodonating groups on the phenyl ring of the nucleophile; lack of correlation of the rate with solvent-ionizing power; and negative activation entropy of the process, ΔS^* -27.9 cal/mol degree. The above mechanism, in contrast to that proposed by Corriu, does not assume formation of an achiral intermediate. An important role is attributed to permutative isomerization, which has been well documented to occur at the silicon center (*289–293*). It should be also mentioned that a spontaneous edge inversion process of tetracoordinate silicon compounds may occur in somewhat similar bicyclic systems (*294*).

In conclusion, studies of nucleophile-catalyzed inversion at the silicon center point to various mechanistic possibilities of this transformation. Results of such studies also support the role of silylonium ions as intermediates.

G. *Alternative Mechanisms of Catalysis by Uncharged Nucleophiles*

1. *Pathways Involving Extended Coordination of Silicon*

These mechanisms have been the subject of several reviews (*10–13*), and there is no room to discuss them in detail. A short comparison with the silylonium intermediate pathway should, however, be included. Both of these conflicting mechanisms assume the formation of a substrate–nucleophile complex in a fast preequilibrium step, and the complex reacts with a nucleophilic reagent to give the product in the rate-limitng step. It is, therefore, appropriate to compare how easily the corresponding complexes are formed and how readily they react toward the product [Eq. (71)]. The

$$
\begin{array}{c}
\text{Nu}-\overset{+}{\text{Si}}R_3 \; X^- \quad \xrightarrow{\substack{\text{nucleophilic} \\ \text{reagent}}} \quad \text{product} \\
\text{\textbf{A}} \\
\\
R_3\text{SiX} + \text{Nu} \\
\\
\underset{\overset{|}{R}}{\text{Nu}-\text{Si}-X} \quad \xrightarrow{\substack{\text{nucleophilic} \\ \text{reagent}}} \quad \text{product} \\
\text{\textbf{B}}
\end{array}
\tag{71}
$$

relative thermodynamic stability of structures **A** and **B** is a complex function of the character of substituents at silicon, nucleophile, and leaving group, as well as of the nature of the solvent. In solvents of good or medium ionizing power like methylene chloride, often explored in silylation reactions, structure **A** is usually more stable for a good leaving group like Cl, Br, I, CF_3SO_3, and ClO_4 and if all substituents R are alkyl or aryl. An acyclic structure and a high steric effect in the silane or nucleophile also favor structure **A**.

The situation may be reversed, however, if more electronegative groups of rather small size are bound to silicon, the solvent is nonpolar, the leaving group is poor like F, RCOO, RS, and H, and if the silicon atom is a part of a strained ring or if such a ring may be formed as a result of the interaction of the Si center with an internal nucleophilic group. All these features favor an extended coordination structure of the intermediate. Numerous compounds of pentacoordinate silicon having these features have been isolated (for reviews see Refs. *247, 290, 295–297*). For example, additional coordination of the internal nucleophile leading to chelate-stabilized trigonal bipyramids is well documented (*247,267,293,298–305*).

No data have been obtained to make a direct comparison of the electrophilic reactivity of **A** and **B** [Eq. (71)]. The high reactivity of silylonium complexes in silylation processes has been already discussed. There is, however, good reason why structure (**B**) [Eq. (72)] may be reactive toward displacement of a ligand by a nucleophile. Donor–acceptor complexes are often represented by formula **A** [Eq. (71)] with the positive charge located on the central atom. Such representation has only formal character and is confusing. In many cases calculations point unequivocally to reverse polarization of the central atom (*84*). Silicon becomes more positive on complexing, while all negative charge from the nucleophile donor is located on the ligands, more precisely on one occupying the apical position. The Si—X bond becomes longer and bond energy poorer. The silylonium ion structure **C** can be considered as the limiting case (*305*).

$$Nu + {\scriptstyle >}Si{\scriptstyle <}_X \longrightarrow \left[Nu^+ \longrightarrow {}^-Si-X \right] \quad \left[Nu-\overset{\delta+}{Si}-\overset{\delta-}{X} \right] \quad Nu-{}^+Si{\scriptstyle <} \; X^-$$

A	**B**	**C**
formal charge		limiting case
distribution	true charge distribution	

(72)

The ability of silicon extended coordination species to react with nucleophiles was well documented by Corriu *et al.* *(247,285,306–308)*. Even negatively charged pentacoordinate silicon anions showed high reactivity toward nucleophilic reagents *(247,306–308)*. For example, $PhMeSiF_3^-$ reacts faster with t-BuMgBr than the parent uncomplexed compound of tetravalent silicon, by a factor of 120 *(247)*. Thus, nucleophilic activation of the Si—X bond by extracoordination is unquestionable. There remains the problem of the detailed mechanism of this activation. Four general routes of transformation of the pentacoordinate silicon intermediate to the product should be distinguished. The Si—X bond may be broken as a result of (1) nucleophilic attack on silicon, (2) attack of electrophile on the departing group, (3) synchronous attack (push-pull), or (4) ligand coupling.[5]

Route 1, attack of the nucleophilic reagent leading to a hexacoordinate intermediate or transition state, has been given preference *(11,12,247,306–308)*. A similar pathway has been also postulated in organophosphorus chemistry *(309–311)*. This mechanism may account for many experimental observations. It does not, however, seem to be generally accepted as the most favorable route (see, for example, Refs. *243* and *290*). Activation toward nucleophilic attack intuitively should be less effective in the extracoordination structure [Eq. (72), **B**] than in the silylonium ion (**C**), because of the higher positive charge and lower steric requirements of the Si center in the latter.

Extracoordination seems to be very effective in activation of the Si—X bond toward electrophilic attack on the leaving group *(311–314)*. The push–pull variant of the mechanism with considerable electrophilic assistance to the leaving group departure thus seems to be more plausible. The push–pull pathway was considered in the HMPA- or Et_3N-catalyzed exchange of halogen between α-NpPhMeSiHF and chlorosilanes *(250)*.

Electrophilic attack may also lead to an elimination–addition reaction involving the silylonium intermediate [Eq. (73)]. Route 2 involving single

$$Nu{-}\overset{\diagdown/}{\underset{|}{Si}}{-}X \; \rightleftharpoons \; Nu\overset{|}{\underset{|}{Si}}{}^+{-}X^- \xrightarrow[-Nu]{Y^-} \overset{\diagdown}{\underset{\diagup}{Si}}\overset{\diagup}{\underset{\diagdown Y}{}} \tag{73}$$

electron transfer (SET) is another possibility. The SET pathway accounts for some reductive processes involving Si—H bond cleavage in the extracoordinate species *(312,315)*. It has also been considered in silicon–alkyl bond cleavage in some hexacoordinate silicon compounds *(313,314)*.

---

[5] Considered by S. Oae in reactions of sulfuranes [*Phosphorus Sulfur* **27**, 13 (1986)].

2. *Brönsted Base Catalysis*

An uncharged nucleophile may also play the role of a Brönsted base if interaction with a protic substrate can increase its nucleophilicity [Eqs. (41) and (42)]. This pathway, formulated by Allen and Modena (*137*), seems to be important in many solvolytic reactions (*7*). There is evidence that more than one solvent molecule participates in the solvolysis of chlorosilanes (*7,137,244*). One functions as the nucleophile attacking silicon, while the others, forming hydrogen bonding complexes, enhance the nucleophilic reactivity.

Brönsted base catalysis is expected to operate when acid–base interaction strongly enhances the substrate nucleophilicity or deeply changes the structure favoring the reaction. The silylation of phosphorous acid with the Me_3SiCl–amine system may serve as a good example [Eq. (74)] (see references cited in Ref. *273*). The first two molecules of silyl chloride enter the

$$
\begin{array}{c}
\underset{HO}{\overset{HO}{\diagdown}} P \overset{O}{\underset{H}{\diagup}} + 2\ Me_3SiCl \longrightarrow \underset{Me_3SiO}{\overset{Me_3SiO}{\diagdown}} P \overset{O}{\underset{H}{\diagup}} \overset{Et_3N}{\longrightarrow}
\end{array}
$$

$$
\underset{Me_3SiO}{\overset{Me_3SiO}{\diagdown}} \overset{-\ +}{POHNEt_3} \xrightarrow{Me_3SiCl} (Me_3SiO)_3P \tag{74}
$$

reaction easily without any catalyst, since they do not change the coordination state of phosphorus. Introduction of the third group requires the action of a base which converts the tetracoordinate phosphorus acid into a highly reactive anion of tricoordinate phosphorus.

It has recently been shown that within the same reaction series the mechanism of catalysis by uncharged Lewis bases may change with variation of the structure (*275*). Tandem kinetic and IR hydrogen bond studies revealed that silylation reaction of a silanol catalyzed with triethylamine is first order with respect to the silanol–amine hydrogen bond complex [Eq. (75).] The same reaction was shown to be effectively catalyzed by

$$
R_2R'SiOH + Et_3N \ \rightleftharpoons \ [Et_3N\text{---}HOSiR_2R']
$$

$$
[Et_3N\text{---}HOSiR_2R'] + RR'R''SiCl \ \rightarrow \ R_2R'SiOSiRR'R'' + Et_3N\cdot HCl
$$

$$
R = Me,\ R' = OSiMe_3,\ R'' = i\text{-Pr} \tag{75}
$$

some nitrogen heterocycles (e.g., *N*-methylimidazole, DMAP), which function as nucleophiles to produce silylonium intermediates.

The strong catalytic effect of amines in silylation with silyl hydrides of some protic substrates, like alcohols, phenols, and carboxylic acids, is well known (for review, see Ref. *316*). In this case, the amine is not likely to ionize the silicon–leaving group bond. General base catalysis has been often postulated (*316–320*). The Si—H bond, however, owing to a high equatorial preference, stabilizes the trigonal bipyramidal structure (*300–302*), thus favoring the extended Si coordination mechanism.

VI
FINAL REMARKS

There are a variety of mechanistic pathways in the chemistry of substitution at the silicon atom. Dissociative routes in which a silylenium ion or a double-bonded tricoordinate silicon compound appear in solution are often postulated. These transient species are considered either as diffusionally equilibrated intermediates or as species of incipient character. Pathways involving the dissociation of a silicon reactant as a result of the interaction with a neutral nucleophile molecule merit particular attention since many important silylation processes are likely to take this route. It should, however, be stressed that any generalization, in terms of a definite dissociative or associative pathway for a broader class of reactions in organosilicon chemistry, is dangerous. The particular route followed by a reaction depends to a considerable extent on the structure of reactants and catalysts and on conditions of the process. The role of polarity and electron donor–acceptor properties of the solvent is often of crucial importance. Progress in understanding of the mechanisms of displacement reactions at silicon will depend to a great extent on systematic studies aiming to determine more closely the role of these factors in determining the mechanistic route.

REFERENCES

1. R. R. Holmes, *Adv. Chem. Res.* **12**, 257 (1979); R. R. Holmes, *Prog. Inorg. Chem.* **32**, 119 (1984).
2. J. C. Martin, *Science* **221**, 501 (1983).
3. R. J. P. Corriu, *Phosphorus Sulfur* **27**, 1 (1986).
4. N. T. Anh and C. J. Minot, *J. Am. Chem. Soc.* **102**, 103 (1980).
5. M. J. S. Dewar and E. Healy, *Organometallics* **1**, 1705 (1982).
6. H. Kwart and K. G. King, "*d*-Orbitals in the Chemistry of Silicon, Phosphorus and Sulfur." Springer-Verlag, Berlin, 1977.
7. C. Eaborn, "Organosilicon Compounds." Butterworth, London, 1960.
8. L. H. Sommer, "Stereochemistry, Mechanism and Silicon." McGraw-Hill, New York, 1965; L. H. Sommer, *Intra-Sci. Chem. Rep.* **7**, 1 (1973).
9. R. H. Prince, *React. Mech. Inorg. Chem.* **9**, 359 (1972).
10. R. J. P. Corriu and M. Henner, *J. Organomet. Chem.* **74**, 1 (1974).
11. R. J. P. Corriu and C. Guerin, *J. Organomet. Chem.* **198**, 231 (1980).
12. R. J. P. Corriu and C. Guerin, *Adv. Organomet. Chem.* **20**, 265 (1982).

13. R. J. P. Corriu, C. Guerin, and J. J. E. Moreau, *Top. Stereochem.* **15**, 43 (1984).
14. F. K. Cartledge, *J. Organomet. Chem. Libr.* **14**, 409 (1984).
15. V. D. Nefedov, T. A. Kochina, and E. N. Sinotova, *Usp. Khim.* **55**, 794 (1986).
16. G. K. S. Prakash, S. Keyaniyan, R. Aniszfeld, L. Heiliger, G. A. Olah, R. C. Stevens, M. H. Choi, and R. Bau, *J. Am. Chem. Soc.* **109**, 5123 (1987).
17. J. B. Lambert and W. J. Schulz, Jr., *J. Am. Chem. Soc.* **105**, 1671 (1983).
18. J. B. Lambert, W. J. Schulz, Jr., J. A. McConnell, and W. Schilf, *J. Am. Chem. Soc.* **110**, 2201 (1988).
19. T. J. Barton, A. K. Hovland, and C. R. Tully, *J. Am. Chem. Soc.* **98**, 5695 (1976).
20. F. P. Lossing, *Can. J. Chem.* **50**, 3973 (1972).
21. Y. Apeloig and P. v. R. Schleyer, *Tetrahedron Lett.*, 4647 (1977).
22. A. C. Hopkinson and M. H. Lien, *J. Mol. Struct.* **104**, 303 (1983).
23. A. C. Hopkinson and M. H. Lien, *J. Chem. Soc., Chem. Commun.*, 107 (1980).
24. A. C. Hopkinson and M. H. Lien, *J. Org. Chem.* **46**, 998 (1981).
25. Y. Apeloig, M. Karni, A. Stanger, H. Schwarz, T. Drewello, and G. Czekay, *J. Chem. Soc., Chem. Commun.*, 989 (1987).
26. Y. Apeloig, S. A. Godleski, D. J. Heacock, and J. M. McKelvey, *Tetrahedron Lett.* **22**, 3297 (1981).
27. T. Truong, M. S. Gordon, and P. Boudjouk, *Organometallics* **3**, 484 (1984).
28. S. A. Godleski, D. J. Heacock, and J. M. McKelvey, *Tetrahedron Lett.* **23**, 4453 (1982).
29. G. G. Hess, F. W. Lampe, and L. H. Sommer, *J. Am. Chem. Soc.* **86**, 3174 (1964).
30. P. Potzinger and F. W. Lampe, *J. Phys. Chem.* **74**, 587 (1970); P. Potzinger and F. W. Lampe, *J. Phys. Chem.* **74**, 719 (1970).
31. I. A. Blair, V. C. Trenerry, and J. H. Bowie, *Org. Mass Spectrom.* **15**, 15 (1980).
32. I. A. Blair, G. Phillipou, and J. H. Bowie, *Aust. J. Chem.* **32**, 59 (1979).
33. W. J. Pietro and W. J. Hehre, *J. Am. Chem. Soc.* **104**, 4329 (1982).
34. M. K. Murphy and J. L. Beauchamp, *J. Am. Chem. Soc.* **99**, 2085 (1977).
35. A. Ding, R. A. Cassidy, L. S. Cordis, and F. W. Lampe, *J. Chem. Phys.* **83**, 3426 (1985).
36. L. Szepes and T. Baer, *J. Am. Chem. Soc.* **106**, 273 (1984).
37. T. M. Mayer and F. W. Lampe, *J. Phys. Chem.* **78**, 2422 (1974); T. M. Mayer and F. W. Lampe, *J. Phys. Chem.* **78**, 2429 (1974).
38. G. W. Goodloe, E. R. Austin, and F. W. Lampe, *J. Am. Chem. Soc.* **101**, 3472 (1979).
39. G. W. Goodloe and F. W. Lampe, *J. Am. Chem. Soc.* **101**, 5649 (1979).
40. G. W. Goodloe and F. W. Lampe, *J. Am. Chem. Soc.* **101**, 6028 (1979).
41. S. N. Senzer and F. W. Lampe, *J. Appl. Phys.* **54**, 3524 (1983).
42. A. Oppenstein and F. W. Lampe, *Rev. Chem. Intermed.* **6**, 275 (1986).
43. G. W. Stewart, J. M. S. Henis, and P. P. Gaspar, *J. Chem. Phys.* **57**, 1990 (1972); G. W. Stewart, J. M. S. Henis, and P. P. Gaspar, *J. Chem. Phys.* **57**, 2247 (1972).
44. J. R. Krause and F. W. Lampe, *J. Phys. Chem.* **81**, 281 (1977).
45. W. N. Allen and F. W. Lampe, *J. Am. Chem. Soc.* **99**, 2943 (1977); W. N. Allen and F. W. Lampe, *J. Am. Chem. Soc.* **99**, 6816 (1977).
46. T. Weiske, S. Akkök, and H. Schwarz, *J. Organomet. Chem.* **336**, 105 (1987).
47. S. K. Shin and J. L. Beauchamp, *J. Am. Chem. Soc.* **111**, 900 (1989).
48. M. K. Murphy and J. L. Beauchamp, *J. Am. Chem. Soc.* **98**, 5781 (1976).
49. F. Cacace and P. Giacomello, *J. Chem. Soc., Perkin Trans. 2*, 652 (1978).
50. W. P. Weber and H. Z. Boettger, *Intra-Sci. Chem. Rep.* **7**, 109 (1973).
51. W. P. Weber, R. A. Felix, and A. K. Willard, *Tetrahedron Lett.*, 907 (1970).
52. D. J. Harvey, M. G. Horning, and P. Vouros, *J. Chem. Soc., Chem. Commun.*, 898 (1970).

53. T. M. Mayer and F. W. Lampe, *J. Phys. Chem.* **78**, 2433 (1974); T. M. Mayer and F. W. Lampe, *J. Phys. Chem.* **78**, 2645 (1974).
54. W. N. Allen and F. W. Lampe, *J. Chem. Phys.* **65**, 3378 (1976).
55. R. Orlando, D. P. Ridge, and B. Munson, *Org. Mass Spectrom.* **23**, 527 (1988).
56. T. J. Odiorne, D. J. Harvey, and P. Vouros, *J. Phys. Chem.* **76**, 3217 (1972).
57. T. J. Odiorne, D. J. Harvey, and P. Vouros, *J. Org. Chem.* **38**, 4274 (1973).
58. D. Clemens and B. Munson, *Org. Mass Spectrom.* **20**, 368 (1985).
59. W. D. Reents, Jr., and A. M. Mujsce, *Int. J. Mass Spectrom. Ion. Phys.* **59**, 65 (1984).
60. J. M. S. Henis, G. W. Stewart, and P. P. Gaspar, *J. Chem. Phys.* **58**, 3639 (1973).
61. J. R. Eyler, G. Silverman, and M. A. Battiste, *Organometallics* **1**, 477 (1982).
62. A. B. Thomas and E. G. Rochow, *J. Inorg. Nucl. Chem.* **4**, 205 (1957); A. B. Thomas and E. G. Rochow, *J. Inorg. Nucl. Chem.* **4**, 1843 (1957).
63. U. Wannagat and F. Brandmair, *Z. Anorg. Allg. Chem.* **280**, 223 (1955).
64. U. Wannagat and W. Liehr, *Angew. Chem.* **69**, 783 (1957).
65. J. B. Lambert and H. Sun, *J. Am. Chem. Soc.* **98**, 5611 (1976).
66. M. E. Peach and T. C. Waddington, *J. Chem. Soc.,* 1238 (1961).
67. F. Brandmair and U. Wannagat, *Z. Anorg. Allg. Chem.* **288**, 91 (1956).
68. G. A. Olah and J. Lukas, *J. Am. Chem. Soc.* **89**, 4739 (1967).
69. G. A. Olah, D. H. O'Brien, and C. Y. Lui, *J. Am. Chem. Soc.* **91**, 701 (1969).
70. G. A. Olah and Y. K. Mo, *J. Am. Chem. Soc.* **93**, 4942 (1971).
71. J. Y. Corey, *J. Am. Chem. Soc.* **97**, 3237 (1975).
72. J. Y. Corey, D. Gust, and K. Mislow, *J. Organomet. Chem.* **101**, C7 (1975).
73. P. Bickart, F. M. Llort, and K. Mislow, *J. Organomet. Chem.* **116**, C1 (1976).
74. J. B. Lambert, W. J. Schulz, Jr., J. A. McConnell, and W. Schilf, *in* "Silicon Chemistry" (E. R. Corey and P. P. Gaspar, eds.), p. 183. Horwood, Chichester, England, 1988.
75. J. B. Lambert, J. A. McConnell, and W. J. Schulz, Jr., *J. Am. Chem. Soc.* **108**, 2482 (1986).
76. J. B. Lambert, J. A. McConnell, W. Schilf, and W. J. Schulz, Jr., *J. Chem. Soc., Chem. Commun.,* 455 (1988).
77. T. J. Barton and C. R. Tully, *J. Org. Chem.* **43**, 3649 (1978).
78. A. R. Bassindale and T. Stout, *Tetrahedron Lett.* **26**, 3403 (1985).
79. B. Lndman and S. Forsen, *NMR: Basic Princ. Prog.* **12**, 326 (1979).
80. G. A. Olah, K. Laali, and O. Farooq, *Organometallics* **3**, 1337 (1984).
81. G. A. Olah and L. D. Field, *Organometallics* **1**, 1485 (1982).
82. A. H. Cowley, M. C. Cushner, and P. E. Riley, *J. Am. Chem. Soc.* **102**, 624 (1980).
83. W. Clegg, U. Klingebiel, J. Neemann, and G. M. Sheldrick, *J. Organomet. Chem.* **249**, 47 (1983).
84. V. Gutmann, "The Donor–Acceptor Approach to Molecular Interactions." Plenum, New York, 1978.
85. M. Zeldin, P. Mehta, and V. D. Vernon, *Inorg. Chem.* **18**, 463 (1979).
86. G. A. Olah and S. C. Narang, *Tetrahedron* **38**, 2225 (1982).
87. F. A. Carey, *Intra-Sci. Chem. Rep.* **7**, 55 (1973).
88. D. N. Kursanov, Z. N. Parnes, and N. M. Loim, *Synthesis,* 633 (1974).
89. W. P. Weber, "Silicon Reagents for Organic Synthesis." Springer-Verlag, Berlin, 1983.
90. J. Y. Corey and R. West, *J. Am. Chem. Soc.* **85**, 2430 (1963).
91. J. D. Austin and C. Eaborn, *J. Chem. Soc.,* 2279 (1964).
92. F. A. Carey and C. L. Wang-Hsu, *J. Organomet. Chem.* **19**, 29 (1969).
93. L. H. Sommer and D. L. Bauman, *J. Am. Chem. Soc.* **91**, 7076 (1969).
94. J. Chojnowski, L. Wilczek, and W. Fortuniak, *J. Organomet. Chem.* **135**, 13 (1977).
95. S. Słomkowski and S. Penczek, *J. Chem. Soc., Perkin Trans. 2,* 1718 (1974).

96. J. Chojnowski, W. Fortuniak, and W. Stańczyk, *J. Am. Chem. Soc.* **109**, 7776 (1987).
97. L. H. Sommer and O. F. Bennett, *J. Am. Chem. Soc.* **81**, 251 (1959).
98. W. Stańczyk and J. Chojnowski, *J. Organomet. Chem.* **117**, 219 (1976).
99. B. G. McKinnie and F. Cartledge, *J. Organomet. Chem.* **104**, 407 (1976).
100. J. Chojnowski, M. Mazurek, and L. Wilczek, *Analyst (London)* **101**, 286 (1976).
101. R. J. Klinger, K. Mochida, and J. K. Kochi, *J. Am. Chem. Soc.* **101**, 6626 (1974).
102. O. Y. Okhlobistin and N. T. Berberova, *Zh. Obshch. Khim.* **17**, 88 (1981).
103. A. A. Khapitcheva, N. T. Berberova, E. S. Klimov, and O. Y. Okhlobistin, *Zh. Obshch. Khim.* **55**, 1533 (1985).
104. Y.-L. Chen and T. J. Barton, *Organometallics* **6**, 2590 (1987).
105. F. C. Whitmore, L. H. Sommer, and J. Gold, *J. Am. Chem. Soc.* **69**, 1976 (1947).
106. R. W. Bott, C. Eaborn, and B. M. Rushton, *J. Organomet. Chem.* **3**, 455 (1965).
107. O. W. Steward, W. J. Uhl, and B. W. Sands, *J. Organomet. Chem.* **15**, 329 (1968).
108. A. G. Brook and K. H. Pannel, *Can. J. Chem.* **48**, 3679 (1970).
109. T. J. Harriston and D. H. O'Brien, *J. Organomet. Chem.* **29**, 79 (1971).
110. K. Tamao and M. Kumada, *J. Organomet. Chem.* **30**, 339 (1971).
111. L. R. Robinson, G. T. Burns, and T. J. Barton, *J. Am. Chem. Soc.* **107**, 3935 (1985).
112. Y. Apeloig and A. Stanger, *J. Am. Chem. Soc.* **109**, 272 (1987).
113. Y. Apeloig, in "Physical Organic Chemistry 1986" (M. Kobayashi, ed.), p. 33. Elsevier, Amsterdam, 1987.
114. C. Eaborn, *J. Organomet. Chem.* **232**, 93 (1982).
115. C. Eaborn, in "Organosilicon and Bioorganosilicon Chemistry" (H. Sakurai, ed.), p. 123. Horwood, Chichester, England, 1985.
116. J. D. Smith, *Pure Appl. Chem.* **58**, 623 (1986).
117. S. S. Dua, C. Eaborn, and F. M. S. Mahmoud, *J. Organomet. Chem.* **192**, 293 (1980).
118. C. Eaborn, D. A. R. Happer, S. P. Hopper, and K. D. Safa, *J. Organomet. Chem.* **170**, C9 (1979).
119. C. Eaborn and F. M. S. Mahmoud, *J. Chem. Soc., Perkin Trans. 2*, 1309 (1981).
120. C. Eaborn, D. A. R. Happer, S. P. Hopper, and K. D. Safa, *J. Organomet. Chem.* **188**, 179 (1980).
121. C. Eaborn, P. D. Lickiss, S. T. Najim, and M. N. Romanelli, *J. Chem. Soc., Chem. Commun.*, 1754 (1985).
122. K. Jones, Ph.D thesis. University of Sussex, Brighton, England, 1988.
123. C. Eaborn, P. D. Lickiss, S. T. Najim, and M. N. Romanelli, *J. Organomet. Chem.* **315**, C5 (1986).
124. C. Eaborn, P. D. Lickiss, and N. A. Ramadan, *J. Chem. Soc., Perkin Trans. 2*, 267 (1984).
125. C. Eaborn, P. D. Lickiss, S. T. Najim, and W. Stańczyk, *J. Chem. Soc., Chem. Commun.*, 1461 (1987).
126. C. Eaborn and D. E. Reed, *J. Chem. Soc., Perkin Trans. 2*, 1687 (1985).
127. S. A. I. Al-Shali, C. Eaborn, F. A. Fattah, and S. T. Najim, *J. Chem. Soc., Chem. Commun.*, 318 (1984).
128. T. W. Bentley and G. E. Carter, *J. Am. Chem. Soc.* **104**, 5741 (1982).
129. D. B. Azarian, C. Eaborn, and P. D. Lickiss, *J. Organomet. Chem.* **328**, 255 (1987).
130. S. A. I. Al-Shali and C. Eaborn, *J. Organomet. Chem.* **246**, C34 (1983).
131. A. I. Al-Wassil, C. Eaborn, and M. N. Romanelli, *J. Chem. Soc., Perkin Trans. 2*, 1363 (1986).
132. G. A. Ayoko and C. Eaborn, *J. Chem. Soc., Chem. Commun.*, 630 (1986).
133. G. A. Ayoko and C. Eaborn, *J. Chem. Soc., Perkin Trans. 2*, 1047 (1987).
134. C. Eaborn, P. D. Lickiss, and A. D. Taylor, *J. Organomet. Chem.* **338**, C27 (1988).

135. C. Eaborn and D. E. Reed, *J. Chem. Soc., Chem. Commun.*, 495 (1983).
136. L. H. Sommer, F. O. Stark, and K. W. Michael, *J. Am. Chem. Soc.* **86**, 5683 (1964).
137. A. D. Allen and G. D. Modena, *J. Chem. Soc.*, 3671 (1957).
138. M. W. Grant and R. H. Prince, *J. Chem. Soc. A.*, 2305 (1968); M. W. Grant and R. H. Prince, *J. Chem. Soc. A.*, 1138 (1969); M. W. Grant and R. H. Prince, *J. Chem. Soc., Chem. Commun.*, 1076 (1968); M. W. Grant and R. H. Prince, *Nature (London)* **222**, 1163 (1969).
139. L. H. Sommer, G. D. Hommer, A. W. Messing, J. L. Kutschinski, F. O. Stark, and K. W. Michael, *J. Am. Chem. Soc.* **93**, 2093 (1971).
140. L. H. Sommer, D. L. Bailey, and F. C. Whitmore, *J. Am. Chem. Soc.* **70**, 2869 (1948).
141. L. H. Sommer and G. A. Baughman, *J. Am. Chem. Soc.* **83**, 3346 (1961).
142. F. Carré, R. Corriu, and B. Henner, *J. Organomet. Chem.* **22**, 589 (1970).
143. A.W. P. Jarvie, A. Holt, and J. Thompson, *J. Chem. Soc. B.*, 852 (1969).
144. M. A. Cook, C. Eaborn, and D. R. M. Walton, *J. Organomet. Chem.* **24**, 301 (1970).
145. T. G. Taylor, W. Hanstein, H. J. Berwin, N. Y. A. Clinton, and R. S. Brown, *J. Am. Chem. Soc.* **93**, 5715 (1971).
146. U. Weidner and A. Schweig, *Angew. Chem., Int. Ed. Engl.* **11**, 146 (1972).
147. A. Guyot, *ACS Polym. Preprint* **26**, 46 (1985).
148. B. N. Dolgov, M. G. Voronkov, and S. N. Borisov, *Zh. Obshch. Khim.* **27**, 709 (1957); B. N. Dolgov, M. G. Voronkov, and S. N. Borisov, *Zh. Obshch. Khim.* **27**, 2069 (1957).
149. S. N. Borisov, M. G. Voronkov, and B. N. Dolgov, *Dokl. Acad. Nauk SSSR* **144**, 93 (1957).
150. J. Chojnowski and W. Stańczyk, *Phosphorus Sulfur* **27**, 211 (1986).
151. D. A. Armitage, *Compr. Organomet. Chem.* **2**, 1 (1982).
152. I. Fleming, *Compr. Org. Chem.* **3**, 540 (1979).
153. L. E. Guselnikov, N. S. Nametkin, and V. M. Vdovin, *Acc. Chem. Res.* **8**, 18 (1975).
154. H. F. Schaefer III, *Acc. Chem. Res.* **15**, 283 (1982).
155. G. Bertrand, G. Trinquier, and P. Mazerolles, *J. Organomet. Chem. Libr.* **12**, 1 (1981).
156. N. Wiberg, *J. Organomet. Chem.* **273**, 141 (1984).
157. D. J. DeYoung, M. J. Fink, R. West, and J. Michl, *Main Group Met. Chem.* **10**, 19 (1987).
158. A. G. Brook and K. M. Baines, *Adv. Organomet. Chem.* **25**, 1 (1986).
159. M. W. Schmidt, M. S. Gordon, and M. Dupuis, *J. Am. Chem. Soc.* **107**, 2585 (1985).
160. K. D. Dobbs and W. J. Hehre, *Organometallics* **5**, 2057 (1986).
161. J. E. Douglas, B. S. Rabinovitch, and F. S. Looney, *J. Chem. Phys.* **23**, 315 (1955).
162. G. Raabe and J. Michl, *Chem. Rev.* **85**, 419 (1985).
163. P. G. Mahaffy, R. Gutowsky, and L. K. Montgomery, *J. Am. Chem. Soc.* **102**, 2854 (1980).
164. G. Olbrich, P. Potzinger, B. Reimann, and R. Walsh, *Organometallics* **3**, 1267 (1984).
165. M. J. S. Dewar, J. Friedheim, G. Grady, E. F. Healy, and J. J. P. Stewart, *Organometallics* **5**, 375 (1986).
166. M. J. S. Dewar and C. Jie, *Organometallics* **6**, 1486 (1987).
167. T. Kudo and S. Nagase, *J. Phys. Chem.* **88**, 2833 (1984).
168. T. Kudo and S. Nagase, *Organometallics* **5**, 1207 (1986).
169. N. Wiberg, K. Schurz, G. Reber, and G. Müller, *J. Chem. Soc., Chem. Commun.*, 591 (1986).
170. B. T. Luke, J. A. Pople, M. B. Krogh-Jespersen, Y. Apeloig, M. Karni, J. Chandrasekhar, and P. v. R. Schleyer, *J. Am. Chem. Soc.* **108**, 270 (1986).
171. E. A. V. Ebsworth, *in* "Organometallic Compounds of the Group IV Elements: The Bond to Carbon" (A. G. MacDiarmid, ed.), Vol. 1, p. 1. Dekker, New York, 1968.

172. A. G. Brook, *J. Organomet. Chem.* **300**, 21 (1986).
173. R. West, *Science* **225**, 1109 (1984).
174. R. West, *Pure Appl. Chem.* **56**, 163 (1984).
175. A. H. Cowley, *Acc. Chem. Res.* **17**, 386 (1984).
176. W. E. Dasent, "Nonexistent Compounds." Dekker, New York, 1965.
177. P. H. Blustin, *J. Organomet. Chem.* **105**, 161 (1976).
178. A. G. Brook, J. W. Harris, J. Lennon, and M. El Sheikh, *J. Am. Chem. Soc.* **101**, 83 (1979).
179. A. G. Brook, F. Abdesaken, B. Gutekunst, and R. K. M.R. Kallury, *J. Chem. Soc., Chem. Commun.*, 191 (1981).
180. A. G. Brook, F. Abdesaken, G. Gutekunst, and N. Plavac, *Organometallics* **1**, 994 (1982).
181. A. G. Brook, S. C. Nyburg, F. Abdesaken, B. Gutekunst, G. Gutekunst, R. K. M. R. Kallury, Y. C. Poon, Y. M. Chang, and W. Wong-Ng, *J. Am. Chem. Soc.* **104**, 5667 (1982).
182. A. G. Brook, S. C. Nyburg, W. F. Reynolds, Y. C. Poon, Y. M. Chang, J. S. Lee, and J.-P. Picard, *J. Am. Chem. Soc.* **101**, 6750 (1979).
183. A. G. Brook, K. D. Safa, P. D. Lickiss, and K. M. Baines, *J. Am. Chem. Soc.* **107**, 4338 (1985).
184. N. Wiberg, G. Wagner, G. Müller, and J. Riede, *J. Organomet. Chem.* **271**, 381 (1984).
185. N. Wiberg, G. Wagner, and G. Müller, *Angew. Chem., Int. Ed. Engl.* **24**, 229 (1985).
186. N. Wiberg, G. Wagner, G. Reber, J. Riede, and G. Müller, *Organometallics* **6**, 35 (1987).
187. R. West, M. J. Fink, and J. Michl, *Science* **214**, 1343 (1981).
188. P. Boudjouk, B.-H. Han, and K. R. Anderson, *J. Am. Chem. Soc.* **104**, 4992 (1982).
189. M. J. Michalczyk, R. West, and J. Michl, *J. Am. Chem. Soc.* **106**, 821 (1984).
190. M. J. Fink, M. J. Michalczyk, K. J. Haller, R. West, and J. Michl, *J. Chem. Soc., Chem. Commun.*, 1010 (1983).
191. S. Masamune, Y. Hanzawa, S. Murakami, T. Bally, and J. F. Blount, *J. Am. Chem. Soc.* **104**, 1150 (1982).
192. H. B. Yokelson, J. Maxka, D. A. Siegel, and R. West, *J. Am. Chem. Soc.* **108**, 4239 (1986).
193. N. Wiberg and K. Schurz, *J. Organomet. Chem.* **341**, 145 (1988).
194. M. Hesse and U. Klingebiel, *Angew. Chem., Int. Ed. Engl.* **25**, 649 (1986).
195. C. N. Smit and F. Bickelhaupt, *Organometallics* **6**, 1156 (1987).
196. H. Bock and R. Dammel, *Angew. Chem., Int. Ed. Engl.* **24**, 111 (1985).
197. A. Sekiguchi, S. Zigler, and R. West, *J. Am. Chem. Soc.* **108**, 4241 (1986).
198. N. S. Nametkin, L. E. Guselnikov, V. M. Vdovin, P. N. Grinberg, V. I. Zavyalov, and V. O. Oppengeim, *Dok. Acad. Nauk SSSR* **171**, 630 (1966).
199. L. Guselnikov, *Chem. Rev.* **79**, 529 (1979).
200. T. J. Barton, *Compr. Organomet. Chem.* **2**, 205 (1982).
201. I. M. T. Davidson, *Annu. Rep. Prog. Chem., Sect. C* **82**, 47 (1985).
202. G. Manuel, G. Bertrand, W. P. Weber, and S. A. Kazoura, *Organometallics* **3**, 1340 (1984).
203. T. J. Barton and S. Bain, *Organometallics* **7**, 528 (1988).
204. T. J. Barton and B. L. Groh, *J. Am. Chem. Soc.* **107**, 7221 (1985).
205. T. J. Barton and G. C. Paul, *J. Am. Chem. Soc.* **109**, 5292 (1987).
206. I. M. T. Davidson and A. Fenton, *Organometallics* **4**, 2060 (1985).
207. R. Damrauer, Ch. H. DePuy, I. M. T. Davidson, and K. J. Hughes, *Organometallics* **5**, 2050 (1986).
208. C. W. Carlson and R. West, *Organometallics* **2**, 1798 (1983).

209. P. Boudjouk, J. R. Roberts, C. M. Golino, and L. H. Sommer, *J. Am. Chem. Soc.* **94**, 7926 (1972).
210. M. Ishikawa and M. Kumada, *Adv. Organomet. Chem.* **19**, 51 (1981).
211. N. Wiberg and G. Preiner, *Angew. Chem., Int. Ed. Engl.* **16**, 328 (1977).
212. N. Wiberg, G. Preiner, O. Schieda, and G. Fischer, *Chem. Ber.* **114**, 3505 (1981).
213. N. Wiberg and G. Wagner, *Angew. Chem.* **95**, 1027 (1983).
214. P. R. Jones and T. F. O. Lim, *J. Am. Chem. Soc.* **99**, 2013 (1977).
215. P. R. Jones, T. F. O. Lim, and R. A. Pierce, *J. Am. Chem. Soc.* **102**, 4970 (1980).
216. P. R. Jones, A. H.-B. Cheng, and T. E. Albanesi, *Organometallics* **3**, 78 (1984).
217. P. R. Jones, M. E. Lee, and L. T. Lin, *Organometallics* **2**, 1039 (1983).
218. A. H.-B. Cheng, P. R. Jones, M. E. Lee, and P. Roussi, *Organometallics* **4**, 581 (1985).
219. G. H. Wiseman, D. R. Wheeler, and D. Seyferth, *Organometallics* **5**, 146 (1986).
220. C. Eaborn, D. A. R. Happer, and K. D. Safa, *J. Organomet. Chem.* **191**, 355 (1980).
221. Z. H. Aiube, J. Chojnowski, C. Eaborn, and W. A. Stańczyk, *J. Chem. Soc., Chem. Commun.*, 493 (1983).
222. R. I. Damja and C. Eaborn, *J. Organomet. Chem.* **290**, 267 (1985).
223. J. Chmielecka, J. Chojnowski, C. Eaborn, and W. A Stańczyk, *J. Chem. Soc., Chem. Commun.*, 1337 (1987).
224. J. Chmielecka, J. Chojnowski, C. Eaborn, and W. A. Stańczyk, *J. Chem. Soc., Perkin Trans. 2*, 865 (1989).
225. C. Eaborn and W. A. Stańczyk, *J. Chem. Soc., Perkin Trans. 2*, 2099 (1984).
226. O. W. Steward and J. L. Williams, *J. Organomet. Chem.* **341**, 199 (1988).
227. J. Chojnowski, K. Kaźmierski, S. Rubinsztajn, and W. Stańczyk, *Makromol. Chem.* **187**, 2039 (1987).
228. J. Chojnowski, S. Rubinsztajn, W. Stańczyk, and M. Ścibiorek, *Makromol Chem. Rapid Commun.* **4**, 703 (1983).
229. A. Streitwieser, Jr., *Chem. Rev.* **56**, 571 (1956).
230. W. P. Jencks, *Acc. Chem. Res.* **13**, 161 (1980).
231. R. A. Sneen, *Acc. Chem. Res.* **6**, 46 (1973).
232. D. J. McLennan, *Acc. Chem. Res.* **9**, 281 (1976).
233. F. H. Westheimer, *Chem. Rev.* **81**, 313 (1981).
234. S. L. Buchwald, J. M. Friedman, and J. R. Knowles, *J. Am. Chem. Soc.* **106**, 4911 (1984).
235. A. E. Pierce, "Silylation of Organic Compounds." Pierce Chem. Co., Rockford, Illinois, 1968.
236. J. F. Klebe, *Adv. Org. Chem.* **8**, 97 (1972).
237. E. Colvin, "Silylation of Organic Compounds." Butterworth, London, 1981; L. Birkofer and O. Stuhl, *Top. Curr. Chem.* **88**, 33 (1981).
238. G. G. Furin, O. A. Vyazankina, B. A. Gostevsky, and N. S. Vyazankin, *Tetrahedron* **44**, 2675 (1988).
239. J. Chojnowski, M. Cypryk, and J. Michalski, *J. Organomet. Chem.* **161**, C31 (1978).
240. A. R. Bassindale and T. Stout, *J. Organomet. Chem.* **238**, C41 (1982).
241. S. C. Chaudhry and O. Hernandez, *Tetrahedron Lett.*, 99 (1979).
242. S. C. Chaudhry and D. Kummer, *J. Organomet. Chem.* **339**, 241 (1988).
243. H. K. Chu, M. D. Johnson, and C. L. Frye, *J. Organomet. Chem.* **271**, 327 (1984).
244. J. R. Chipperfield and R. H. Prince, *J. Chem. Soc.*, 3567 (1963).
245. G. Scott, M. Kelling, and R. Schild, *Chem. Ber.* **99**, 291 (1966).
246. J. R. Gibson and A. F. Janzen, *Can. J. Chem.* **50**, 3087 (1972).
247. R. Corriu, *in* "Silicon Chemistry" (E. R. Corey, J. Y. Corey, and P. P. Gaspar, eds.), p. 225. Horwood, Chichester, England, 1988.

248. J. P. Corriu, G. Dabosi, and M. Martineau, *J. Chem. Soc., Chem. Commun.*, 649 (1977).
249. R. J. P. Corriu, G. Dabosi, and M. Martineau, *J. Organomet. Chem.* **150**, 27 (1978).
250. R. J. P. Corriu, F. Larcher, and G. Royo, *J. Organomet. Chem.* **129**, 299 (1977).
251. R. J. P. Corriu, G. Dabosi, and M. Martineau, *J. Organomet. Chem.* **154**, 33 (1978).
252. A. R. Bassindale and T. Stout, *J. Chem. Soc., Perkin Trans. 2*, 221 (1986).
253. A. R. Bassindale, J. C. Y. Lau, T. Stout, and P. G. Taylor, *J. Chem. Soc., Perkin Trans. 2*, 227 (1986).
254. H. J. Campbell-Ferguson and E. A. V. Ebsworth, *J. Chem. Soc. A*, 1508 (1966); H. J. Campbell-Ferguson and E. A. V. Ebsworth, *J. Chem. Soc. A*, 705 (1967).
255. M. Cypryk, J. Chojnowski, and J. Michalski, *Tetrahedron* **41**, 2471 (1985); M. Cypryk, J. Chojnowski, and W. Fortuniak, unpublished observations.
256. B. Borecka, J. Chojnowski, M. Cypryk, J. Michalski, and J. Zielińska, *J. Organomet. Chem.* **171**, 17 (1979).
257. V. D. Romanenko, V. I. Tovstenko, and L. N. Markovski, *Zh. Obshch. Khim.* **49**, 1907 (1979).
258. H. Schmidbaur and R. Seeber, *Chem. Ber.* **107**, 1731 (1974).
259. G. D. Hommer and L. H. Sommer, *J. Organomet. Chem.* **67**, C10 (1974).
260. P. M. Nowakovski and L. H. Sommer, *J. Organomet. Chem.* **178**, 95 (1979).
261. C. Eaborn and F. M. S. Mahmoud, *J. Organomet. Chem.* **220**, 139 (1981).
262. C. L. Frye and J. M. Klosovski, *J. Am. Chem. Soc.* **94**, 7186 (1972).
263. R. J. P. Corriu, G. Dabosi, and M. Martineau, *J. Organomet. Chem.* **186**, 25 (1980).
264. K. Hensen, T. Zengerly, P. Pickel, and G. Klebe, *Angew. Chem., Int. Ed. Engl.* **22**, 725 (1983).
265. K. Hensen, T. Zengerly, T. H. Müller, and P. Pickel, *Z. Anorg. Allg. Chem.* **558**, 21 (1988).
266. A. E. Williams and J. D. Cargioli, *Annu. Rep. NMR Spectrosc.* **9**, 221 (1979); H. Marsman, *NMR: Basic Princ. Prog.* **17**, 67 (1981).
267. C. H. Yoder, C. M. Ryan, G. F. Martin, and P. S. Ho, *J. Organomet. Chem.* **190**, 1 (1980); K. D. Onan, A. T. McPhail, C. H. Yoder, and R. W. Hillyard, Jr., *J. Chem. Soc., Chem. Commun.*, 209 (1978).
268. W. J. Stec, A. Okruszek, B. Uznański, and J. Michalski, *Phosphorus* **2**, 97 (1972).
269. A. MacDiarmid, *Intra-Sci. Chem. Rep.* **7**, 83 (1973).
270. J. R. Beattie and F. W. Parrett, *J. Chem. Soc. A*, 1784 (1966).
271. A. R. Bassindale and T. Stout, *J. Chem. Soc., Chem. Commun.*, 1387 (1987).
272. V. D. Romanenko, V. I. Tovstenko, and L. N. Markovski, *Synthesis*, 823 (1980).
273. L. Woźniak and J. Chojnowski, *Tetrahedron* **45**, 2465 (1989).
274. T. A. Bryson, G. H. Bonitz, C. J. Reichel, and R. E. Dardis, *J. Org. Chem.* **45**, 524 (1980).
275. S. Rubinsztajn, M. Cypryk, and J. Chojnowski, *J. Organomet. Chem.* **367**, 27 (1989); M. Cypryk, S. Rubinsztajn, and J. Chojnowski, *J. Organomet. Chem.* (in press).
276. J. Pikies, K. Przyjemska, and W. Wojnowski, *Z. Anorg. Allg. Chem.* **551**, 209 (1987).
277. T. J. Barton and C. R. Tully, *J. Organomet. Chem.* **172**, 11 (1979).
278. E. J. Corey and A. Venkateswarlu, *J. Am. Chem. Soc.* **94**, 6190 (1972).
279. H. H. Hergott and G. Simchen, *Liebigs Ann. Chem.*, 1718 (1980).
280. R. J. P. Corriu and M. Leard, *J. Organomet. Chem.* **15**, 25 (1968).
281. R. Corriu, M. Leard, and J. Masse, *Bull. Soc. Chim. Fr.*, 2555 (1968).
282. F. Carré, R. Corriu, and M. Leard, *J. Organomet. Chem.* **24**, 101 (1970).
283. R. Corriu and M. Henner-Leard, *J. Organomet. Chem.* **64**, 351 (1974); *Bull. Soc. Chim. Fr.*, 1447 (1974).

284. J. Chojnowski, M. Cypryk, J. Michalski, and L. Wozniak, *J. Organomet. Chem.* **288**, 275 (1985).
285. R. J. P. Corriu, F. Larcher, and G. Royo, *J. Organomet. Chem.* **104**, 293 (1976).
286. F. K. Cartledge, B. G. McKinnie, and J. M. Wolcott, *J. Organomet. Chem.* **118**, 7 (1976).
287. W. H. Stevenson III and J. C. Martin, *J. Am. Chem. Soc.* **107**, 6352 (1985).
288. A. R. Bassindale, J. C. Y. Lau, and P. G. Taylor, *J. Organomet. Chem.* **341**, 213 (1988).
289. W. B. Farnham and R. L. Harlow, *J. Am. Chem. Soc.* **103**, 4608 (1981).
290. W. H. Stevenson III, S. Wilson, J. C. Martin, and W. B. Farnham, *J. Am. Chem. Soc.* **107**, 6340 (1985).
291. R. Damrauer and S. E. Danahey, *Organometallics* **5**, 1490 (1986).
292. R. J. P. Corriu, A. Kpoton, M. Poirier, G. Royo, and J. Corey, *J. Organomet. Chem.* **277**, C25 (1984).
293. R. J. P. Corriu, M. Mazhar, M. Poirier, and G. Royo, *J. Organomet. Chem.* **306**, C5 (1986).
294. A. J. Arduengo, D. A. Dixon, D. C. Roe, and M. Kline, *J. Am. Chem. Soc.* **110**, 4437 (1988).
295. R. Krebs, Ph.D. thesis. Technical University, Braunschweig, Federal Republic of Germany.
296. S. N. Tandura, N. W. Alekseev, and M. G. Voronkov, *in* "Structural Chemistry of Boron and Silicon," p. 99. Akademie-Verlag, Berlin, 1985.
297. V. Chvalovsky, *in* "Organosilicon Compounds: Advances in Organosilicon Chemistry" (V. Bazant, M. Horak, V. Chvalovsky, and J. Schraml, eds.), Vol. 3, p. 80. Inst. Chem. Process Fundam., Prague, Czechoslovakia, 1973.
298. G. Klebe, M. Nix, and K. Hensen, *Chem. Ber.* **117**, 797 (1984).
299. R. J. P. Corriu, G. Royo, and A. de Saxce, *J. Chem. Soc., Chem. Commun.,* 892 (1980).
300. C. Brelière, F. Carré, R. J. P. Corriu, M. Poirier, and G. Royo, *Organometallics* **5**, 388 (1986).
301. J. Boyer, R. J. P. Corriu, A. Kpoton, M. Mazhar, M. Poirier, and G. Royo, *J. Organomet. Chem.* **301**, 131 (1986).
302. J. Boyer, C. Brelière, R. Corriu, A. Kpoton, M. Poirier, and G. Royo, *J. Organomet. Chem.* **311**, C39 (1986).
303. B. J. Helmer, R. West, R. J. P. Corriu, M. Poirier, and G. Royo, *J. Organomet. Chem.* **251**, 295 (1983).
304. G. Klebe, J. W. Bats, and K. Hensen, *J. Chem. Soc., Dalton Trans.,* 1 (1985).
305. A. A. Macharashvili, V. E. Shklover, Y. T. Struchkov, G. I. Oleneva, E. P. Kramarova, A. G. Shipov, and Y. I. Baukov, *J. Chem. Soc., Chem. Commun.,* 683 (1988).
306. A. Boudin, G. Cerveau, C. Chuit, R. J. P. Corriu, and C. Reyé, *Organometallics* **7**, 1165 (1988).
307. A. Boudin, G. Cerveau, C. Chuit, R. J. P. Corriu, and C. Reyé, *Angew. Chem., Int. Ed. Engl.* **25**, 473 (1986).
308. R. J. P. Corriu, C. Guerin, B. J. L. Henner, and W. W. C. Wong Chi Man, *Organometallics* **7**, 237 (1988).
309. R. Corriu, J. P. Dutheil, and G. Lanneau, *J. Am. Chem. Soc.* **106**, 1060 (1984).
310. G. Lanneau, *Phosphorus Sulfur* **27**, 43 (1986).
311. A. Hosomi, S. Iijima, and H. Sakurai, *Chem. Lett.,* 243 (1981).
312. K. Tamao, *in* "Organosilicon and Bioorganosilicon Chemistry" (H. Sakurai, ed.), p. 231. Horwood, Chichester, England, 1985.
313. K. Tamao, M. Akita, H. Kato, and M. Kumada, *J. Organomet. Chem.* **341**, 165 (1988).

314. M. Kumada, K. Tamao, and J.-I. Yoshida, *J. Organomet. Chem.* **239**, 115 (1982).
315. D. Yang and D. D. Tanner, *J. Org. Chem.* **51**, 2267 (1986).
316. E. Lukevics and M. Dzintara, *J. Organomet. Chem.* **271**, 307 (1984).
317. G. Schott, B. Schneider, and H. Kelling, *Z. Anorg. Allg. Chem.* **398**, 293 (1973).
318. V. A. Ivanov, V. O. Reikhsfeld, and I. E. Saratov, *Zh. Obshch. Khim.* **47**, 1781 (1977).
319. E. Lukevics and M. Dzintara, *Zh. Obshch. Khim.* **51**, 2043 (1981); E. Lukevics and M. Dzintara, *Zh. Obshch. Khim.* **52**, 1176 (1982).
320. E. Lukevics, O. A. Pudova, and M. Dzintara, *Zh. Obshch. Khim.* **54**, 339 (1984).

Index

Cumulative List of Contributors

Abel, E. W., **5**, 1; **8**, 117
Aguilo, A., **5**, 321
Albano, V. G., **14**, 285
Alper, H., **19**, 183
Anderson, G. K., **20**, 39
Angelici, R. J., **27**, 51
Aradi, A. A., **30**, 189
Armitage, D. A., **5**, 1
Armor, J. N., **19**, 1
Ash, C. E., **27**, 1
Ashe III, A. J., **30**, 77
Atwell, W. H., **4**, 1
Baines, K. M., **25**, 1
Barone, R., **26**, 165
Bassner, S. L., **28**, 1
Behrens, H., **18**, 1
Bennett, M. A., **4**, 353
Birmingham, J., **2**, 365
Blinka, T. A., **23**, 193
Bogdanović, B., **17**, 105
Bottomley, F., **28**, 339
Bradley, J. S., **22**, 1
Brinckman, F. E., **20**, 313
Brook, A. G., **7**, 95; **25**, 1
Brown, H. C., **11**, 1
Brown, T. L., **3**, 365
Bruce, M. I., **6**, 273; **10**, 273; **11**, 447; **12**, 379; **22**, 59
Brunner, H., **18**, 151
Buhro, W. E., **27**, 311
Cais, M., **8**, 211
Calderon, N., **17**, 449
Callahan, K. P., **14**, 145
Cartledge, F. K., **4**, 1
Chalk, A. J., **6**, 119
Chanon, M., **26**, 165
Chatt, J., **12**, 1
Chini, P., **14**, 285
Chisholm, M. H., **26**, 97; **27**, 311
Chiusoli, G. P., **17**, 195
Chojnowski, J., **30**, 243
Churchill, M. R., **5**, 93
Coates, G. E., **9**, 195
Collman, J. P., **7**, 53
Connelly, N. G., **23**, 1; **24**, 87
Connolly, J. W., **19**, 123
Corey, J. Y., **13**, 139

Corriu, R. J. P., **20**, 265
Courtney, A., **16**, 241
Coutts, R. S. P., **9**, 135
Coyle, T. D., **10**, 237
Crabtree, R. H., **28**, 299
Craig, P. J., **11**, 331
Csuk, R., **28**, 85
Cullen, W. R., **4**, 145
Cundy, C. S., **11**, 253
Curtis, M. D., **19**, 213
Darensbourg, D. J., **21**, 113; **22**, 129
Darensbourg, M. Y., **27**, 1
Davies, S. G., **30**, 1
Deacon, G. B., **25**, 237
de Boer, E., **2**, 115
Deeming, A. J., **26**, 1
Dessy, R. E., **4**, 267
Dickson, R. S., **12**, 323
Dixneuf, P. H., **29**, 163
Eisch, J. J., **16**, 67
Emerson, G. F., **1**, 1
Epstein, P. S., **19**, 213
Erker, G., **24**, 1
Ernst, C. R., **10**, 79
Evans, J., **16**, 319
Evans, W. J., **24**, 131
Faller, J. W., **16**, 211
Faulks, S. J., **25**, 237
Fehlner, T. P., **21**, 57; **30**, 189
Fessenden, J. S., **18**, 275
Fessenden, R. J., **18**, 275
Fischer, E. O., **14**, 1
Ford, P. C., **28**, 139
Forniés, J., **28**, 219
Forster, D., **17**, 255
Fraser, P. J., **12**, 323
Fritz, H. P., **1**, 239
Fürstner, A., **28**, 85
Furukawa, J., **12**, 83
Fuson, R. C., **1**, 221
Gallop, M. A., **25**, 121
Garrou, P. E., **23**, 95
Geiger, W. E., **23**, 1; **24**, 87
Geoffroy, G. L., **18**, 207; **24**, 249; **28**, 1
Gilman, H., **1**, 89; **4**, 1; **7**, 1
Gladfelter, W. L., **18**, 207; **24**, 41
Gladysz, J. A., **20**, 1